YEAST

PRODUCTION AND APPLICATIONS

YEAST
PRODUCTION AND APPLICATIONS

Turibio Kuiate
Guranda Avkopashvili

Kruger Brentt
Publishers

2025

Kruger Brentt Publishers UK. LTD.
Company Number 9728962
Regd. Office: 68 St Margarets Road, Edgware, Middlesex HA8 9UU

ISBN: 978-1-78715-298-4

For information on all our publications visit our website at http://krugerbrentt.com/

PREFACE

"Yeast: Production and Applications" offers a comprehensive exploration of the biology, production processes, and diverse applications of yeast in various industries. This book serves as an essential resource for researchers, biotechnologists, engineers, and students interested in harnessing the potential of yeast for industrial, agricultural, and medical purposes.

Yeast, a single-celled microorganism belonging to the fungi kingdom, has been utilized for millennia in various human endeavors, including baking, brewing, and fermentation. In recent decades, yeast has emerged as a versatile workhorse in biotechnology, serving as a platform for the production of biofuels, pharmaceuticals, enzymes, and other valuable products.

As editors of this volume, we have endeavored to compile a diverse array of perspectives, methodologies, and case studies from leading experts and practitioners in the field of yeast production and applications. Through a blend of theoretical foundations, practical insights, and real-world examples, this book covers a wide range of topics, including yeast genetics and physiology, fermentation processes, metabolic engineering, and biotechnological applications in food, agriculture, healthcare, and environmental remediation.

Our overarching goal in presenting this book is to provide readers with a comprehensive understanding of the biology, production methodologies, and applications of yeast in various industries. By offering insights into the latest advancements, best practices, and emerging trends in yeast biotechnology, we aim to inspire innovation, collaboration, and excellence in harnessing the potential of yeast for addressing global challenges and improving human well-being.

We extend our sincere gratitude to all the contributors who have generously shared their expertise, experiences, and insights in the field of yeast production and applications. It is our hope that this book will serve as a valuable reference and guide for researchers, biotechnologists, engineers, and students as they explore the multifaceted world of yeast biotechnology and its myriad applications.

Turibio Kuiate

Guranda Avkopashvili

CONTENTS

SECTION-1

INTRODUCTION

CHAPTER 1

THE VERSATILITY OF YEAST IN BIOTECHNOLOGY

1. INTRODUCTION: AN OVERVIEW

1.1. Yeasts: commercial applications

Yeasts are extensively used in various applications, mostly in the food industry for activities like as wine making, brewing, distilled spirits manufacturing, and baking. They are also utilized in biomass production, namely for the synthesis of single-cell protein (SCP). In recent times, yeast has also found applications in the biofuel business and in the synthesis of heterologous chemicals. Their primary function stems from their metabolic ability to convert carbohydrates into ethyl alcohol and carbon dioxide in the absence of oxygen. Furthermore, a significant quantity of secondary taste compounds are formed, which directly affects the sensory characteristics of specific food products. Nevertheless, it would be erroneous to underestimate their metabolic capabilities only based on fermentative activity. The primary determinants that impact yeast metabolism are the level of oxygen accessibility and the specific carbon substrate utilized. Several yeast strains are capable of functioning in both anaerobic and aerobic environments, effortlessly transitioning between different metabolic types [1]. Clearly, the fundamental metabolic pathways remain same, however several regulatory mechanisms are noteworthy for their exceptional metabolic flexibility [2]. Saccharomyces cerevisiae is the predominant genus of yeast utilized in the food business, whereas Candida, Endomycopsis, and Kluyveromyces play a vital role in the creation of single-cell protein (SCP). Yeast strains with industrial significance are meticulously chosen from the vast natural diversity and their characteristics are enhanced to optimize the desired output of the procedure. Both traditional methodologies and contemporary techniques of genetic manipulation are utilized to produce variations that are pertinent to functioning in distinct industrial environments [3, 4]. Strain modifications impact not only the process of fermentation and its immediate output, but also the ease of product recovery methods and ultimately the optimal quality of the specific final product. These nonpathogenic strains, which exhibit both genotypic and phenotypic stability, should possess a short generation time and low nutritional demands. In order to limit the possibility of contamination, it is advisable to expedite the fermentation process. In addition, they should exhibit tolerance to various physiological challenges, including low pH, high ethanol concentration, and high osmotic stress.

The enzymatic activity of yeast plays a crucial role in the process of beer production (Figure 1). Industrial species are specifically chosen polyploids that do not undergo sexual reproduction and do not produce spores [5]. Brewer's yeasts are categorized into two distinct groups: top-fermenting yeast, also known as ale yeast, and bottom-fermenting yeast, also known as lager yeast. Both varieties of yeast exhibit comparable cellular appearance, but they diverge in certain physiological and metabolic characteristics [6], which closely align with the specific process conditions and the type of final product. Ale yeast undergoes fermentation at ambient temperature, leading to the production of beers that possess a distinctive fruity fragrance. When it comes to lager yeast, the fermentation temperature is lower, resulting in a longer fermentation process compared to ale strains [7].

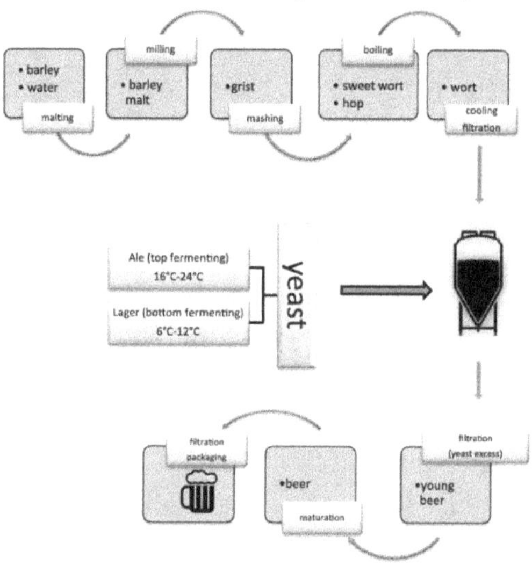

Figure 1. Beer production process.

Ale strains of yeast exhibit a greater proportion of aerobic respiration in their metabolism compared to lager yeast. Both top-fermenting and bottom-fermenting yeast are classified under the genus Saccharomyces. Ale-brewing yeasts exhibit greater genetic diversity and are categorized as Saccharomyces cerevisiae, while the taxonomy of lager strains has undergone multiple revisions. Originally, bottom-fermenting yeast was categorized as Saccharomyces carlsbergensis. However, with the use of modern taxonomic methods, it was determined that the species

S. carlsbergensis should be placed within the Saccharomyces pastorianus group. Genetic studies have determined that S. pastorianus strains are now recognized as allopolyploid interspecies hybrids of S. cerevisiae and S. bayanus [8]. From a technological perspective, in addition to their high fermentative activity and ability to tolerate various environmental stresses, the yeast cells' capacity for asexual aggregation, known as flocculation, is particularly important. This ability allows the yeast to settle at the bottom of the fermenter at the end of the fermentation process, making downstream processing simpler [9, 10].

The predominant microbial group in natural fruit ecosystems is comprised of several yeast species. This is the reason why juices or musts undergo quick and spontaneous fermentation during wine production. The yeast that colonizes different fruits belongs to several genera, including Saccharomyces, Brettanomyces (also known as Dekkera in its sexual form), Candida, Cryptococcus, Debaromyces, Hanseniaspora (also known as Kloeckera in its anamorph form), Hansenula, Kluyveromyces, Pichia, Rhodotorula, Torulaspora, Schizosaccharomyces, and Zygosaccharomyces [11]. It is universally understood that the majority of fermented items are produced by the combination of several microorganisms. The microbial consortia are responsible for different biological activities that contribute to the nutritional, sanitary, and aromatic properties of the product [12]. Undoubtedly, yeast plays a crucial part in the process of wine production, both in small-scale home settings and on a larger industrial scale (Figure 2). Non-Saccharomyces species of yeast, which are not traditional, are the dominant ones in the initial and intermediate stages of the fermentation process. However, the later phase of the natural process is controlled by Saccharomyces cerevisiae. Grapes, collected at specific stages of ripeness, are a crucial raw material in the production of wine worldwide, with the timing of harvest determined by the desired type of the wine.

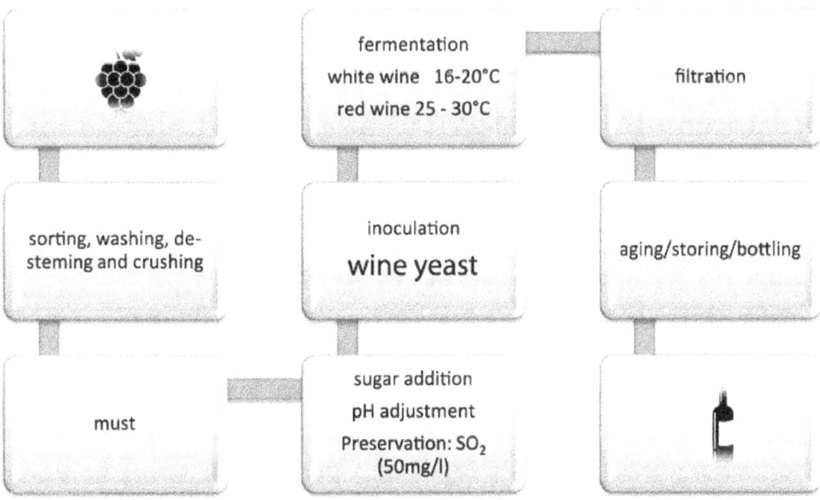

Figure 2. Wine production process.

The enzymatic activity of yeast is once again essential for determining the chemical characteristics of the product. In addition to producing ethanol, yeast also plays a key role in creating flavor compounds and perfumes from the copious substrates found in fruits, which contribute to the sensory qualities of the specific wine product. During the breakdown of sugars in grapes, various substances such as amino acids, fatty acids, terpenes, and thiols are produced. This process also results in the formation of by-products including glycerol, carboxylic acids, aldehydes, higher alcohols, esters, and sulfides. The specific strains utilized in the process greatly influence the synthesis of these by-products [13, 14]. In 1965, a big Californian winery produced the first two commercially available dry wine yeasts, known as Montrachet and Pasteur Champagne [15]. In contemporary times, wine producers commonly employ specifically chosen yeast cultures as a regular procedure. These bacteria outcompete indigenous yeast species and control the specific chemical changes that occur in musts, resulting in a product of consistent and predictable quality [16]. Currently, commercially accessible starters have been developed by a combination of natural genome renewal and deliberate genetic improvement procedures [17], followed by meticulous selection based on their superior fermentation performance. The genome renewal theory, as described in the standard version, proposes that occasional sexual cycles with a high level of self-fertilization can eliminate harmful genetic variations and promote the spread of advantageous variations. This process aids in the adaptability of yeast. [18] This theory needed to be reassessed [19] because numerous environmental isolates showed significant levels of genomic heterozygosity [20, 21]. Currently, the main type of yeast used in commercial wine production is Saccharomyces cerevisiae, which includes strains previously referred to as S. bayanus. However, recent research has shown that these strains are actually a kind of S. cerevisiae [22, 23]. The increasing desire for a wider variety of wines or for specific attributes has prompted the investigation of novel species for the production of wine [24–26]. This atypical yeast strain can enhance the sensory characteristics of the wine by creating a diverse array of distinct secondary compounds and releasing specific enzymes or displaying other significant attributes (such as the release of mannoproteins and contributions to color stability) [27]. The wine business currently suggests using unconventional yeast strains such as Torulaspora delbrueckii, Metschnikowia pulcherrima, Pichia kluyveri, Lachancea thermotolerans, etc. as starters [28]. Yeast strains utilized as starters must meet the requirements for commercial use. The list of desired properties is extensive and encompasses both fermentation characteristics and flavor characteristics (such as minimal production of sulphide, dimethyl sulphide, and thiols, release of glycosylated flavor precursors, and limited production of higher alcohols), metabolic properties with health implications (such as low formation of sulphite, biogenic amines, and urea), and technological properties (such as high tolerance to sulphite, minimal foam formation, and effective flocculation) [29].

Another instance of commercially significant Saccharomyces cerevisiae strains are distiller's yeasts, which are utilized in industries where alcohol production is then followed by distillation. Spontaneous fermentations are not employed in this particular industry, instead a specific strain of yeast is deliberately introduced into the fermenter. Starters utilized in the commercial production of distilled beverages such as aquavit, gin, vodka, whisky, rum, tequila, cognac, brandy, and kirsch should demonstrate a highly efficient anaerobic metabolism. This metabolic process should not only lead to a high production of ethanol, but also result in a significant level of cellular tolerance to this substance. Furthermore, there is a higher need for thermostable versions that can tolerate acids and increased osmotic pressure among distillers [30]. Over the years, there has been a growing fascination with novel variations that have the ability to break down starch. Several efforts have been undertaken to generate ethanol from starch by utilizing genetically modified haploid strains that produce amylolytic enzymes, owing to the ease of genetic manipulation of these strains. Nevertheless, the utilization of laboratory haploid strains in practical applications is challenging due to their inferior fermentation characteristics compared to industrial polyploidy strains [31, 32].

Yeasts play a crucial role in the baking industry by providing the necessary enzymatic power for dough fermentation. Concentrated yeast biomass is used as a starter to manufacture bread and other bakery products, in addition to its application in alcoholic drinks production. There are three varieties of baker's yeast that can be purchased commercially: fresh compressed biomass, dehydrated cells (dry yeast), and lyophilized cells (instant). The fermentation of dough substrates results in the synthesis of ethanol, as well as several volatile and nonvolatile chemicals that significantly contribute to the flavor of bread [33]. Carbohydrates, primarily maltose, undergo fermentation, resulting in the production of carbon dioxide. This leads to a rise in the volume of the dough and is responsible for the texture of the crumb. Baker's yeast is brewery yeast that is created through a submerged fermentation process conducted in the presence of oxygen (Figure 3). Yeast cell production is favored under aerobic circumstances, which is not desirable for ethanol producers. However, it becomes crucial when there is a need to create a big number of cell mass.

The primary component utilized in yeast production factories for industrial production medium is beet or cane molasses, mostly due to the economical nature of these byproducts and their high sucrose concentration. Typically, industrial production involves a series of steps that are carried out in batches or fed-batch settings. These stages vary in the size of the fermenter used and are completed with controlled and strong aeration [34]. Aeration is widely recognized as the primary element for enhancing yeast yield, and extensive research has been conducted to explore the optimization of certain technical approaches [35]. The distinctive feature of yeast metabolism is its dual requirement for efficient respiratory metabolism during yeast manufacturing to achieve high biomass yield, while also possessing strong fermentative potential to produce high-quality bakery products. During the fermentation of dough, yeast cells are subjected to several environmental challenges, including freeze-thaw cycles, high sugar concentrations, air drying, and oxidative stresses [36]. It is not unexpected that industrial starters would possess suitable stress tolerance. Yeasts have indeed developed adaptive mechanisms, but excessive stress can limit their enzymatic capacity. Research has shown that two molecules, trehalose and proline, play a crucial role in enhancing yeast's ability to withstand stress. Therefore, manipulating the metabolism of these molecules is a promising strategy for creating yeast strains that are more resistant to stress, which is particularly significant for industrial applications [37].

Figure 3. Baker's yeast production process.

Another enduring industrial operation utilizing yeast is the manufacture of single-cell protein (SCP), which serves as an alternate source of highly nutritious proteins used in food or as feed supplements. The concept of producing protein concentrate arose as a response to the increasing global population and the widespread lack of protein globally [38]. SCP production is the cultivation of nonpathogenic microorganisms (such as bacteria, fungus, and algae) using inexpensive and waste source materials. This is done through submerged fermentation, but solid-state fermentation is occasionally used. In addition to its substantial protein composition (about 60-82% of dry matter), SCP also includes lipids, carbs, nucleic acids, vitamins, and minerals [39]. Various technologies were assessed and multiple products were brought to market. Currently, the primary use of SCP is in animal nutrition. Furthermore, the aforementioned baker's yeast can also be regarded as an exemplification of a certain SCP formulation. Several fungal species are utilized as a source of protein-rich diet. The yeast species Candida, Hansenula, Pichia, Torulopsis, and Saccharomyces [40] are the most widely recognized. Additionally, marine yeast is now recognized as a valuable protein source [41, 42]. The yeast-derived SCP possesses the appropriate nutritional content and is rich in critical amino acids, particularly those that include sulfur. Furthermore, the historical effect of yeast is a crucial factor in people's acceptance of it as a food source. The primary drawback of this preparation, which hinders its use as food, is the presence of nucleic acids. These nucleic acids require additional purification steps in downstream processing [43] and are also connected with an exterior mannoprotein layer on the yeast cell wall, which hampers digestion.

Yeast can be regarded as an alternate lipid source. Certain species have the ability to produce and store more than 20% of their biomass as neutral lipids, which is why they are referred to as "oleaginous." Under ideal growth conditions and/or because to genetic enhancements, the amount of fat accumulation can potentially reach up to 70%. The group of oleaginous yeast comprises many species such as Candida, Cryptococcus, Pichia (Hansenula), Lipomyces, Pseudozyma, Rhodosporidium, Rhodotorula, Trichosporon, Trigonopsis, Yarrowia, and Saccharomycopsis [44]. The yeast species possess the capacity to amass substantial amounts of lipids, which presents a promising opportunity for the commercial production of single-cell oil (SCO). Microbial oils have the potential to be used as alternative edible oils in the food industry and can also be used as raw materials for the production of various oleochemicals, including fuels, soaps, plastics, paints, detergents, textiles, rubber, surfactants, lubricants, additives for the food and cosmetic industry, and other chemicals [45]. Microorganisms, which are considered as an alternative source of oils, are currently unable to directly compete with plants. However, their utilization offers several advantages, such as shorter process cycles, independence from seasonal and climatic conditions, the ability for genetic enhancement, and the potential to produce lipids with unique structures that are not synthesized by plants. It is worth noting that yeast oils have a composition that is comparable to vegetable products, with the main component being triacylglycerols (TGA) that are abundant in polyunsaturated fatty acids [46].

Lipids are mostly stored intracellularly in the form of granular structures known as "lipid bodies." The composition and abundance of fatty acids in these lipid bodies vary across different species [47]. The primary fatty acids produced by oleaginous yeast include myristic acid (C14:0), palmitic acid (C16:0), stearic acid (C18:0), oleic acid (C18:1), palmitoleic acid (C16:1), and linoleic acid (C18:2) [48]. Oleaginous yeast utilizes two distinct mechanisms for lipid accumulation: de novo lipid production and ex novo lipid accumulation. There are significant metabolic distinctions between the process of de novo lipid accumulation from hydrophilic substrates and ex novo lipid accumulation from hydrophobic substrates. On the other hand, ex novo lipid creation is a process that is linked to cell development and happens at the same time, regardless of whether the culture medium has run out of nitrogen or not. The production of ex novo lipids refers to the process of altering fats and oils using oleaginous bacteria [49]. The significance of microbial oil production has increased in relation to the generation of biodiesel. The oleaginous yeast is regarded as a promising candidate for producing "2nd generation" biodiesel. This biodiesel is derived from lipids produced by oleaginous microorganisms that grow on various waste or agro-industrial residues such as sewage sludge, hemicelluloses, hydrolysates, waste glycerol, and cheese whey [50, 51]. Saccharomyces cerevisiae (SCO) can be generated using submerged fermentation conducted in the presence of oxygen, using either batch, fed batch, or continuous operation methods. In order to attain optimal biomass development and appropriate lipid accumulation, yeast cultivation necessitates meticulous evaluation of the circumstances. Several elements have been identified as impacting the process of lipid formation, and it appears that the composition of the medium is particularly significant [52]. Once biomass has been generated, stress conditions must be produced to enhance the process of lipid accumulation. Lipid accumulation is induced by a deficiency in nutrients coupled with an abundance of carbon. The primary method employed to initiate lipid accumulation is nitrogen limitation, while other nutrients like as phosphorus and sulfur have also been demonstrated to induce lipid storage. Despite the potential for practical use, the commercialization of microbial products is now restricted to oils that include polyunsaturated fatty acids such as docosahexanoic acid (DHA), arachidonic acid (ARA), and eicosapentaenoic acid (EPA) [53, 54]. The production of microbial oils, whether for human consumption, industrial usage, or biofuel generation, is currently more expensive than the production of vegetable oils, making it cost-prohibitive [55]. In order to further the industrialization of yeast as an alternative source of oils, it is crucial to concentrate all efforts on enhancing the genetic capabilities of cell factories, optimizing the provision of nutrients and culture conditions, and making improvements to downstream technologies [56].

2. YEASTS USED AS WHOLE-CELL BIOCATALYSTS

Various types of yeasts are useful biocatalysts utilized in numerous areas of chemistry, particularly for the production of chiral building blocks and high-quality compounds. These catalytic tools are intriguing due to their diverse enzymatic activity as well as their microbiological characteristics, including easy cultivation, minimal nutritional needs, and adaptability. Their adaptable metabolism allows them to possess these capabilities, as it reacts to the effects of the environment. Consequently, the physical and chemical parameters of the process can influence the direction (including stereoselectivity) and efficiency of the biotransformations. Additionally, they can be influenced by the manipulation of the reaction environment, such as the choice of reaction media (e.g., water, organic solvents), as well as the physical shape of the biocatalysts (e.g., permeabilization, immobilization). This greatly expands the range of their applicability by overcoming restrictions such as the low solubility of the bioconversion substrates. Yeasts have been utilized for many years and are among the earliest whole-cell biocatalysts employed in industrial processes.

Literature evidence has demonstrated that yeasts possess exceptional reductive abilities, which are closely linked to their fundamental uses. Due to its demonstrated chemoselectivity, regioselectivity, and stereoselectivity, as well as its ability to regenerate dehydrogenase cofactors under biocatalytic conditions, whole-cell biocatalysis has been thoroughly investigated as a reductive catalyst for synthesizing chiral building blocks. Yeasts, in particular, have been extensively studied for the synthesis of chiral alcohols with a specific absolute configuration. The reactivity of several yeast genera has been examined with respect to various ketones, resulting in the production of pure enantiomeric alcohols in the majority of cases [57]. The alcohols depicted in Figure 4 encompass both simple and complex structures, which are the result of whole-cell biocatalysis facilitated by yeasts.

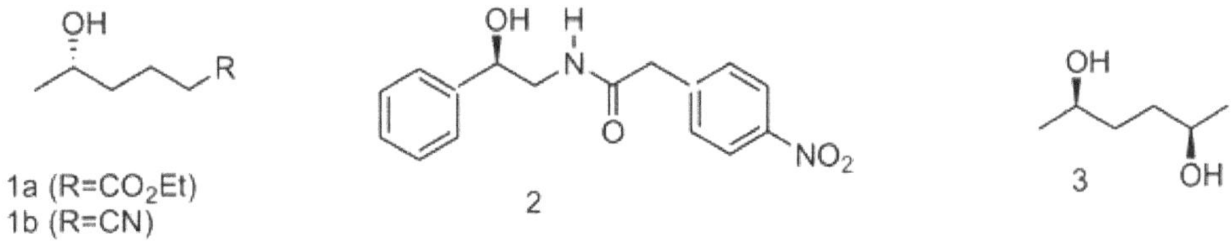

Figure 4. Alcohols synthesized using whole yeasts cells.

Pichia methanolica is highly valued in the industry due to its exceptional reductive capacity and its ability to work with a wide range of substrates. Therefore, Pichia methanolica facilitated the reduction of ethyl-5-oxo-hexanoate and 5-oxohexanenitrile, resulting in the production of S-alcohols: 1a and 1b (as shown in Figure 4). These alcohols were obtained with a conversion degree of up to 90% and 97% enantiomeric excess (ee). Additionally, the bioreduction of compound

1a (Figure 4) was successfully carried out on a large scale, yielding similar results. The reduction process at Bristol-Mayers Squibb Company (USA) utilizes Pichia methanolica. Alcohol 2 (Figure 4) plays a crucial role in the pharmaceutical business. This molecule is recognized as a therapeutic agent in the treatment of type 2 diabetes by enhancing metabolic rates. The utilization of Candida sorbophila MY 1833 enabled the production of a high-purity product on a large scale, achieving a conversion rate of up to 60% and an enantiomeric excess of 98%. Optically pure diols are crucial for their role as chiral precursors in the production of medicines and fine chemicals. For instance, compound 3 (Figure 4) can be obtained by reducing a diketone using Saccharomyces cerevisiae on a large scale. This process results in complete reduction of the substrate and a distereomeric excess (de) of 96% for 2S, 5S-hexanediol. In addition to the stated diol, this process can also be used to generate 5-hydroxyhexane-2-one, which serves as a substrate for synthesizing chiral tetrahydrofuranes. These compounds are of great importance for producing biodegradable polymers, medicines, and perfumes. Within the category of chiral compounds containing hydroxyl groups, there are certain molecules that are particularly noteworthy. These compounds serve as foundational structures for the production of several other compounds. The (S)-4-chloro-3-hydroxybutanoate ester (S)-CHBE (Figure 5) is of significant importance [58]. Several fungal catalysts, such as Geotrichum sp., Candida sp., and Aureobasidium sp., exhibit activity towards suitable ketones. However, the reaction occurs with only moderate stereoselectivity. Pichia stipites CBS 6054 was one of the organisms that enabled for the production of a chiral product with an enantiomeric excess (ee) of 80%. Yeast-mediated enantioselective bioreduction produces chiral bicyclic alcohols, such as (S)-pramipexole (an anti-Parkinson medication) and its (R) isomer, which is considered an anti-amyotrophic lateral sclerosis (ALS) agent [59]. The molecules acquire chirality through the reduction of the prochiral bicyclic ketone (Figure 6) mediated by Sachcaromyces cerevisiae. This reduction is the critical step in the reaction sequence that leads to the formation of the final enantiomers of pramipexole. The discussed objective exemplifies the yeast biotransformation of compounds including heteroatoms in their structures, specifically a sulfur atom within a bicyclic molecule.

The reductive activity of Saccharomyces cerevisiae is also utilized for non-typical reactions, such as the hydrogenation of geraniol into citronellol. This process is done by using resting yeast cells on a preparative scale, as shown in Figure 7 [60]. The process productivity for the reaction conducted in the continuous-closed-gas-loop bioreactor has reached 2.38 g/L.

Recently, yeasts have been proven to have practical applications as biocatalysts in a completely different activity. The studies with Saccharomyces cerevisiae were successful, as shown in Figure 8[61]. The figure demonstrates the addition of several 3-substituted indol derivatives to nitroolefins. The chemical synthesis of substituted indoles necessitates the use of harmful organic solvents, whereas biological synthesis greatly mitigates this issue and provides a convenient and efficient technique (yielding reaction rates ranging from 72 to 90%).

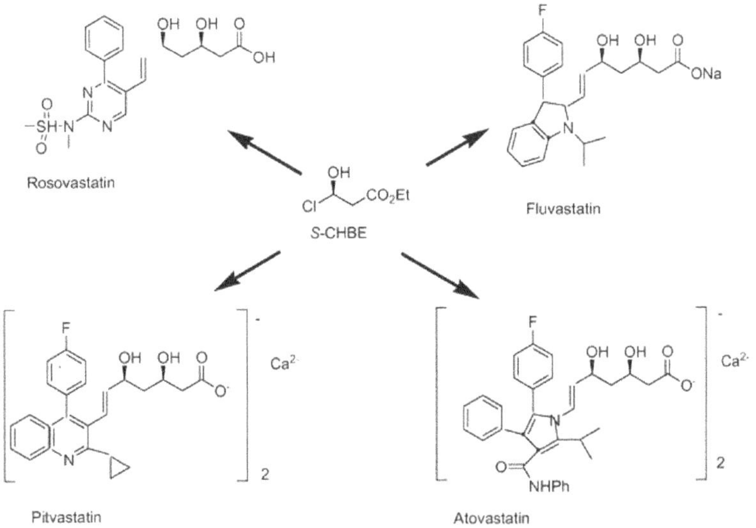

Figure 5. (S)-CHBE as chiral building platform.

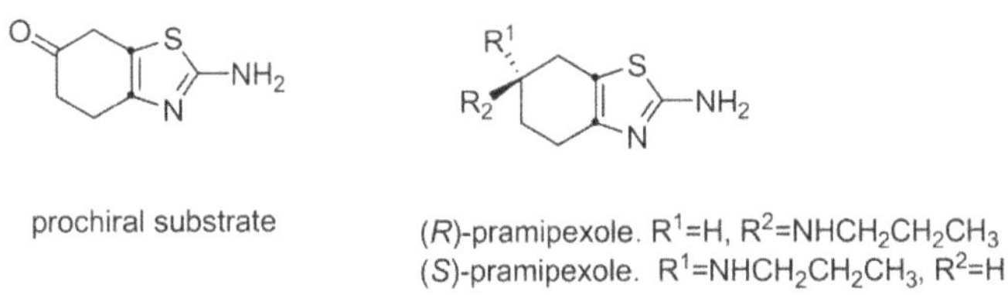

Figure 6. Pramipexole (2-amin-6-propylamino-4,5,6,7-tetrahydrobenzothioazole).

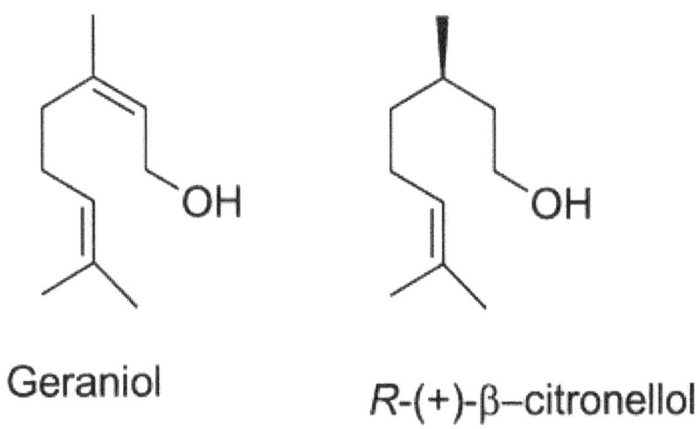

Figure 7. Structures of geraniol and citronellol.

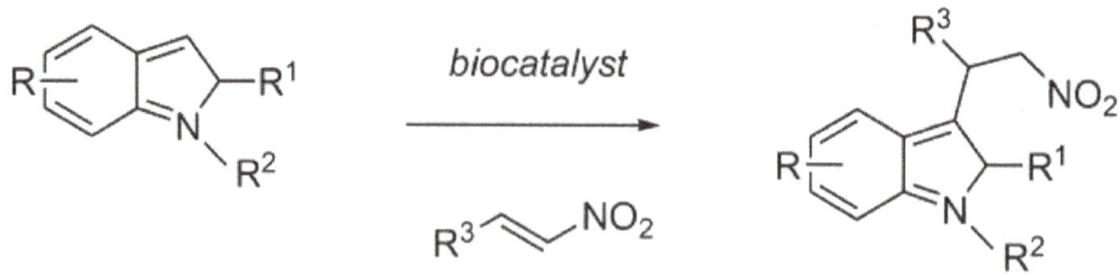

Figure 8. Biocatalyzed addition of indoles to nitroolefins.

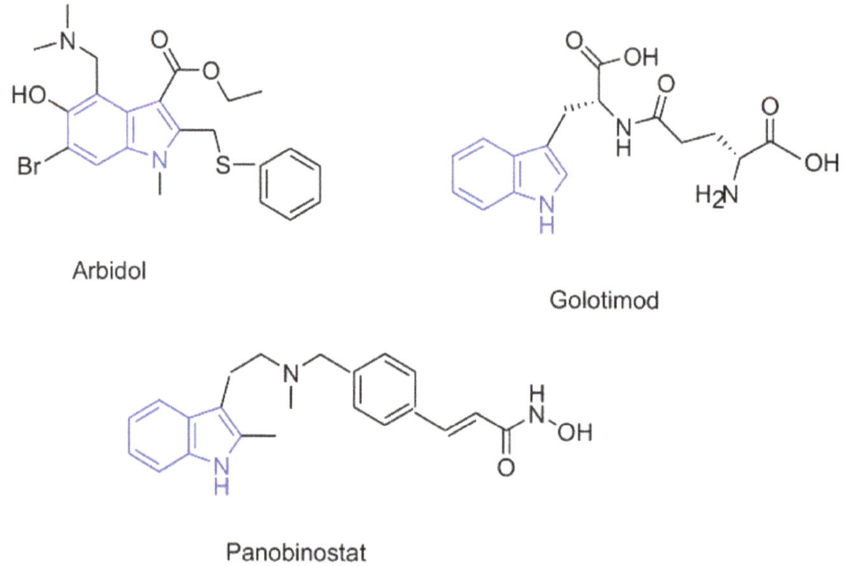

Figure 9. Compounds with the indol-motif.

Indole motifs play a vital role in the structure of biologically active substances including arbidol (used for treating and preventing influenza A and B viruses), golotimod (an agent that stimulates the immune system, fights against microbes, and has anti-cancer properties), and panobinostat (used for treating acute myeloid leukemia; see Figure 9).

3. YEAST'S ENZYMES APPLICATIONS

Biocatalysis encompasses two processes: biotransformations, which involve the conversion of xenobiotics utilizing entire cells or resting cell systems, and enzyme catalysis, which involves the conversion of xenobiotics using cell-free extracts or purified enzymes [62]. While both whole cells and isolated enzymes can serve as biocatalysts, whole cells are generally preferred due to their greater stability and cost-effectiveness compared to purified enzymes. Whole cells eliminate the requirement for purification and the inclusion of coenzymes. However, in the context of single-step biotransformation, isolated enzymes are a more favorable option and can be utilized as a free or immobilized biocatalyst in either aqueous or organic environments [63]. Yeasts, particularly species of Saccharomyces, are primarily recognized for their whole cell reductive activity [64, 65]. They are commonly employed in the food industry for the manufacturing of alcoholic drinks and for fermenting bread [3]. Yeasts serve as a reservoir of enzymes, including lipases, dehydrogenases, and invertase.

3.1. Yeast's lipases

Lipases have a broad distribution in the natural world and are synthesized by plants, animals, and microbes. Microbial enzymes are superior to other enzymes due to their wide range of catalytic activity, ease of manipulation, and cost-effective production (via extracellular production during rapid growth on inexpensive conditions) [66]. Moreover, microbial enzymes are devoid of issues related to contamination by hormones or viruses, making them suitable for use in the manufacturing of food and pharmaceuticals for vegetarian or kosher diets [67]. Microbial lipases (EC 3.1.1.3) are effective enzymes for organic synthesis due to their ability to function on a wide range of nonphysiological substrates and their stability in biphasic systems or pure organic media. Lipases have versatile applications in lipid modifications and the manufacture of specialized chemicals such as medicines, polymers, biodiesels, and biosurfactants [68]. Lipases under normal bodily conditions facilitate the hydrolysis of ester bonds in triacylglycerol, resulting in the production of glycerol and free fatty acids. In nonaqueous environments, they act as catalysts for the reverse process of esterification. Transesterification is the process of exchanging a group between an ester and acid, ester and alcohol, or at least between two esters (Figure 10) [69].

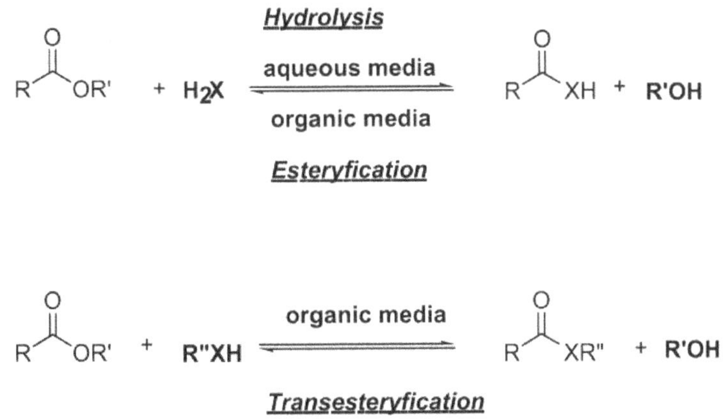

Figure 10. Types of reactions carried out by lipases.

Mentioned features make them significant biocatalyst for various applications. There are a certain number of yeast species able to produce lipases, most of them belong to Candida genus: Candida utilis, C. rugosa (cylindracea), C. antarctica, C. viswanathii, and additionally, Yarrowia lipolytica [70].

3.1.1. Hydrolysis

In the pharmaceutical sector, lipases are employed to separate racemic mixtures of alcohols or carboxylic acids by means of asymmetric hydrolysis of acyl derivatives. Candida antarctica lipase, isoform B (CAL-B) is widely used as a psychrophilic lipase in several applications such as kinetic resolutions, desymmetrization, and aminolysis [71]. Commercial CAL-B is offered in three different forms: free, lyophilized, and immobilized. The immobilized form is attached to Lewatit VP OC 1600 (poly(methyl methacrylate-co-divinylbenzene)—Novozyme 435, Chirazyme L2-C2). Novozym 435 is an appropriate catalyst for both small organic compounds [72, 73] and polymerization processes [74, 75]. Furthermore, the attachment of CAL-B to various supports can lead to variations in its activity and enantioselectivity (as shown in Table 1), making it a useful means of controlling the selectivity of hydrolysis. This property was employed to separate the racemic mixture of 2-O-butyryl-2-phenylacetic acid, which is a precursor for both enantiomers of mandelic acid (as shown in Figure 11). It was also used for the selective hydrolysis of 3-phenylglutaric dimethyl diester, which is a precursor in the manufacture of drugs such as HIV inhibitors [76]. Table 1 demonstrates that the enantioselectivity of the biocatalyst can be altered by replacing one support material with another (Figure 11).

Table 1. The influence of different methods of immobilization on the activity and enantioselectivity of CAL-B applied for hydrolytic resolution of 2-O-butyryl-2-phenylacetic acid.

Biocatalyst	Activity $\mu mol g_{cat}^{-1} h^{-1}$	*ee* p (%)[1]	Stereochemical preference	E[2]
Novozym 435	0.75	>99	S	>100
NOVO-CAL-B	0.56	>99	S	>100
Lewatit-CAL-B	0.55	>99	S	>100
Octyl-agarose-CAL-B	0.19	95	R	49
Octadecyl-sepabeads-CAL-B	0.35	90	R	23
Butyl-agarose-CALB-B	0.16	72	R	7.5

Optical purity of product.

Enantioselectivity of the enzyme towards substrate.

An alternative method to modify the enantioselectivity of hydrolysis involves introducing an organic co-solvent into the reaction media. The conformation of the enzyme in organic media is more stiff, which can potentially affect the enantioselectivity of the process. The enantioselectivity of the lipase-catalyzed hydrolysis of several substituted phenoxypropionates in Candida rugosa was enhanced by the inclusion of 30-70% dimethyl sulfoxide (DMSO) or sodium dodecyl sulfate (SDS), resulting in an improvement from E = 4 to >100 [77, 78]. The lipase (C. rugose) exhibited a full reversal of enantioselectivity in the hydrolysis of 1,4-dihydropyridines when various organic solvents saturated with water were used. This enabled the production of both enantiomers of 1,4-dihydropyridine (Figure 12).

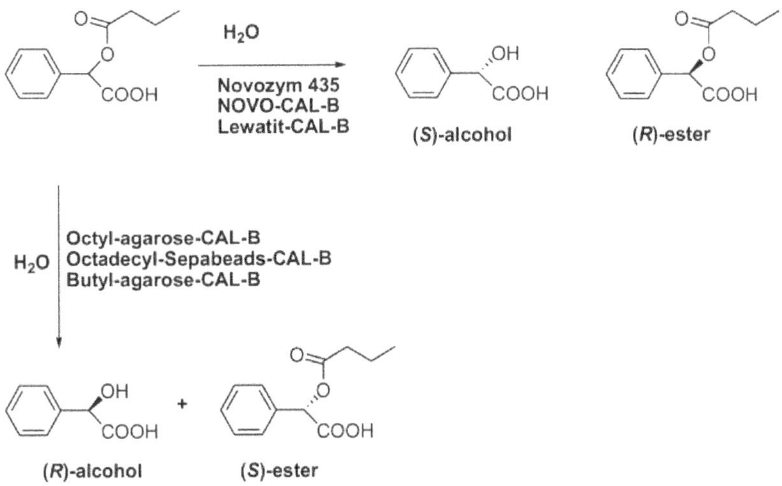

Figure 11. Kinetic resolution of 2-O-butyryl-2-phenylacetic acid by immobilized CAL-B on different supports.

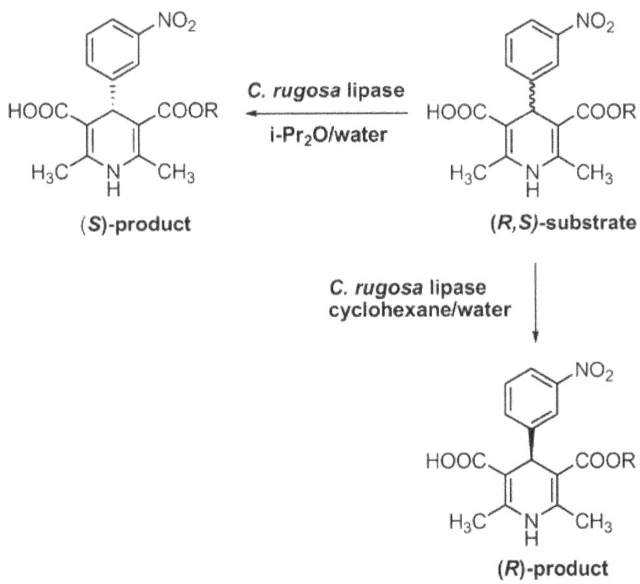

Figure 12. Reversal of enantioselectivity in hydrolysis catalyzed by Candida rugosa lipase in organic solvent saturated with water [79].

Candida antarctica synthesizes two variants of lipases, namely isoforms A and B. Recently, there has been a greater focus on the use of CAL-B. However, in the past several years, CAL-A has also been applied in notable ways. CAL-A exhibits a remarkable biochemical characteristic of being highly thermally stable, with a temperature tolerance exceeding 90°C [71, 80]. It is peculiar that a bacterium adapted to cold temperatures develops an enzyme that remains stable at high temperatures. This property enables the use of CAL-A in distinctive reaction circumstances [81], particularly for sterically hindered tertiary alcohols and their derivatives, as well as bulky cyclic compounds. The bioconversion of tertiary alcohols can effectively remove the tert-butyl protecting group, even when exposed to high temperatures (Table 2) [82]. Table 2 provides a summary of many examples of these applications. Another intriguing usage of CAL-A is its ability to selectively break down cyclic diacetates, which are important components in the production of vitamin D3 derivatives [83], as well as sterically hindered carboxylic acids [84].

3.1.2. Esterification and transesterification

Lipases play a crucial role in chemical synthesis, particularly in esterification, which is essential for separating racemic mixtures of secondary alcohols and carboxylic acids. Chiral secondary alcohols are used as intermediates in the production of pharmaceuticals [85–87]. There are three lipase-catalyzed procedures for preparing enantiopure compounds: kinetic resolution (KR), dynamic kinetic resolution (DKR), and desymmetrization. Enzymatic kinetic resolution relies on the disparity in reaction rates between the enantiomers of a racemate in the presence of a chiral catalyst, which is an enzyme. Dynamic kinetic resolution is a process that involves the simultaneous racemization of the unreacted enantiomer and the separation of the enantiomers based on their different reaction rates. Racemization can be accomplished either by chemical means or enzymatic processes. The process of separating secondary alcohols and esters into their individual components is conducted in organic solvents using lipase-catalyzed acylation and alcoholysis. It results in the production of one enantiomer in the form of an alcohol and the other enantiomer in the form of an ester. The maximum potential yield for each enantiomer is 50%. Enzymes from C. antarctica are frequently employed for the separation of secondary alcohols, particularly those with large substituents [88, 89]. They are also utilized for the separation of aliphatic compounds [90]. For instance, CAL-B was used to separate a racemic combination of 2-pentanol in heptane solvent, resulting in a yield of 49.6% and an enantiomeric excess (ee) of over 99% (Figure 13). S-(+)-2-pentanol is an important compound used in the production of medications that treat Alzheimer's disease. These drugs work by preventing the release or synthesis of β-amyloid peptide, which is associated with the disease [91, 92].

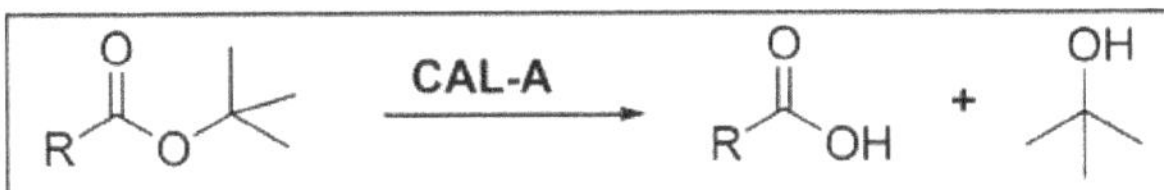

R	Yield (%)
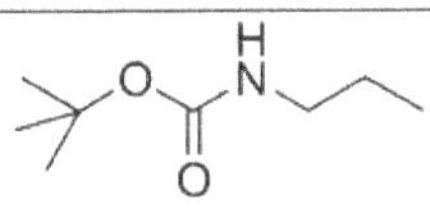	50
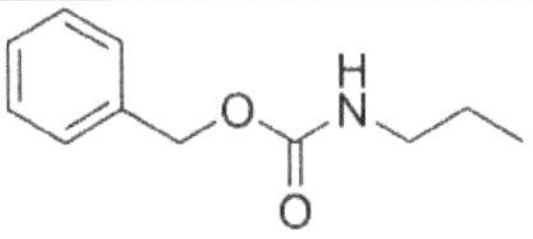	80
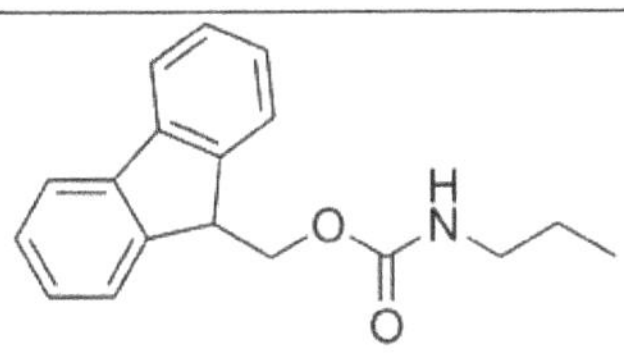	40
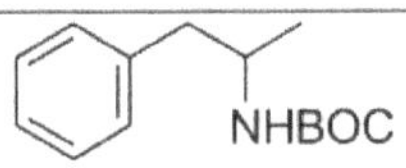	68

Table 2. Examples of substrate structures accepted by CAL-A–protecting group removal.

C. rugosa produces lipase that is used to resolve profens (2-aryl propionic acids) in an enantioselective transesterification process. Profens are a significant category of nonsteroidal anti-inflammatory medicines, and their biological effectiveness relies on the optical purity of the molecules, primarily the (S)-enantiomer [93]. For example, (S)-ibuprofen, also known as (S)-2(4-isobutylphenyl) propionic acid, is 160 times more potent than the (R)-isomer in suppressing the production of prostaglandins. The synthesis of optically pure profens can be achieved through many methods such as asymmetric chemical synthesis, catalytic kinetic resolution, and chiral chromatography [69]. However, enzymatic enantioselective esterification appears to be the most effective approach (Figure 14). The reaction was conducted in saturated cyclohexane using 1-propanol or 2-propanol as acyl agents. The reaction achieved a high conversion rate and produced (S)-ibuprofen with excellent enantiomeric excess [94].

Enantiomerically pure amines are a group of chemicals that have potential biological qualities and can be used in various industrial applications [95]. Candida antarctica lipase B is a highly efficient catalyst for synthesizing nitrogenated chemicals with high enantiomeric purity, such as amines, amides, amino acids, and amino alcohols. This is accomplished through enantioselective acetylation [96]. An experiment was conducted to determine the most successful method for resolving aminoalkylpyridines. It was found that using CAL-B and ethyl acetate as an acyl donor in the tert-butyl methyl ether (TBME-medium) resulted in a 50% conversion rate, a reaction duration of 4 hours, and an enantiomeric excess of the product and substrate greater than 99%. This information is illustrated in Figure 15 of reference [97].

The CAL-A isoform of C. antarctica lipase has the ability to preferentially attach acyl groups to cyclic, sterically hindered structures through a process called kinetic resolution (as shown in Figure 16). Alicyclic β-aminocarboxylic acid esters, which serve as fundamental components for creating diverse medicinal heterocycles of significant importance, are produced using this method [81]. The highest level of activity and selectivity was reported when using diethyl ether or diisopropyl ether as solvents and 2,2,2-trifluoroethyl hexanoate as the acyl donor.

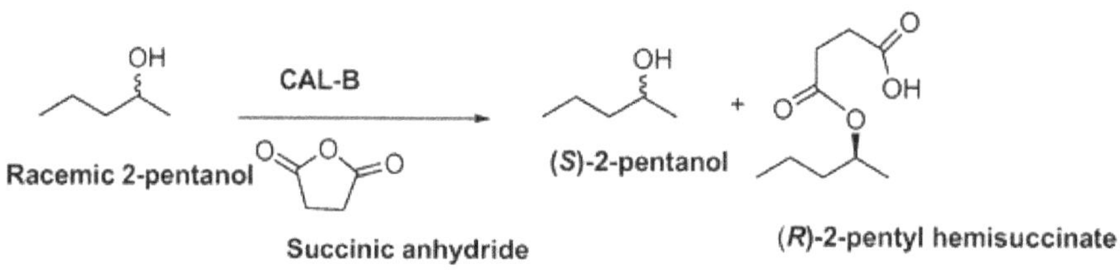

Figure 13. Kinetic resolution of 2-pentanol by esterification.

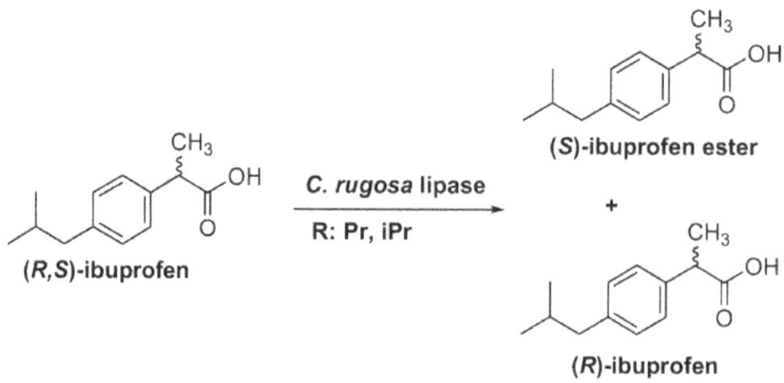

Figure 14. Enantioselective esterification of (R,S)-ibuprofen.

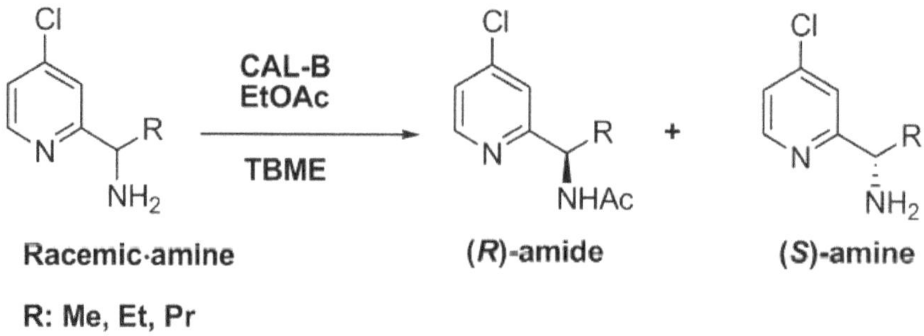

Figure 15. Kinetic resolution chloro-aminopyridines.

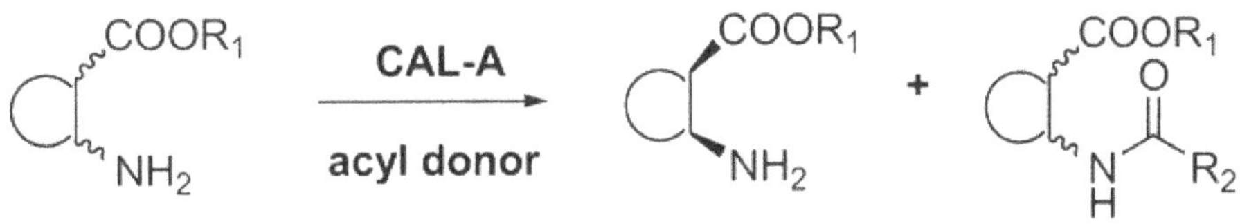

Figure 16. Kinetic resolution of alicyclic amino acids.

CAL-A exhibits reactivity towards sterically hindered tertiary alcohols. This characteristic is highly distinctive among hydrolases. The initial instance of enantioselective kinetic resolution of a racemic mixture of tertiary alcohol involved the acylation of 2-phenylbut-3-yn-2-ol. The reaction exhibited significant enantioselectivity, however, the yield was very modest (25%) due to steric hindrance. Another intriguing use of CAL-A is the targeted acylation of sterols [98], furyl substituted allyl alcohol [99], or cyanohydrins [100, 101].

3.2. Yeast's invertase

Invertase, also known as β-fructofuranosidase (EC 3.2.1.26), facilitates the breakdown of the glycoside bond in disaccharides by targeting the nonreducing beta-fructofuranoside side at the terminal end [102]. Furthermore, it has a broad distribution in the environment, primarily found in plants and microbes. An essential usage of invertase is in the manufacturing of invert syrup, which is a combination of fructose and glucose in equal amounts. This syrup is derived from sucrose (as shown in Figure 17) and is widely employed in the food and beverage sectors. A blend of monosaccharides is both sweeter than sucrose and hygroscopic. It is mostly utilized in the manufacturing of soft-centered sweets and fondants. Invertase is utilized in the production of synthetic honey, as well as in the creation of plasticizing agents for cosmetics, pharmaceuticals, and paper sectors. Additionally, it is employed in the development of enzyme electrodes for the detection of sucrose [103, 104]. Furthermore, it can be utilized for the production of probiotic oligosaccharides such as non-digestible oligosaccharides (NDO), for example, lactosucrose [105]. In the commercial setting, invertase is primarily manufactured using Saccharomyces cerevisiae (Baker's yeast) or Saccharomyces carlsbergensis. In yeast cells, invertase is synthesized in either an intracellular or extracellular form [106].

3.3. Yeast's oxidoreductases

Enantiomerically pure alcohols, such as α- and β-hydroxyesters, are crucial and valuable compounds used as intermediates in the production of medicines and other high-quality chemicals [107]. Enantioselective ketone reductions are widely used for the manufacture of optically pure alcohols. Using whole cells instead of isolated enzymes is favored for reactions catalyzed by dehydrogenases/reductases because it reduces the expenses associated with enzyme purification and cofactor regeneration, as these processes require cofactors such as NADH or NADPH. Nevertheless, the use of isolated dehydrogenases resulted in a reduction of issues related to product purification (Figure 18).

In general, α-ketoesters are reduced by entire yeast cells with lower enantioselectivities. Therefore, a pair of purified reductases is chosen to create both enantiomers of (S)-ethyl-3-hydroxybutyrate, which is a pharmaceutical building block, in optically pure forms on a preparative scale [108]. The reduction of β-ketoesters is influenced by both the structure of the substrate and the specificity of the enzyme, typically resulting in the desired enantiomer with a high level of optical purity [109].

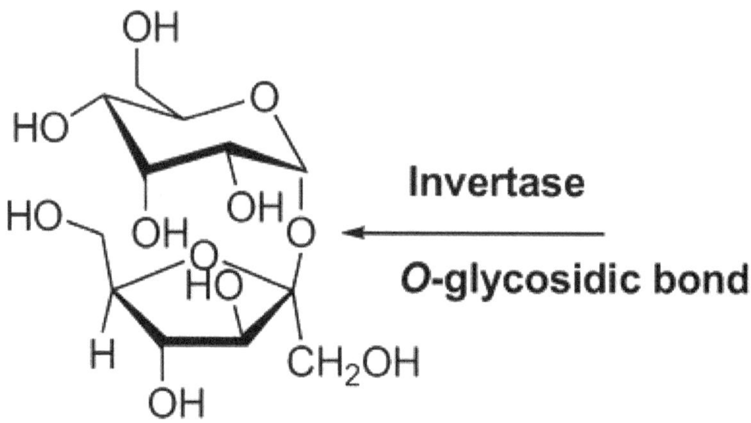

Figure 17. Hydrolysis of sucrose by invertase.

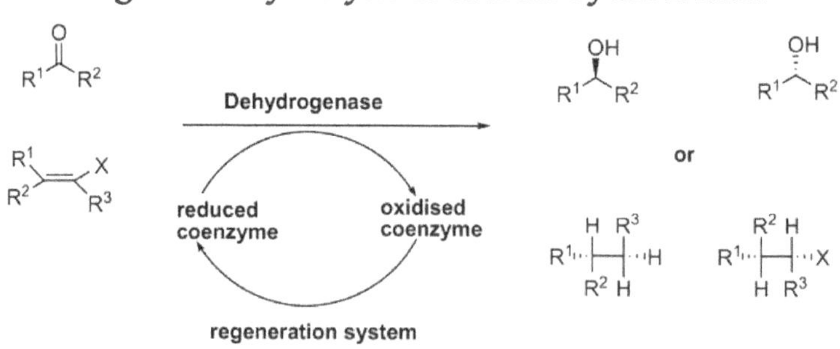

Figure 18. Reduction reactions catalyzed by dehydrogenases. Cofactor regeneration circle.

Saccharomyces, also known as producers of old yellow enzyme (OYE), are the first flavoproteins to be discovered and characterized. OYE can be utilized for double bond reduction (Figure 19), dismutation reactions towards cyclic substrates (Figure 20), and the reduction of nitrate esters (Figure 112) when coenzyme NADPH is added.

Another example of a reductase is the selective carbonyl reductase from Candida magnolia. This enzyme is capable of reducing structurally diverse ketones to produce optically pure (R)-aryl and aliphatic alcohols. This enzyme also catalyzes the reduction of ketones with anti-Prelog enantioselectivity, which is an atypical characteristic of bioreductions (Figure 21) [107]. The configuration of the aryl alcohols obtained is predominantly in the R-enantiomer form, but it is also significantly influenced by the structure of the R group (Figure 21).

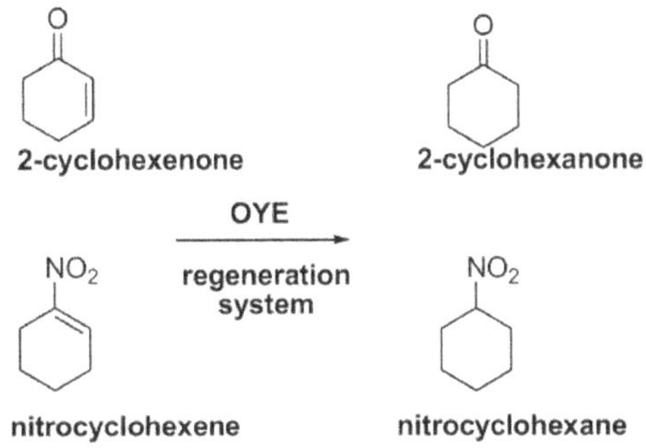

Figure 19. Example of reduction of double bonds catalysed by OYE.

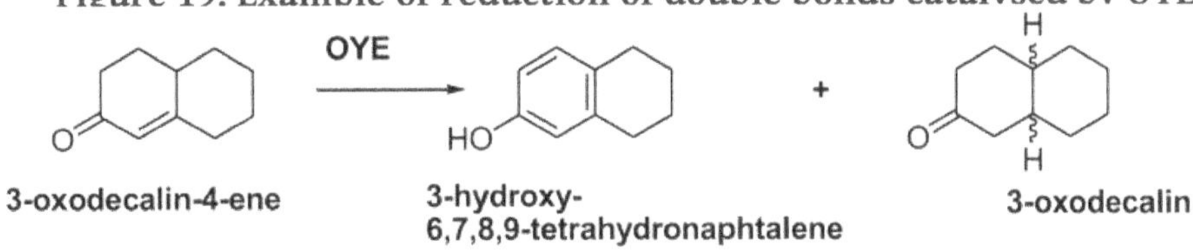

Figure 20. Dismutation reaction calalysed by OYE.

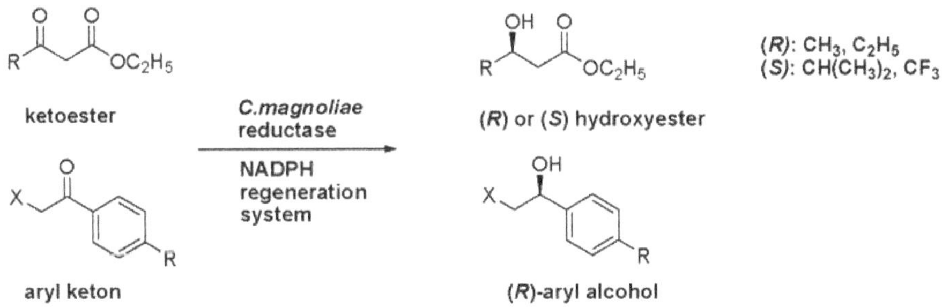

Figure 21. Anti-Prelog reduction of ketones by C. magnoliae reductase.

4. THE APPLICATIONS OF YEAST IN MOLECULAR BIOLOGY

Yeasts belonging to the Saccharomyces genus, specifically S. cerevisiae, are well recognized as essential models for eukaryotic organisms and are frequently employed in genetic and molecular biology research. Saccharomyces cerevisiae is a single-celled organism that may be cultivated on certain media, allowing for precise manipulation of its chemical and physical surroundings. Yeast cultivation is a straightforward, cost-effective, and speedy process that may be carried out in both aerobic and anaerobic environments. Being nonpathogenic and harmless, these organisms are acceptable to use in laboratory settings without the need for any specific precautions. The yeast organism possesses enormous biotechnological capabilities due to its large accessibility and ease of culture on both liquid and solid mediums. This makes yeast a cost-effective and convenient organism to work with.

Despite having evolved independently for 1 billion years, yeast and humans nevertheless share a significant number of genes that show a strong similarity to those found in mammals. Due to the similarity in basic cellular processes such as replication, recombination, cell division, and metabolism between yeast and larger eukaryotes, yeast serves as a suitable model for studying various processes including aging, gene expression regulation, signal transduction, cell cycle, metabolism, apoptosis, neurodegenerative disorders, and more [113]. Moreover, the protein expression systems of this organism share more similarities with higher organisms rather than prokaryotes. This is primarily because of the posttranscriptional and posttranslational processing, which makes it highly suitable for producing various biomolecules that are of industrial or medical importance, such as recombinant proteins used in pharmaceutical applications [114].

The life cycle of Saccharomyces cerevisiae strains consists of two phases: haploid and diploid. Both phases usually undergo asexual reproduction through budding. The cell cycle comprises four distinct stages (G1, S, G2, and M) and is regulated in a comparable manner to that of the cell cycle in bigger eukaryotes [115]. Yeast cells in their haploid state can exist as either mating type an or α. Under typical circumstances, these cells can engage in mating to produce a/α diploid cells. The diploid cells are incapable of mating, but they can undergo asexual reproduction through a process called budding, similar to haploid cells. Nevertheless, in certain situations, such as when the environment is unfavorable due to a shortage of nutrients, a diploid cell has the ability to perform meiosis, resulting in the production of haploid spores. Afterwards, the recently formed haploid nuclei are enclosed in four spores with altered cell walls, creating structures that are highly resilient to external pressures [116]. Each individual haploid spore, formed as a result of meiosis and grouped in a tetrad, can be separated and examined using different micromanipulation techniques. It offers a distinct opportunity to examine the interconnection between genes, among various other possibilities. Haploid cells can be utilized for studying recessive mutations, whilst diploid strains can be employed for conducting complementation assays.

Saccharomyces cerevisiae was the first eukaryote to have its genome entirely sequenced and published in 1996. The nucleus genome consists of 12,068 kilobases (kb) and is structured into a haploid set of 16 chromosomes, which range in size from 200 to 1600 kb. An distinguishing property is that the yeast genome is significantly more condensed compared to other eukaryotic organisms, with genes accounting for 70% of the entire sequence. The organism contains roughly 5885 genes that have the capacity to encode proteins. It also has around 140 genes that specify ribosomal RNA, 40 genes for small nuclear RNA molecules, and 275 genes for transfer RNA. An worldwide, multidisciplinary team is currently using synthetic biology technologies to produce 16 chromosomes of S. cerevisiae. The team anticipates obtaining the results by the end of the year [117]. Another distinctive and atypical characteristic of the S. cerevisiae genome, specifically for eukaryotes, is the existence of DNA plasmids. These plasmids play a crucial role in various genetic manipulations and hold significant value in contemporary molecular biology [118].

The methods employed for yeast transformation and targeted selection have been thoroughly documented. Shuttle vectors are frequently employed for this purpose because they have the ability to transform both yeast and bacteria, including Escherichia coli. Different yeast strains possess distinct auxotrophic indicators that can be created using genetic engineering techniques, such as deleting genes involved in amino acid production pathways. Researchers have created a variety of bifunctional vectors that can be easily separated and have the ability to independently reproduce in both yeast and bacterium cells.

4.1. Yeast's plasmid vectors

Saccharomyces cerevisiae is a highly significant microorganism in contemporary biotechnology, not only for its role in the brewing and bread-making industry, but also for its remarkable potential in the domains of molecular biology and biomedicine, owing to its distinctive genetic material and protein expression systems. S. cerevisiae is one of the rare eukaryotic organisms that possess circular DNA in the form of plasmids. The majority of yeast strains possess the 2-μm plasmid, which serves as an excellent foundation for cloning vectors (Figure 22). This plasmid is 6 kb in size and contains four essential elements: an origin of replication, genes REP1 and REP2 that encode proteins involved in the replication process, the FLP gene which allows the plasmid to switch between isoforms, and the function of gene D is still unknown [119]. To ensure the 2-μm plasmid functions as an effective cloning vector, it is necessary to include a selected marker element. This element enables the identification of transformed cells in the cloning experiment.

The majority of bacterial vectors possess genes that encode resistance to different types of antibiotics, such as ampicillin (ampR) or tetracycline (tetR). Consequently, when transformed cells were cultured in a medium containing the antibiotic, the colonies that were incapable of growing did not possess the resistance gene. This indicates that the vector was not taken up by these colonies. Therefore, the only cells that were able to survive were the transformed ones that carried the vector correctly inserted.

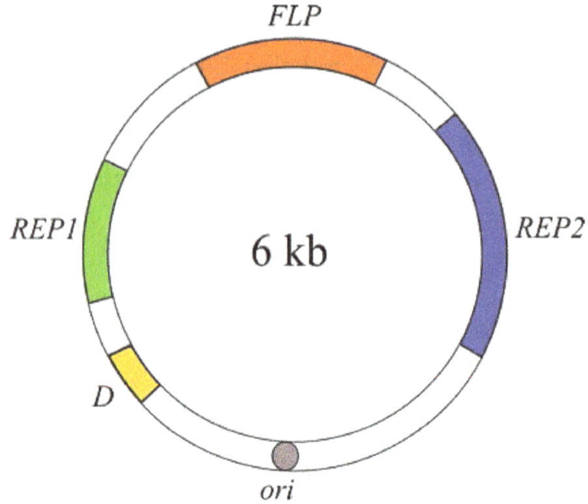

Figure 22. Genetic structure of 2 µm yeast plasmid.

The variations in cloning techniques using yeast primarily involve the strategic utilization of selected markers. Typically, a certain type of organism is needed as the host in this situation. Specifically, an auxotrophic mutant is required, which is unable to acquire or produce a crucial component necessary for one of its metabolic pathways. An illustrative instance is leu2- yeast, which possesses a non-functional variant of the LEU2 gene, rendering it incapable of producing leucine. Consequently, it can only thrive in a medium supplemented with this particular amino acid. In order to effectively utilize that organism as the host, it is necessary to build a vector containing an active LEU2 gene. Subsequently, the cells are genetically modified using the plasmid and cultivated in a nutrient-deficient medium devoid of leucine. Only the altered cells [120] will proliferate in this manner.

There exist a limited number of yeast cloning vectors, all of which are referred to as shuttle vectors. This designation indicates that they have the ability to reproduce and be chosen in both bacteria and yeast. Shuttle vectors were primarily created due to the inefficiency of plasmid manufacture solely from yeast. As a result, large-scale DNA propagation and convenient genetic manipulation are carried out in bacterial organisms, specifically Escherichia coli.

Yeast cloning vectors that rely on the 2-µm plasmid are referred known as yeast episomal plasmids (YEps) (Figure 23(a)). YEp can vary in their composition, either containing the entire 2-µm plasmid or only the origin of replication. Typically, their structure is based on E. coli vectors, such as pBR322, which include genes that confer resistance to antibiotics such as ampicillin or tetracycline. YEps have the ability to reproduce autonomously or be incorporated into a yeast chromosome. YEps integration is primarily driven by the presence of selective marker genes that closely resemble the mutant chromosomal DNA of the host organism. Examples of such genes include LEU2, URA3, HIS3, TRP1, and LYS2, which are all involved in the biosynthesis pathways of pyrimidine nucleotides, histidine, tryptophan, or lysine, respectively. Yeast episomes (YEps) are regarded as vectors with a high number of copies, producing up to 200 copies per cell. The transformation frequency is estimated to be between 10,000 and 100,000 transformed cells per microgram. Regrettably, the primary disadvantage is that the recombinants produced by YEps exhibit a high degree of instability, hence posing significant challenges and requiring much time in obtaining definitive and dependable outcomes.

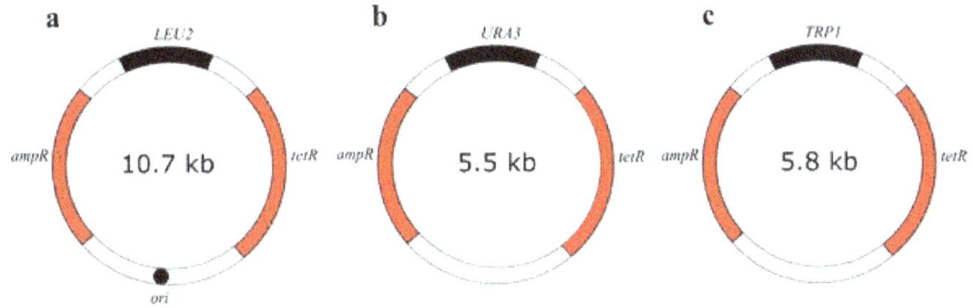

Figure 23. Genetic structure of exemplary yeast vectors. AmpR and tetR are the antibiotic resinstance genes from E. coli pBR322 plasmid, whereas LEU2, URA3, TRP1 fragments represent yeast chromosomal DNA: (a) yeast episomal plasmid YEp13, (b) yeast integrative plasmid YIp5, (c) yeast replicative /plasmid YRp7.

Yeast integrative plasmids (YIps) are another significant category of yeast vectors, as seen in Figure 23(b). They mostly comprise E. coli plasmids, such as pBR322, along with a selection marker (often one of those listed before). The crucial aspect is that they lack yeast origin of replication, therefore rendering them incapable of replicating by any means other than integration with chromosomal DNA. YIps have the lowest transformation frequency, with significantly fewer than 1000 transformed cells per µg and typically only one copy per cell. However, their recombinants exhibit high stability, making them the preferred choice for experimental reasons.

Yeast replicative plasmids (YRps), shown in Figure 23(c), are a distinct category of cloning vectors used in yeast. Their composition includes a DNA sequence derived from the E. coli vector, with the yeast origin of replication located near the selection marker. The structure indicates that YRps have the ability to multiply on their own with a rather high frequency of transformation, ranging from 1000 to 10,000 transformed cells per µg. Additionally, each cell can have between 5 and 100 copies of YRps. YRps recombinants exhibit similar instability as YEps recombinants, rendering them a less preferable option for laboratory work.

As time progressed, there was an increasing need for larger DNA fragments to be modified via genetic engineering techniques. The yeast artificial chromosomes (YACs) were produced at this stage as the final type of yeast cloning vectors.

The underlying concept of these constructs was to mimic the natural DNA by incorporating several hundred kilobases of genetic information, which is typically found in yeast chromosomes. Yeast Artificial Chromosomes (YACs) were believed to possess the three essential components of a chromosome:

- Centromeres required for the proper chromosome positioning during the cell division;
- Origins of replication, which are the places on chromosome where the replication of genetic material starts;
- Telomers as the defenders of the chromosomes against exonucleases.

Over the years, various types of YACs have been created. Typically, they are composed primarily of an E. coli pBR322 plasmid containing certain yeast genes. These genes include selective markers (usually at least two located in opposite directions), an origin of replication, the gene CEN4 which codes for the centromere region, two telomere fragments TEL, and an extra selective marker with a restriction enzyme site. One example of such a marker is the SUP4 gene, which counteracts a mutation that leads to the buildup of a red pigment (Figure 24).

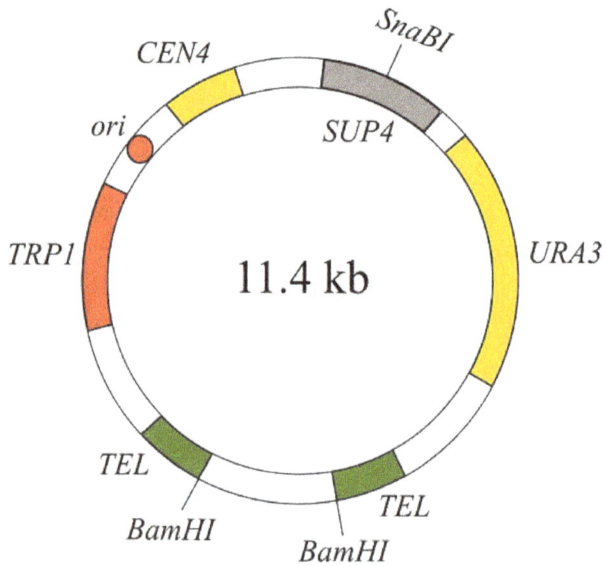

Figure 24. Genetic structure of yeast artificial chromosome pYAC3.

Yeast Artificial Chromosomes (YACs) contain a restriction enzyme site located between two TEL sites. When this site is cleaved, it generates a "artificial" linear chromosome. This chromosome is then further divided by cutting the SUP4 gene, resulting in two arms. The DNA fragment to be cloned is then inserted between these arms and ligated, effectively restoring the original single-line chromosome structure. The subsequent procedure involves introducing double auxotrophic mutant organisms that are incapable of surviving in the minimal medium unless they acquire the appropriate YAC. Further experimental control can be assessed using straightforward optical examination. Colonies exhibiting a white coloration indicate successful transformation due to disruption of the SUP4 gene, while any other color indicates a lack of correct transformation [121].

4.2. Yeast expression system

Recombinant proteins hold significant importance as biomolecules due to their ability to imitate the functions of natural proteins. As a result, they are extensively researched in the fields of biotechnology and biopharmaceuticals. The key factor in achieving successful synthesis of the target protein is selecting an effective expression system that allows for the generation of a functioning product with a high yield. The yeast expression system is a viable alternative to commonly used bacterial and higher eukaryote expression systems. The organisms possess a clear genetic makeup and are capable of carrying out various posttranslational changes, such as protein folding, disulfide bond formation, and glycosylation [122]. The cultivation of yeast is also simple, fast, and cost-effective, giving them a significant edge over insect or mammalian cells. Yeasts are very amenable to genetic manipulation and can readily adapt to fermentation processes. This makes them a suitable choice for use as a cell factory, allowing for the production of a significant quantity of target protein. Unlike bacteria, yeast expression systems produce recombinant proteins that are devoid of endotoxins, making this system safer for medicinal and food applications [114]. Approximately 20% of all biopharmaceuticals are manufactured using S. cerevisiae. The most prevalent substances include insulin, human serum albumin, hepatitis vaccines, and virus-like particles utilized for immunization against human papillomavirus [123].

Yeast cells are unable to create complex N-linked glycans, which limits their ability to produce glycoproteins that are similar to those found in humans. Furthermore, S. cerevisiae is capable of synthesizing hypermannosylated N-linked glycans, where the mannose residues are linked to the chitobiose core, which is a dimer of β-1,4-linked glucosamine units. Hypermannosylation leads to a reduced lifespan in living organisms, which in turn undermines the effectiveness of several therapeutic glycoproteins [124]. In order to address this problem, a significant amount of work has been dedicated to modifying the glycosylation pathways in P. pastorsis. The goal is to create strains that exhibit N-glycosylation patterns similar to those found in humans [125, 126]. This accomplishment has enhanced the utility of yeast in the industrial manufacture of stable and recombinant glycoproteins. There is no universal protocol for yeast expression systems that can be equally effective in producing all types of proteins. The optimization of the entire process is a crucial step in order to achieve adequate quantities of pure, correctly folded, and secreted protein of interest. Although small and simple proteins are readily available, larger and multi-domain proteins may require specific chaperones to assist in the folding process [127]. An advantage of the yeast expression system is that it enables the release of generated proteins outside the cell when a suitable signaling sequence is connected to the structural gene [114]. It greatly simplifies the process of purifying recombinant proteins from the culture medium and enables the optimization of culturing conditions. Several ways have been devised to enhance the level of protein secretion. One approach is protein engineering to produce a desired product. This can be achieved by changing the coding regions and signaling sequences of proteins [128–130]. Due to the extreme specificity of this technology towards each protein, the optimum settings for one protein may not be effective for another. An alternative method involves modifying the host strains and optimizing the folding and secretory machinery by increasing or removing genes that play a crucial role in protein secretion [131, 132]. Furthermore, research has demonstrated that the expression of genes is increased in colder temperatures, resulting in a higher degree of secretion [133].

There is a wide range of expression vectors that can be used to produce foreign proteins in yeast. These vectors are derived from YIp, YEp, and YRp plasmids, which have been previously characterized. The DNA encoding the desired protein is introduced into the vector. The choice of selective marker and the strength of the promoter are crucial elements that dictate the plasmid copy number and the quantity of mRNA expression of the recombinant protein. In the past, different types of promotors, both inducible and constitutive, have been utilized to regulate gene expression in yeasts. The first of them enables the regulation of gene expression. The majority of inducible promoters are susceptible to catabolite repression or are influenced by other environmental factors, such as intracellular iron levels, stress, or deficiency of critical amino acids. The GAL promoter, which is activated by the addition of galactose, offers a simple mechanism for controlling the expression of the foreign gene that has been cloned [134]. Another notable example is the CUP1 promoter, which is activated in response to copper [135], or the heat shock factor promoter, which is activated in response to heat stress at 39°C [136]. Additionally, there exist other categories of promoters that induce robust and constant expression. The TEF1 promoter, exemplifying S. cerevisiae, is an often employed representative of this category due to its ability to facilitate robust gene expression in both high and low glucose environments [137]. The choice of an appropriate promoter is contingent upon the unique process demands and the characteristics of the desired protein for production.

Furthermore, yeasts are acknowledged as a generally recognized as safe (GRAS) organism, thus solidifying its status as the most commonly employed microbial eukaryote for the production of recombinant proteins.

4.3. Yeast's two hybrid system

Following the Field and Song discovery in 1989, a novel method for studying protein-protein interactions was introduced, known as the yeast two-hybrid system. This technique enables the identification of protein interactions within yeast cells and can be utilized to identify a partner protein that interacts with a known protein. This approach exploits the fact that most eukaryotic transcription factors, like Gal4p, have two separate and functional DNA domains: a binding domain (BD) and a transcription activation domain (AD). Although the two domains are often found on the same polypeptide chain, the transcription factor can also operate when these two domains are brought together through noncovalent protein-protein interactions. In the yeast hybrid system, each of these domains is linked to the corresponding domain from the protein under study. Consequently, the process yields two fusion proteins: one that is linked to the DNA-binding domain (BD) and the other that is attached to the activation domain (AD) [138]. The genes encoding both fusion proteins are present on separate plasmids, however each plasmid is expressed within the same yeast cell. If there is a connection between the proteins being studied, the BD and AD domains are in close proximity, which leads to the activation of transcription for a reporter gene that is controlled by the transcription factors.

The yeast two-hybrid system involves the interaction between transcriptional factors and a gene coding for β-galactosidase. This interaction has the potential to result in the expression of the reporter gene, which codes for β-galactosidase, in the E. coli lacZ reporter gene. The existence of this enzyme can be subsequently confirmed by a straightforward reaction with X-gal, resulting in the formation of a blueish insoluble product. This reaction serves to either validate or refute the connection of the proteins under investigation. The yeast two-hybrid system has undergone substantial research and development since 1989, resulting in numerous new applications. A summary of these applications may be found in reference [139].

5. CONCISE SUMMARY

The distinctive characteristics of yeasts, which are crucial for their diverse applications, are currently being studied. Once the "Synthetic Yeast Genome Project (Sc2.0)" is completed, new possibilities for utilizing these yeasts will emerge, both in the field of molecular biology and in industrial applications.

AT A GLANCE

Yeasts are a highly diverse collection of microorganisms, and even strains that are categorized as the same species can exhibit significant genetic variation. The biodiversity of yeasts is intimately linked to their suitability for various applications. The biotechnological significance of yeast is nearly incalculable. Throughout the millennia, individuals have used the enzymatic capacity of this substance to create fermented food items such as bread and alcoholic beverages. Undoubtedly, the use of yeasts was initially based on instinct, but with the advancement of science and technology, these microorganisms became the subject of extensive scientific research. Yeast is a highly valuable scientific model due to its eukaryotic nature and extensive understanding of yeast cell genetics and metabolism. The elucidation of the genome of the baker's yeast Saccharomyces cerevisiae in 1996 provided an opportunity for the comprehensive study of the expression and functioning of the eukaryotic genome. Currently, a multinational team is utilizing synthetic biology technologies to synthesize the 16 yeast chromosomes. The team anticipates obtaining the results by the end of the year. In modern times, yeast is considered a versatile instrument for biotechnological applications.

REFERENCES

1. Otterstedt K, Larsson C, Bill RM, Stahlberg A, Boles E, Hohmann S, Gustafsson L. Switching the mode of metabolism in the yeast Saccharomyces cerevisiae. EMBO reports. 2004;5:532-537. DOI: 10.1038/sj.embor.7400132
2. Barnett JA, Entian K-D. A history of research on yeasts 9: Regulation of sugar metabolism. Yeast. 2005;22:835-894. DOI: 10.1002/yea.1249
3. Steensels J, Snoek T, Meersman E, Nicolino MP, Voordeckers K, Verstrepen KJ. Improving industrial yeast strains: Exploiting natural and artificial diversity. FEMS Microbiology Reviews. 2014;38:947-995. DOI: 10.1111/1574-6976.12073
4. Hahn-Hägerdal B, Karhumaa K, Larsson CU, Gorwa-Grauslund M, Görgens J, van Zyl WH. Role of cultivation media in the development of yeast strains for large scale industrial use. Microbial Cell Factories. 2005;4(31):1-16. DOI: 10.1186/1475-2859-4-31
5. Krogerus K, Magalhães F, Vidgren V, Gibson B. Novel brewing yeast hybrids: Creation and application. Applied Microbiology and Biotechnology. 2017;101:65-78. DOI: 10.1007/s00253-016-8007-5
6. Piskur J, Rozpedowska E, Polakova S, Merico A, Compagno C. How did Saccharomyces evolve to become a good brewer? Trends in Genetics. 2006;22(4):183-186. DOI: 10.1016/j.tig.2006.02.002
7. Berlowska J, Kregiel D, Rajkowska K. Biodiversity of brewery yeast strains and their fermentative activities. Yeast. 2015;32:289-300. DOI: 10.1002/yea.3041
8. Bond U. Chapter 6: The genomes of lager yeasts. Advances in Applied Microbiology. 2009;69:159-182. DOI: 10.1016/S0065-2164(09)69006-7
9. Soares EV. Flocculation in Saccharomyces cerevisiae: A review. Journal of Applied Micro-biology. 2010;110:1-18. DOI: 10.1111/j.1365-2672.2010.04897.x
10. Verstrepen KJ, Derdelinckx G, Verachtert H, Delvaux FR. Yeast flocculation: What brewers should know. Applied Microbiology and Biotechnology. 2003;61:197-205. DOI: 10.1007/s00253-002-1200-8
11. Pretorius IS, Van der Westhuizen TJ, Augustyn OPH. Yeast biodiversity in vineyards and wineries and its importance to the South African wine industry. South African Journal of Enology and Viticulture. 1999;20:61-74
12. Liu Y, Rousseaux S, Tourdot-Marechal R, Sadoudi M, Gougeon R, Schmitt-Kopplin P, Alexandre H. Wine microbiome: A dynamic world of microbial interactions. Critical Reviews in Food Science and Nutrition. 2017;57(4):56-873. DOI: 10.1080/10408398.2014.983591

13. Cordente AG, Curtin CD, Varela C, Pretorius IS. Flavor-active wine yeasts. Applied Microbiology an Biotechnology. 2012;96:601-618. DOI: 10.1007/s00253-012-4370-z

14. Styger G, Prior B, Bauer FF. Wine flavor and aroma. Journal of Industrial Microbiology and Biotechnology. 2011;38:1145-1159. DOI: 10.1007/s10295-011-1018-4

15. Degre AR. Selection and commercial cultivation of wine yeast and bacteria. In: Fleet GH, editor. Wine Microbiology and Biotechnology. Reading: Harwood Academic; 1993. pp. 421-447

16. Ciani M, Capece A, Comitini F, Canonico L, Siesto G, Romano P. Yeast interactions in inoculated wine fermentation. Frontiers in Microbiology. 2016;7(555):1-7. DOI: 10.3389/fmicb.2016.00555

17. Perez-Torrado R, Querol A, Guillamon JM. Genetic improvement of non-GMO wine yeasts: Strategies, advantages and safety. Trends in Food Science & Technology. 2015;45:1-11. DOI: 10.1016/j.tifs.2015.05.002

18. Mortimer RK, Romano P, Suzzi G, Polsinelli M. Genome renewal: A new phenomenon revealed from a genetic study of 43 strains of Saccharomyces cerevisiae derived from natural fermentation of grape musts. Yeast. 1994;10(12):1543-1552. DOI: 10.1002/yea.320101203

19. Magwene PM. Revisiting Mortimer's Genome renewal hypothesis: Heterozygosity, homothallism, and the potential for adaptation in yeast. Advances in Experimental Medicine and Biology. 2014;781:37-48. DOI: 10.1007/978-94-007-7347-9_3

20. Magwene PM, Kaykç Ö, Granek JA, Reininga JM, Scholl Z, Murray D. Outcrossing, mitotic recombination, and life-history trade-offs shape genome evolution in Saccharomyces cerevisiae. Proceedings of the National Academy of Sciences of the USA. 2011;108(5):1987-1992. DOI: 10.1073/pnas.1012544108

21. Borneman AR, Desany BA, Riches D, Affourtit JP, Forgan AH, Pretorius IS, Egholm M, Chambers PJ. Whole-genome comparison reveals novel genetic elements that characterize the genome of industrial strains of Saccharomyces cerevisiae. PLoS Genetics. 2011;7(2):e1001287: 1-10. DOI: 10.1371/journal.pgen.1001287

22. Moreno-Arribas MV, Polo MC. Winemaking biochemistry and microbiology: Current knowledge and future trends. Critical Reviews in Food Science and Nutrition. 2005;45:65-286. DOI: 10.1080/10408690490478118

23. Eldarov MA, Kishkovskaia SA, Tanaschuk TN, Mardanov AV. Genomics and biochemistry of Saccharomyces cerevisiae wine yeast strains. Biochemistry (Moscow). 2016;81(13):1650-1668. DOI: 10.1134/S0006297916130046

24. Hong Y-A, Park H-D. Role of non-Saccharomyces yeasts in Korean wines produced from campbell early grapes: Potential use of Hanseniaspora uvarum as a starter culture. Food Microbiology. 2013;34:207-214. DOI: 10.1016/j.fm.2012.12.011

25. Wang C, Mas A, Esteve-Zarzoso B. Interaction between Hanseniaspora uvarum and Saccharomyces cerevisiae during alcoholic fermentation. International Journal of Food Microbiology. 2015;206:67-74. DOI: 10.1016/j.ijfoodmicro.2015.04.022

26. Steensels J, Daenen L, Malcorps P, Derdelinckx G, Verachtert H, Verstrepen KJ, Kevin J. Brettanomyces yeasts—From spoilage organisms to valuable contributors to industrial fermentations. International Journal of Food Microbiology. 2015;206:24-38. DOI: 10.1016/j.ijfoodmicro.2015.04.005

27. Padilla B, Gil JV, Manzanares P. Past and future of non-Saccharomyces yeasts: From spoilage microorganisms to biotechnological tools for improving wine aroma complexity. Frontiers in Microbiology. 2016;7(411):1-20. DOI: 10.3389/fmicb.2016.00411

28. Masneuf-Pomarede I, Bely M, Marullo P, Albertin W. The genetics of non-conventional wine yeasts: Current knowledge and future challenges. Frontiers in Microbiology. 2016;6(1563):1-15. DOI: 10.3389/fmicb.2015.01563

29. Pretorius IS. Tailoring wine yeast for the new millennium: Novel approaches to the ancient art of winemaking. Yeast. 2000;16:675-729. DOI: 10.1002/1097-0061(20000615)16:8<675::AID-YEA585>3.0.CO;2-B

30. Walker GM, Hill AE. Saccharomyces cerevisiae in the production of whisk(e)y. Beverages. 2016;2(38):1-15. DOI: 10.3390/beverages2040038

31. Kim HR, Im Y-M, Ko H-M, Chin J-E, Kim I-C, Lee HB, Bai S. Raw starch fermentation to ethanol by an industrial distiller's yeast strain of Saccharomyces cerevisiae expressing glucoamylase and α-amylase genes. Biotechnology Letters. 2011;33:1643-1648. DOI: 10.1007/s10529-011-0613-9

32. Stewart GG, Hill AE, Russell I. 125th anniversary review: Developments in brewing and distilling yeast strains. Journal of the Institute of Brewing. 2013;119(4):202-220. DOI: 10.1002/jib.104

33. Cho IH, Peterson DG. Chemistry of bread aroma: A review. Food Science and Bio-technology. 2010;19(3):575-582. DOI: 10.1007/s10068-010-0081-3

34. Perez-Torrado R, Gamero E, Gomez-Pastor R, Garre E, Aranda A, Matallana E. Yeast biomass, an optimized product with myriad applications in the food industry. Trends in Food Science & Technology. 2015;46:167-175. DOI: 10.1016/j.tifs.2015.10.008

35. Gelinas P. Aeration and foam control in Baker's yeast production: Mapping patents. Comprehensive Reviews in Food Science and Food Safety. 2015;15:371-391. DOI: 10.1111/1541-4337.12188

36. Attfield PV. Stress tolerance: The key to effective strains of industrial baker's yeast. Nature Biotechnology. 1997;15:1351-1357. DOI: 10.1038/nbt1297-1351

37. Shima J, Takagi H. Stress-tolerance of baker's-yeast (Saccharomyces cerevisiae) cells: Stress-protective molecules and genes involved in stress tolerance. Biotechnology and Applied Biochemistry. 2009;53:155-164. DOI: 10.1042/BA20090029

38. Suman G, Nupur M, Anuradha S, Pradeep B. Single cell protein production: A Review. International Journal of Current Microbiology and Applied Sciences. 2015;4(9):251-262

39. Patelski P, Berlowska J, Dziugan P, Pielech-Przybylska K, Balcerek M, Dziekonska U, Kalinowska H. Utilisation of sugar beet bagasse for the biosynthesis of yeast SCP. Journal of Food Engineering. 2015;167:32-37.DOI: 10.1016/j.jfoodeng.2015.03.031

40. Nasseri AT, Rasoul-Amini S, Morowvat MH, Ghasemi Y. Single cell protein: Production and process. American Journal of Food Technology. 2011;6:103-116. DOI: 10.3923/ajft.2011.103.116

41. Chi Z, Yan K, Gao L, Li J, Wang X,Wang L. Diversity of marine yeasts with high protein content and evaluation of their nutritive compositions. Journal of the Marine Biological Association of the United Kingdom. 2008;88(7):1347-1352. DOI: 10.1017/S0025315408001938

42. Sarlin PJ, Philip R. Marine yeasts as feed supplement for Indian white prawn Fenneropenaeus indicus: Screening and Testing the Efficacy. International Journal of Current Microbiology and Applied Sciences. 2016;5(1):55-70. DOI: 10.20546/ijcmas.2016.501.005

43. Anupamaa, Ravindra P. Value-added food: Single cell protein. Biotechnology Advances. 2000;18:459-479. DOI: 10.1016/S0734-9750(00)00045-8

44. Liang MH, Jiang JG. Advancing oleaginous microorganisms to produce lipid via metabolic engineering technology. Progress in Lipid Research. 2013;52:395-408. DOI: 10.1016/j.plipres.2013.05.002

45. Probst KV, Schulte LR, Durrett TP, Rezac ME, Vadlan PV. Oleaginous yeast: A value-added platform for renewable oils. Critical Reviews in Biotechnology, 2016;36(5):942-955. DOI: 10.3109/07388551.2015.1064855

46. Papanikolaou S, Chevalot I, Komaitis M, Aggelis G, Marc I. Kinetic profile of the cellular lipid composition in an oleaginous Yarrowia lipolytica capable of producing a cocoa-butter substitute from industrial fats. Antonie van Leeuwenhoek. 2001;80:215-224. DOI: 10.1023/A:1013083211405

47. Beopoulos A, Nicaud JM, Gaillardin C. An overview of lipid metabolism in yeasts and its impact on biotechnological processes. Applied Microbiology and Biotechnology. 2011;90:1193-1206. DOI: 10.1007/s00253-011-3212-8

48. Uemura H. Synthesis and production of unsaturated and polyunsaturated fatty acids in yeast: Current state and perspectives. Applied Microbiology and Biotechnology. 2012;95(1):1-12. DOI: 10.1007/s00253-012-4105-1

49. Papanikolaou S, Aggelis G. Lipids of oleaginous yeasts. Part I: Biochemistry of single cell oil production. European Journal of Lipid Science and Technology. 2011;113(8):1031-1051. DOI: 10.1002/ejlt.201100014

50. Papanikolaou S, Aggelis G. Lipids of oleaginous yeasts. Part II: Technology and potential applications. European Journal of Lipid Science and Technology. 2011;113:1052-1073. DOI: 10.1002/ejlt.201100015

51. Leiva-Candia DE, Pinzi S, Redel-Macías MD, Koutinas A, Webb C, Dorado MP. The potential for agro-industrial waste utilization using oleaginous yeast for the production of biodiesel. Fuel. 2014;123:33-42. DOI: 10.1016/j.fuel.2014.01.054

52. Ageitos JM, Vallejo JA, Veiga-Crespo P, Villa TG. Oily yeasts as oleaginous cell factories. Applied Microbiology and Biotechnology. 2011;90:1219-1227. DOI: 10.1007/s00253-011-3200-z

53. Groenewald M, Boekhout T, Neuveglise C, Gaillardin C, van Dijck PWM, Wyss M. Yarrowia lipolytica: Safety assessment of an oleaginous yeast with a great industrial potential. Critical Reviews in Microbiology. 2014;40(3):187-206. DOI: 10.3109/1040841X.2013.770386

54. Ratledge C. Fatty acid biosynthesis in microorganisms being used for single cell oil production. Biochimie. 2004;86:807-815. DOI: 10.1016/j.biochi.2004.09.017

55. Koutinas AA, Chatzifragkou A, Kopsahelis N, Papanikolaou S, Kookos IK. Design and techno-economic evaluation of microbial oil production as a renewable resource for biodiesel and oleochemical production. Fuel. 2014;116:566-577. DOI: 10.1016/j.fuel.2013.08.045

56. Huang C, Chen X, Xiong L, Chen X, Ma L, Chen Y. Single cell oil production from low-cost substrates: The possibility and potential of its industrialization. Biotechnology Advances. 2013;31:129-139. DOI: 10.1016/j.biotechadv.2012.08.010

57. Goldberg K, Schroer K, Lutz S, Liese A. Biocatalytic ketone reduction-a powerful tool for the production of chiral alcohols-part II: Whole cell reductions. Applied Microbiology and Biotechnology. 2007;76: 249-255. DOI: 10.1007/s00253-007-1005-x

58. Ye Q, Ouyang P, Ying H. A review-biosynthesis of optically pure ethyl (S)-4-chloro-3-hydroxybutanoatw ester: Recent advances and future perspectives. Applied Microbiology and Biotechnology. 2011;89:513-522. DOI: 10.1007/s00253-010-2942-3

59. Ferraboschi P, Ciceri S, Ciuffreda O, De Mieri M, Romano D, Grisenti P. Baker's yeast catalysed preparation of a new enantiomerically pure synthon of (S)-pramipexole and its enantiomer (dexpramipexole). Tetrahedron: Asymmetry. 2014;25:1239-1245. DOI: 10.1016/j.tetasy.2014.07.011

60. Arifin A, Don M, Uzir M. Baker's yeast mediated biotransformation geraniol into citronellol using a continuous-closed-glass-loop bioreactor (CCGLB) system. Bichemical Engineering Journal. 2011;56:219-224. DOI: 10.1016/j.bej.2011.07.001

61. Mane A, Lohar T, Salunkhe R. Baker's yeast as an efficient biocatalyst for regioselective 1,4-conjugate addition of indoles to nitroolefines in aqueous medium. Tetrahedron Letters. 2016;57:2341-2346. DOI: 10.1016/j.tetlet.2016.04.057

62. Straathof AJJ, Panke S, Schmid A. The production of fine chemicals by biotransformations. Current Opinion in Biotechnology. 2002;13:548-556. http://dx.doi.org/10.1016/S0958-1669(02)00360-9

63. Forti L, Di Mauro S, Cramarossa MR, Filippucci S, Turchetti B, Buzzini P. Non-conventional yeasts whole cells as efficient biocatalysts for the production of flavors and fragrances. Molecules. 2015;20:10377-10398. DOI: 10.3390/molecules200610377

64. Schaefer CA, Silva VD, Stambuk BU, Nascimento M. Use of Saccharomyces cerevisiae yeasts in the chemoselective bioreduction of (1E,4E)-1,5-Bis(4-Methoxyphenyl)-1,4 pentadien-3-one in biphasic system. Journal of the Brazilian Chemical Society. 2103;24:1116-1122. http://dx.doi.org/10.5935/0103-5053.20130140

65. Katz M, Frejd T, Hahn-Hagerdal B, Gorwa-Grauslund MF. Efficient anaerobic whole cell stereoselective bioreduction with recombinant Saccharomyces cerevisiae. Biotechnology and Bioengineering. 2003;84(5):573-582. DOI: 10.1002/bit.10824

66. Hasan F, Shah AA, Hameed A. Methods for detection and characterization of lipases: A comprehensive review. Biotechnology Advanced. 2009;27(6):782-798. http://dx.doi.org/10.1016/j.biotechadv.2009.06.001

67. Vakhlu J, Kour A. Yeast lipases: Enzyme purifications, biochemical properties and gene cloning. Electronic Journal of Biotechnology. 2006;9:69-81. DOI: 10.2225/vol9-issue1-fulltext-9

68. Kazlauskas RJ, Bornscheuer UT. Biotransformations with lipases. In: Rehm J, Reed G, Puchler A, Stadler P, editors. Biotechnology. Weinheim: Wiley-VCH; 1998. pp. 37-191

69. Sharma S, Kanwar SS. Organic solvent tolerant lipases and applications. The Scientific World Journal. 2014;2014:15p. http://dx.doi.org/10.1155/2014/625258

70. Gonçalves FAG, Colen G, Takahashi JA. Yarrowia lipolytica and its multiple applications in the biotechnological industry. The Scientific World Journal. 2014;2014:14. p. http://dx.doi.org/10.1155/2014/476207

71. Anderson EM, Larsson KM, Kirk O. One biocatalyst–many applications: The use of Candida antarctica B-lipase in organic synthesis. Biocatalysis and Biotransformations. 1998;16:181-204. http://dx.doi.org/10.3109/10242429809003198

72. Weber N, Weitkamp P, Mukherjee KD. Steryl and stanyl esters of fatty acids by solvent-free esterification and transesterification in vacuo using lipases from Rhizomucor miehei, Candida antarctica, and Carica papaya. Journal of Agriculture and Food Chemistry. 2001;49(11):5210-5216. DOI: 10.1021/jf0107407

73. Hamam F, Shahidi F. Synthesis of structured lipids containing medium-chain and Omega-3 fatty acids. Journal of Agriculture and Food Chemistry. 2006;54:4390-4396. DOI: 10.1021/jf052540r

74. Gross RA, Kalra B. Biodegradable polymers for the environment. Science. 2002;297(5582):803-807. DOI: 10.1126/science.297.5582.803

75. Gross RA, Kumar A, Kalra B. Polymer synthesis by in vitro enzyme catalysis. Chemical Reviews. 2001;10(7):2097-2124. DOI: 10.1021/cr0002590

76. Cabrera Z, Fernandez-Lorente G, Fernandez-Lafuente R, Palomo JM, Guisan JM. Novozym 435 displays very different selectivity compared to lipase from Candida antarctica B adsorbed on other hydrophobic supports. Journal of Molecular Catalysis B: Enzymatic. 2009;57:171-176. http://dx.doi.org/10.1016/j.molcatb.2008.08.012

77. Ueji SI, Nishimura M, Kudo R, Matsumi R, Watanabe K, Ebara Y. A dramatic improvement of enantioselectivity of lipase in organic solvents by addition of aqueous SDS: A close correlation between enantioselectivity and conformational flexibility of lipase. Chemical Letters. 2001;30:912-913. DOI: 10.1002/chin.200206031.

78. Watanabe K, Ueji SI. Dimethyl sulfoxide as a co-solvent dramatically enhances the enantioselectivity in lipase-catalyzed resolutions of 2-phenoxypropionic acyl derivatives. Journal of the Chemical Society, Perkin Transactions 1. 2001;I(12):1386-1390. DOI: 10.1039/B100182P

79. Hirose Y, Kariya K, Sasaki I, Kurono Y, Ebiike H, Achiwa K. Drastic solvent effect on lipase-catalysed enantioselective hydrolysis of prochiral 1,4-dihydropyridines. Tetrahedron Letters. 1992;33:7157-7160. http://dx.doi.org/10.1016/S0040-4039(00)60861-4

80. Kirk O, Christensen MW. Lipases from Candida antarctica: Unique biocatalysts from a unique origin. Organic Process Research & Development. 2002;6(4):446-451. DOI: 10.1021/op0200165

81. Domingues de Maria P, Carboni-Oerlemans C, Tuin B, Bargeman G, van der Meer A, van Gemert R. Biotechnological applications of Candida antarctica lipase A: State-of-the-art. Journal of Catalysis B: Enzymatic. 2005;37:36-46. http://dx.doi.org/10.1016/j.molcatb.2005.09.001

82. Schmidt M, Barbayianni E, Fotakopoulou I, Höhne M, Constantinou-Kokotou V, Bornscheuer UT, Kokotos G. Enzymatic removal of carboxyl protecting groups. 1. Cleavage of the tert-butyl moiety. Journal of Organic Chemistry. 2005;70(9):3737-3740. DOI: 10.1021/jo050114z

83. Gotor-Fernandez V, Brieva R, Gotor V. Lipases: Useful biocatalysts for the preparation of pharmaceuticals.Journal of Molecular Catalysis. 2006;40:111-120. DOI: 10.1016/j.molcatb.2006.02.010

84. Kingery-Wood J, Johnson JS. Resolution of anti-3-oxotricyclo[2.2.1.0]heptane-7-carboxylic acid by Candidia antartica lipase A. Tetrahedron Letters. 1996;37(23):3975-3976. http://dx.doi.org/10.1016/0040-4039(96)00759-9

85. Davies HG, Green RH, Kelly DR, Roberts SM. Recent advances in the generation of chiral intermediates using enzymes. Biotechnology. 1990;10:129-152. http://dx.doi.org/10.3109/07388559009068264

86. Sih CJ, Gu Q-M, Holdgrun X, Harris K. Optically-active compounds via biocatalytic methods. Chirality. 1992;4:91-97. DOI: 10.1002/chir.530040205

87. Margolin AL. Enzymes in the synthesis of chiral drugs. Enzyme and Microbial Technologies. 1993;15:266-280. http://dx.doi.org/10.1016/0141-0229(93)90149-V

88. Salzar L, Bermudez JL, Ramirez C, Llama EF, Sinisterra JV. Resolution of 3-α-Naphthoxy-1,2-propanediol Using Candida antarctica Lipase. Tetrahedron. 1999;10:3507-3514. http://dx.doi.org/10.1016/S0957-4166(99)00366-3

89. Orrenius C, Oehrner N, Rotticci D, Mattson A, Hult K, Norin T. Candida antarctica lipase B catalyzed kinetic resolutions: Substrate structure requirements for the preparation of enantiomerically enriched secondary alcohols. Ibid. 1995;6:1217-1220. 10.1016/0957-4166(95)00147-h

90. Patel RN, Banerjee A, Nanduri V, Goswami A, Comezoglu FT. Enzymatic resolution of racemic secondary alcohols by lipase B from Candida antarctica. Journal of the American Oil Chemists' Society. 2000;Vol. 77:Issue (10). 1015-1019. DOI: 10.1007/s11746-000-0161-y

91. Sauerberg P, Olesen PH. Heterocyclic Compounds and their Preparation and Use. 1995; U.S. Patent 5, 418:240

92. Hamilton GS, Li J-H, Steiner J-P. Method of Using Neutrotrophic Sulfonamide Compounds. 1998; U.S. Patent 5, 721:256

93. Hutt AJ, Caldwell J. The importance of stereochemistry in the clinical pharmacokinetics of the 2-arylpropionic acid non-steroidal anti-inflammatory drugs. Clinical Pharmacokinetics. 1984;9(4):371-373. DOI: 10.2165/00003088-198409040-00007

94. Siódmiak T, Rumiński JK, Marszał M. Application of lipases from Candida rugosa in the enantioselective esterification of (R,S)-ibuprofen. Current Organic Chemistry. 2012;16(8):972-977. DOI: 10.2174/138527212800194728

95. Breuer M, Ditrich K, Habicher T, Hauer B, Keßeler M, Stürmer R, Zelinski T. Industrial methods for the production of optically active intermediates. Angewandte Chemie International Edition. 2004;43(7):788-824. DOI: 10.1002/anie.200300599

96. Gotor-Fernandez V, Busto E, Gotor V. Candida antarctica lipase B: An ideal biocatalyst for the preparation of nitrogenated organic compounds. Advanced Synthesis & Catalysis. 2006;348:797-812. DOI: 10.1002/adsc.200606057

97. Torre O, Busto E, Gotor-Fernández V, Gotor V. Enzymatic preparation of novel aminoalkylpyridines using lipases in organic solvents. Advanced Synthesis & Catalysis. 2007;349:1481-1488. DOI: 10.1002/adsc.200600634

98. Borch K, Christensen M. WO 01/53511 A1; 2001

99. Pihko AJ, Lundell K, Kanerva L, Koskinen AMP. Enantioselective synthesis of a hindered furyl substituted allyl alcohol intermediate: A case study in asymmetric synthesis. Tetrahedron: Asymmetry. 2004;15:1637-1643. http://dx.doi.org/10.1016/j.tetasy.2004.04.004

100. Paizs C, Tähtinen P, Toşa M, Majdik C, Irimie FD, Kanerva LT. Biocatalytic enantioselective preparation of phenothiazine-based cyanohydrin acetates: Kinetic and dynamic kinetic resolution. Tetrahedron. 2004;60(46):10533-10540. http://dx.doi.org/10.1016/j.tet.2004.06.136

101. de Gonzalo G, Lavandera I, Brieva R, Gotor V. Enzymatic acylation reactions on ω-hydroxycyanohydrins. Tetrahedron. 2004;60(46):10525-10532. http://dx.doi.org/10.1016/j.tet.2004.06.134

102. Romero-Gomez S, Augur C, Viniegra-Gonzalez G. Invertase production by Aspergillus niger in submerged and solid-state fermentation. Biotechnology Letters. 2000;22:1255-1258. DOI: 10.1023/A:1005659217932

103. Kotwal SM, Shankar V. Immobilized invertase. Biotechnology Advanced. 2009;27:311-322. http://dx.doi.org/10.1016/j.biotechadv.2009.01.009

104. Kulshrestha S, Tyagi P, Sindhi V, Yadavilli KS. Invertase and its applications—A brief review. Journal of Pharmacy Research. 2013;7(9):792-797. http://dx.doi.org/10.1016/j.jopr.2013.07.014

105. Mussatto SI, Mancilha IM. Non-digestible oligosaccharides: A review. Carbohydrate Polymers. 2007;68:587-597. DOI: 10.1016/j.carbpol.2006.12.011

106. Lahiri S, Basu A, Sengupta S, Banerjee S, Dutta T, Soren D, Chattopadhyay K, Ghosh AK. Purification and characterization of a trehalase-invertase enzyme with dual activity from Candida utilis. Archives of Biochemistry and Biophysics. 2012;522(2):90-99. http://dx.doi.org/10.1016/j.abb.2012.03.026

107. Zhu D, Yang Y, Hua L. Stereoselective enzymatic synthesis of chiral alcohols with the use of a carbonyl reductase from Candida magnoliae with anti-Prelog enantioselectivity. Journal of Organic Chemistry. 2006;71:4202-4205. DOI: 10.1021/jo0603328

108. Seebach D, Sutter MA, Weber RH, Züge MF. Yeast reduction of ethyl acetoacetate: (S)-(+)-ethyl 3-hydroxybutanoate. Organic Synthesis. 1985;63:1-7. DOI: 10.15227/orgsyn.063.0001

109. Kawai Y, Tsujimoto M, Kondo S, Takanobe K, Nakamura K, Ohno A. Asymmetric reduction of β-keto esters with an enzyme from Bakers' yeast. Bulletin of the Chemical Society of Japan. 1994;67:524-528. http://dx.doi.org/10.1246/bcsj.67.524

110. Williams RE, Bruce NC. New uses for an old enzyme—the old yellow enzyme family of flavoenzymes. Microbiology. 2002;148:1607-1614. http://dx.doi.org/10.1099/00221287-148-6-1607

111. Świderska MA. Application of Saccharomyces carlsbergensis old yellow enzyme in synthesis of chiral ketones and building blocks for β-amino acids [thesis]. University Of Florida; 2007, Gainesville, Florida 32611, United Stated of America

112. Meah Y, Massey V. Old yellow enzyme: Stepwise reduction of nitroolefins and catalysis of aci-nitro tautomerization. Proceedings of the National Academy of Science of the United States of America. 2000;97:10733-10738. DOI: 10.1073/pnas.190345597

113. Karathia H, Vilaprinyo E, Sorribas A, Alves R. Saccharomyces cerevisiae as a model organism: A comparative study. PLoS One. 2011;6(2):e16015. DOI: 10.1371/journal.pone.0016015

114. Demain AL, Vaishnav P. Production of recombinant proteins by microbes and higher organisms. Biotechnology Advances. 2009;27(3):297-306. DOI: 10.1016/j.biotechadv.2009.01.008

115. Li F, Long T, Lu Y, Ouyang Q, Tang C. The yeast cell-cycle network is robustly designed. Proceedings of the National Academy of Sciences of the United States of America. 2004;101(14):4781-4786. DOI: 10.1073/pnas.0305937101

116. Duina AA, Miller ME, Keeney JB. Budding yeast for budding geneticists: A primer on the saccharomyces cerevisiae model system. Genetics. 2014;197(1):33-48. DOI: 10.1534/genetics.114.163188

117. http://syntheticyeast,org/

118. Goffeau A, Barrell BG, Bussey H, Davis RW, Dujon B, Dujon B, Feldmann H, Galibert F, Hoheisel JD, Jacq C, Johnston M, Louis EJ, Mewes HW, Murakami Y, Philippsen P, Tettelin H, Oliver SG. Life with 6000 genes. Science. 1996;274:546-567. DOI: 10.1126/science.274.5287.546

119. Kojo H, Greenberg BD, Sugino A. Yeast 2-micrometer plasmid DNA replication in vitro: Origin and direction. Proceedings of the National Academy of Sciences of the United States of America. 1981;78(12):7261-7265. DOI: 10.1073/pnas.78.12.7261

120. Stearns T, Ma H, Botstein D. Manipulating yeast genome using plasmid vectors. Methods in Enzymology. 1990;185:280-297

121. Brown TA. Gene cloning and DNA analysis: An introduction. 6th ed. Manchester: Wiley; 2010. pp. 105-112

122. Hou J, Tyo KEJ, Liu Z, Petranovic D, Nielsen J. Metabolic engineering of recombinant protein secretion by Saccharomyces cerevisiae. FEMS Yeast Research. 2012;12(5):491-510. DOI: 10.1111/j.1567-1364.2012.00810.x

123. Nielsen J. Production of biopharmaceutical proteins by yeast. Bioengineered. 2013;4(4):207-211. DOI: 10.4161/bioe.22856

124. Burda P, Aebi M. The dolichol pathway of N-linked glycosylation. Biochimica et Biophysica Acta (BBA)—General Subjects. 1999;1426(2):239-257. DOI: 10.1016/S0304-4165(98)00127-5

125. Wildt S, Gerngross TU. The humanization of N-glycosylation pathways in yeast. Nature Reviews Microbiology. 2005;3(2):119-128. DOI: 10.1038/nrmicro1087

126. Hamilton SR, Gerngross TU. Glycosylation engineering in yeast: The advent of fully humanized yeast. Current Opinion in Biotechnology. 2007;18(5):387-392. DOI: 10.1016/j.copbio.2007.09.001

127. Payne T, Finnis C, Evans LR, Mead DJ, Avery SV, Archer DB, et al. Modulation of chaperone gene expression in mutagenized Saccharomyces cerevisiae strains developed for recombinant human albumin production results in increased production of multiple heterologous proteins. Applied and Environmental Microbiology. 2008;74(24):7759-7766. DOI: 10.1128/AEM.01178-08

128. Fukuda K, Jensen MH, Haser R, Aghajari N, Svensson B. Biased mutagenesis in the N-terminal region by degenerate oligonucleotide gene shuffling enhances secretory expression of barley α-amylase 2 in yeast. Protein Engineering, Design and Selection. 2005;18(11):515-526. DOI: 10.1093/protein/gzi057

129. Kjeldsen T, Ludvigsen S, Diers I, Balschmidt P, Sørensen AR, Kaarsholm NC. Engineering-enhanced protein secretory expression in yeast with application to insulin. Journal of Biological Chemistry. 2002;277(21):18245-18248. DOI: 10.1074/jbc.C200137200

130. Rakestraw JA, Sazinsky SL, Piatesi A, Antipov E, Wittrup KD. Directed evolution of a secretory leader for the improved expression of heterologous proteins and full-length antibodies in Saccharomyces cerevisiae. Biotechnology and Bioengineering. 2009;103(6):1192-1201. DOI: 10.1002/bit.22338

131. Harmsen MM, Bruyne MI, Raué HA, Maat J. Overexpression of binding protein and disruption of the PMR1 gene synergistically stimulate secretion of bovine prochymosin but not plant thaumatin in yeast. Applied Microbiology and Biotechnology. 1996;46(4):365-370

132. Smith JD, Tang BC, Robinson AS. Protein disulfide isomerase, but not binding protein, overexpression enhances secretion of a non-disulfide-bonded protein in yeast. Biotechnology and Bioengineering. 2004;85(3):340-350. DOI: 10.1002/bit.10853

133. Homma T, Iwahashi H, Komatsu Y. Yeast gene expression during growth at low temperature. Cryobiology. 2003;46(3):230-237

134. Hovland P, Flick J, Johnston M, Sclafani RA. Galactose as a gratuitous inducer of GAL gene expression in yeasts growing on glucose. Gene. 1989;83(1):57-64

135. Koller A, Valesco J, Subramani S. The CUP1 promoter of Saccharomyces cerevisiae is inducible by copper in Pichia pastoris. Yeast (Chichester, England). 2000;16(7):651-656. DOI: 10.1002/(SICI)1097-0061(200005)16:7<651::AID-YEA580>3.0.CO;2-F

136. Li N, Zhang L-M, Zhang K-Q, Deng J-S, Prändl R, Schöffl F. Effects of heat stress on yeast heat shock factor-promoter binding in vivo. Acta Biochimica Et Biophysica Sinica. 2006;38(5):356-362

137. Gatignol A, Dassain M, Tiraby G. Cloning of Saccharomyces cerevisiae promoters using a probe vector based on phleomycin resistance. Gene. 1990;91(1):35-41. DOI: 10.1016/0378-1119(90)90159-O

138. Fields S, Song O. A novel genetic system to detect protein-protein interactions. Nature. 1989;340(6230):245-246. DOI: 10.1038/340245a0

139. Stynen B, Tournu H, Tavernier J, Dijck PV. Diversity in genetic in vivo methods for protein-protein interaction studies: From the yeast two-hybrid system to the mammalian split-luciferase system. Microbiology and Molecular Biology Reviews. 2012;76(2):331-382. DOI: 10.1128/MMBR.05021-11

CHAPTER 2

DETECTION OF ORTHOLOGS IN YEASTS USING SUPERVISED PAIRWISE APPROACH FOR ANALYZING BIG DATA

1. INTRODUCTION: AN OVERVIEW

Orthologs are genes found in various species that have evolved from a shared ancestor, while paralogous genes are sequences that have evolved through a duplication event from a single sequence [1]. The differentiation between orthologs and paralogs is vital as orthologous genes serve as indicators of evolution and aid in deducing the function of unfamiliar proteins.

The orthology relationship is defined by either one-to-one pairs or by groups with one-to-many or many-to-many connections. Therefore, the ability to scale both gene/protein pairwise comparisons and the classification process is a difficulty because of the continuously growing number of sequenced genomes. The topic of unsupervised ortholog categorization is introduced in reference [2], beginning with the comparison of pairwise sequences. Several algorithms, including the Reciprocal Best Hit (RBH) heuristic [3], Inparanoid [4], OrthoMCL [5], the Comprehensive, Automated Project for the Identification of Orthologs from Complete Genome Data (OMA) [6], and the Reciprocal Smallest Distance (RSD) algorithm [7], construct a similarity graph based on these comparisons. Additionally, under this unsupervised learning methodology, there are tree-based and hybrid algorithms [8]. Alternatively, a supervised strategy is introduced in [9], which utilizes pairwise classification. However, this approach does not address the issue of managing the imbalance between the rare ortholog pairs (minority class) and the bulk of non-ortholog pairs. In addition, this strategy does not address the issue of generalization in the model, as it does not entail learning from external genome pairs. A crucial aspect to address in ortholog detection (OD) algorithms is the fundamental knowledge of gene/protein characteristics utilized for classifying the sequences. Classification algorithms that rely solely on sequence similarity often struggle due to the negative impact of genetic and evolutionary events, such as duplications, gene losses, and changes between genome segments. These events include duplications, deletions, horizontal gene transfers, fusions, fissions, inversions, and transpositions, all of which can adversely affect the accuracy of classification based on sequence similarity. Despite the significant promise of OD methods mentioned in both the "omics" [1] tools site and the Quest for Ortholog[2]s Consortium site, literature reports some of their limitations.

- They tend to include paralogs in the orthogroups increasing false positives [8]. In the presence of lots of gene losses caused by whole genome duplications (WGD) in Saccharomycete yeasts, a big amount of paralogs can be included in orthogroups, or in orthologous pairs [10].
- They can reduce specificity due to the presence of short sequences or those that is evolved in a convergent way. This kind of affection may arise when genes inherited by horizontal transfers brought about failure in the inference of near phylogenetic relationships between species that are distantly related and have recently exchanged a gene [11].
- They can fail in ortholog detection mainly in the twilight zone (less than the 30% of sequence identity), that is, they can raise false negatives in the presence of distantly related divergent sequences. They can also fail in front of mosaics of proteins [12] since orthogroups not considering hybrid proteins may contain proteins without a common ancestor.

In order to overcome these restrictions and enhance the effectiveness of ortholog discovery, algorithms are increasingly incorporating various gene information alongside alignment-based characteristics (AB). Ortholog recognition algorithms incorporate several factors such as synteny [13–15], global rearrangements [16, 17], protein interactions [9], protein mosaic architecture [18], and evolutionary distances [6, 7]. Thus far, these endeavors have proven insufficient in enhancing the effectiveness of classification [19]. The incorporation of AB, together with additional factors such as physicochemical characteristics, sequence length, and synteny, can potentially enhance the identification of paired orthologs. Furthermore, leveraging curated classifications in benchmark datasets during the learning process may lead to a more efficient approach.

Salichos and Rokas [10] created a benchmark dataset of Saccharomycete yeast to assess OD methods. The dataset consists of orthogroups that do not contain any paralogs. These orthogroups consist of orthologs from yeast species that

1 http://omictools.com/
2 http://questfororthologs.org/orthology_databases

have experienced both pre- and post-whole genome duplications (WGD), and are considered to be curated or genuine. [10] When evaluating this benchmark dataset, Multiparanoid [20], OrthoMCL, and enhanced versions of RBH and the RSD were found to include paralogs in the orthogroups [10]. This data confirms that the discipline of OD, which stands for bioinformatics, need a continuous supply of new and efficient methods, mostly focused on yeasts. When it comes to efficiency, OD techniques that rely on sequence alignments can require a significant amount of CPU time, ranging from weeks to years, to compute orthologs of sequenced species. This is owing to the quadratic scaling capacity of these comparisons. During the Quest for Orthologs Consortium meeting [19], the focus was on addressing challenges related to the efficiency and handling of large datasets. Various efforts in these areas were presented. According to them, the growing need for computational power in sequence analysis cannot be met by simply expanding the capacity of computation. However, it can be fulfilled by adopting new methodologies or implementing new algorithms. This chapter introduces a novel classification method that utilizes paired protein characteristics in a supervised decision system. The method takes into account the significant imbalance between ortholog and non-ortholog pairs in the benchmark dataset presented by [10]. Several novel gene pair similarity metrics, which are regarded as pairwise protein features, are established by utilizing protein physicochemical profiles, their inclusion in conserved regions within related genomes, and their respective lengths. An analysis is conducted to evaluate the efficiency and scalability of calculating these indicators, with the aim of recommending their application for big data. Overall, the performance of Mahout [21] and MLlib [22] in handling large and imbalanced data in scalable machine learning libraries surpassed RBH, RSD, and OMA [6] in identifying orthologs from complete genome data in Saccharomycete yeast genomes.

2. METHODS

2.1. Yeast genomes in experiments

Yeast genomes selected for the experiments are well-studied eukaryote species that shared a common ancestor with human millions of years ago. Saccharomyces cerevisiae belongs to Saccharomycete yeasts and Schizosaccharomyces pombe is a unicellular organism from Archiascomycete fungus. These species bring about experimental models for many essential processes of eukaryotes since most of their genes have been predicted as homologs to multicellular eukaryotes ones. Specifically, S. pombe genes are more similar to mammals' genes than S. cerevisiae ones [23]. However, S. pombe is a distant relative of S. cerevisiae, accordingly, ortholog detection between them is a difficult task due to the divergence of sequences [13]. On the other hand, other close Saccharomycete species as Kluyveromyces (Kluyveromyces lactis and Kluyveromyces waltii) and Candida glabrata have been included in one of the ortholog detection benchmarking datasets [10] since some of them underwent a whole genome duplication (WGD) process causing a rampant paralogy and a lot of gene losses. Precisely, the genome pair of C. glabrata and S. cerevisiae that are post-WGD species constitutes a target study case for OD algorithms. For the studies reported below we have identified each genome pair as follows:

- S. cerevisiae – K. lactis: ScerKlac
- K. lactis – K. waltii: KlacKwal
- C. glabrata – K. lactis: CglaKlac
- S. cerevisiae – C. glabrata: ScerCgla
- S. cerevisiae – S. pombe: ScerSpombe

2.2. Similarity measures as gene pair features in ortholog detection

Five normalized pairwise sequence similarities (global and local alignment scores, protein length, synteny and physicochemical profile) were previously specified in [24]. Table 1 summarizes the calculations for protein pairs (xi,yj) from the two proteome representations P1={x1, x2,..., xn} and P2={y1, y2, ..., ym} with n and m sequences, respectively.

Table 1. Protein pair features.

Measure	Description of the calculation
S_1 – local and global alignment	The sequence alignment measure averages positives local and global protein alignment scores from the Smith Waterman [32] and the Needleman-Wunsch [33] algorithms calculated with a specified scoring matrix, and "gap open" (GOP) and "gap extended" (GEP) parameters. The scores are normalized by using the maximum score of all (x_i, y_j) protein pairs
S_2 – protein length	Measure S_2 represents the normalized difference for continuous values [34] of the protein sequence lengths
S_3 – membership to locally collinear blocks	This measure is calculated from the distance between pairs of sequences considering their membership to Locally Collinear Blocks (LCBs) which represent truly homologous regions obtained with the Mauve software [35]. The normalized difference is selected for the comparison of the total number of codons in each block
S_4 – physicochemical profile	This measure is based on the spectral representation of sequences from the global protein pairwise alignment by using the linear predictive coding [34]. First, each amino acid that lies in a matching region without "gaps" between two aligned sequences is replaced by its contact energy [36]. The moving average for each spectrum, that is, the average of this physicochemical feature in the predefined window size W is then calculated. Next, the Pearson correlation coefficient with the corresponding significance level between the two spectral representations in a matching region is obtained. Finally, the significant similarities of the regions without "gaps" are aggregated considering the length of each region. From our previous studies presented in [37, 38], three features for the physicochemical profile with W values of 3, 5 and 7 have been considered

2.3. From the computation of similarity measures in parallel to the processing of large-scale data.

Time complexity analysis of the sequential algorithms for the calculation of pairwise similarity measures in terms of N gene/protein pairs has brought about the need of a parallel version with the scalability target. The protein physicochemical profile turns to be the one with the highest quadratic cost $O(N \times (m + n)2)$ since it operates over the aligned sequences with m and n as the maximum sequence lengths for two genomes, respectively. The parallel proposal is sketched in Figure 1 with a fork control flow to distribute the calculation over different processors and a join control flow to terminate the parallel executions.

For the scalability analysis of the parallel physicochemical profile calculation, we estimated the TS sequential execution time in expression (1) by using constants.

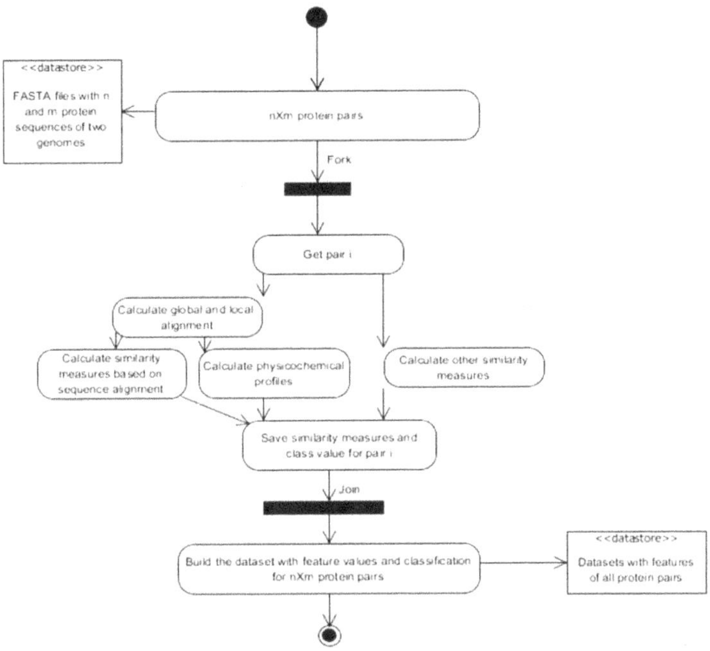

Figure 1. Parallel calculation of protein pair similarity measures.

$$TS=N \times (n \times m + (m+n)^2) \times tc \qquad (1)$$

The constant tc indicates the required time for the arithmetic operations. We assume tc as optimum value, tc = 1 [25]. Consequently, the parallel execution time TP is defined in terms of the calculation time and the communication time among processors. In the parallel version, calculation time depends on the distribution of N cycles among P processors. With this distribution: (i) all the processors execute $\lfloor N/P \rfloor$ iterations or (ii) $p < P$ processors execute $\lfloor N/P \rfloor + 1$ iterations. Thus, when we consider the worst case, the calculation time TCalc can be estimated as in expression (2).

$$T_{Calc} = (N / P + 1) \times \left(n \times m + (m+n)^2\right) \times t_c \quad (2)$$

The communication time among processors is directly related to the information accessed in the sequential section of the algorithm. In each iteration, each processor P needs to receive sequence information. Hence, communication time can be defined as in expression (3).

$$T_{Com} = (N / P + 1) \times \left(t_s + t_w \times P \times (m+n)\right) \quad (3)$$

where ts represents the time required to establish the communication and to prepare the sending information and tw is the required time to send a numeric value. These values can be considered as optimum ones, that is, ts = 0 and tw = 1 [25]. In this way, we combine expressions (2) and (3) to obtain an estimation of the parallel execution time for the physicochemical profile as in expression (4).

$$T_P = (N / P + 1) \times \left(n \times m + (m+n) \times (m+n+P)\right) \quad (4)$$

Time complexity of the parallel version of the algorithm can be estimated from the parallel execution time TP. Expression (⌊N/P⌋ + 1) can be taken as N/P.

It holds that m×n ≤ (m + n)×(m+n+P). Therefore, time complexity is $O\left(\frac{N \times (n+m)^2}{P} + N \times (m+n)\right)$, that is lower than O(N×(m+n)2) as demonstrated in expression (5).

$$\frac{\frac{N \times (n+m)^2}{P} + N \times (m+n)}{N \times (n+m)^2} = \frac{1}{P} + \frac{1}{m+n} \leq 1 \quad (5)$$

By using expressions TS and TP, we can analyze the speed-up and the efficiency performance for different samples obtained from N protein pairs and a range of 1–100 processors. Figure 2A and B shows the speed-up and efficiency values, respectively, in a genome comparison dataset of S. cerevisiae and S. pombe. The 100% sample represents 5006 sequence pairs; the 75% sample represents 3754 pairs, the 50% represents 2503 pairs, and the 25% represents 1251 pairs. An increased number of processors lead to a saturation in the speed-up. Specifically, when the problem size increase, thus the speed-up and the efficiency values increase for a single number of processors, although both values tend to decrease with an increasing number of processors. The efficiency tends to keep constant when we simultaneously increase both the problem size and the number of processors, consequently, contributing to the scalable conception of the algorithm [26].

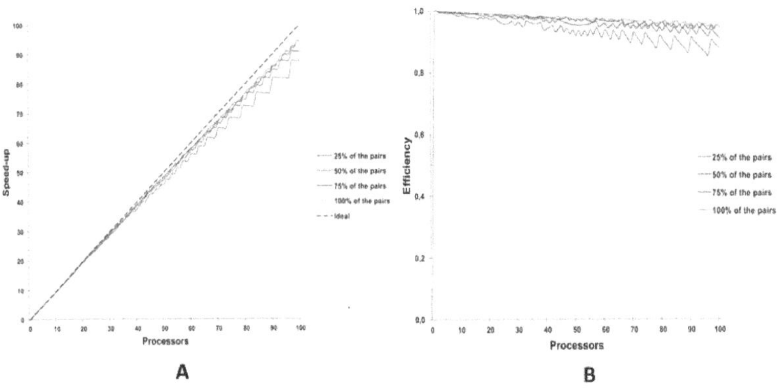

Figure 2. Estimated performance of the speed-up (A) and efficiency (B) in the parallel calculation of the physicochemical profile of proteins for the genome comparison dataset of S. cerevisiae and S. pombe.

With the sequential and parallel MATLAB (2010) implementations, we calculated real execution time for one, two, three and four processors in a personal computer Intel® Core™ i3 CPU, 2.53 GHz, RAM DDR3 de 4.0 Gb, 64Bits Windows7. With 1921 pairs of S. cerevisiae and S. pombe representing 100% of the sample, Figure 3A and B shows the real and estimated values of the speed-up and efficiency, respectively. We can observe a similar performance of estimated and real values for the total and 50% of the sample. The best speed-up was obtained for the execution in four processors, but the efficiency is best with two processors.

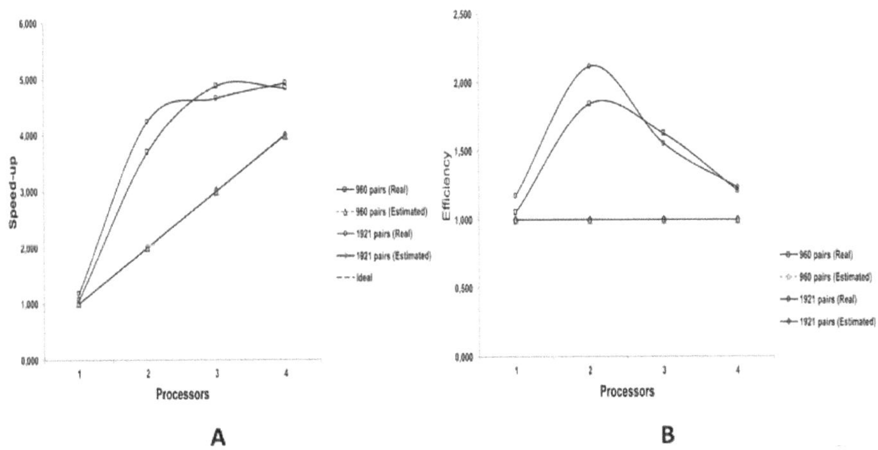

Figure 3. Real and estimated values for speed-up (A) and efficiency (B) in a S. cerevisiae and S. pombe sample.

Theoretically, the speed-up value is bounded by the number of processors and the efficiency value is in the interval [0, 1]. However, the real speed-up obtained is higher than the number of processors in each execution and the efficiency is higher than one. This event is known as super-linear speed-up. It is a consequence of the fact that each processor consume less time than $\frac{T_S}{P}$. It can be possible since the parallel version may execute less work than the sequential version or may take advantage of resources such as the cache memory [25].

In a further scalability analysis with 133,666,445 pairs of sequences from different yeast genomes, the increased number of processors from 1 to 100 saturates the speed-up. Table 2 shows the genome data used for this scalability analysis. Datasets are conformed by accumulating gene pairs from different genome pairs with different maximum sequence lengths. Figure 4A illustrates the effects of the gene pair accumulation in the speed-up of the parallel physicochemical profile similarity measure calculation while Figure 4B depicted the effect in the efficiency of this calculation. When the problem size increases, the speed-up and the efficiency increase for a specific number of processors, although both values tend to decrease when the number of processors increases. Efficiency tends to be constant when both the problem size and the number of processors increase. This is an essential issue to achieve scalability, specifically, the horizontal one.

Table 2. Genome data used in the scalability analysis.

Datasets	Total of gene pairs	Maximum length for the first genome *m*	Maximum length for the second genome *n*
ScerSpombe	16,324,500	4910	4924
ScerSpombe + ScerKlac	47,546,047	4910	4924
ScerSpombe + ScerKlac + ScerCgla	78,111,162	4910	4924
ScerSpombe + ScerKlac + ScerCgla + CglaKlac	105,891,467	4910	4924
ScerSpombe + ScerKlac + ScerCgla + CglaKlac + KlacKwal	133,666,445	4915	4924

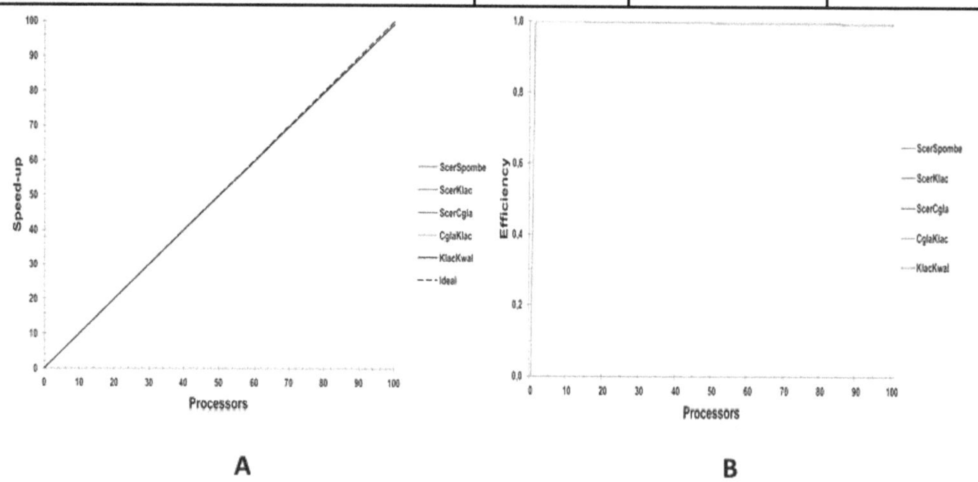

Figure 4. Estimated speed-up (A) efficiency (B) and when the input data and processors increase.

In sum, the parallel algorithm time complexity considering P processors is lower than the cost of the sequential version and the scalability may be achieved with the parallel proposal. Nevertheless, considering implementation hazards of the parallel models, scalability can be improved when these models are executed in a cloud within a big data framework as MapReduce [27, 28]. This framework guarantees reliability, availability, and a good execution performance in a cloud with a distributed file system.

In a MapReduce design, each mapper process should build a subset of the resulting dataset with calculated features of all protein pairs of its partition. Figure 5 shows three steps such as Initial, Map and Final. The Initial phase divides the protein pair set into independent Hadoop distributed file system (HDFS) blocks; it also duplicates and transfers them to other nodes of the cloud. Then, in the Map step, each mapper calculates the protein features of all the pairs in its partition. Finally, in the Final step, the files created by each mapper would be concatenated to build the final dataset file. In this procedure the Reduce step is omitted since the mappers output are directly combined following the operation reported in [29].

Algorithm 1 shows el pseudo-code of the Map function in the MapReduce process for the protein feature calculations. Step 2 calculates the features for a pair of proteins and Step 5 save the previously calculated data.

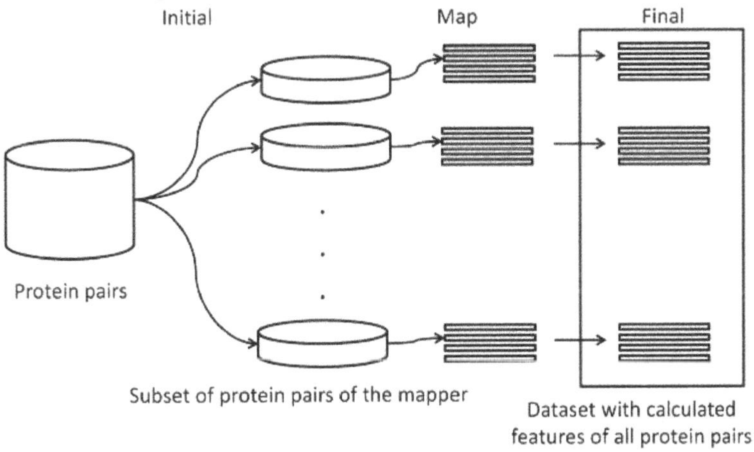

Figure 5. MapReduce design of protein pair feature calculations.

MAP (key, value)

Input: <key; value> % key represents the location in bytes, and value, the contents of a protein pair.

Output: <key'; value'> % key' indicates the identification of the pair, and value', the calculated features.

- 1 pair ← PAIR_REPRESENTATION(value)% features variable contains the values of the feature for the pair

- 2. features ← CALCULATE_FEATURES(pair)
- 3. lkey ← lkey:set(pair:get_id)
- 4. lvalue ← lvalue:set(features)
- 5. EMIT (lkey, lvalue)

Algorithm 1. Map phase for feature calculation.

2.4. Big data supervised classification with imbalance management

Given a set A = {Sr(xi, yj)} of gene pair features as discrete or continuous values of r gene pair similarity measure functions, previously specified, we represent a pairwise ortholog detection decision system DS = (U, A ∪ {d}), where U = {(xi, yj)}, ∀ xi ∈ G1, ∀ yj ∈ G2 is the universe of the gene pairs, and d ∉ A is the binary decision feature obtained from a curated classification. This decision feature defines the low ratio of orthologs to the total number of possible gene pairs. The big data supervised classification divides DS into train and test sets to build a learning/training model and to classify the instances by means of a big data supervised algorithm managing the imbalance between classes. The built model can be generalized to related pairs of genomes/yeast species not used in the learning/training process following the generalization concept in [30]. Thus, the test set is used to compare both supervised and classical unsupervised algorithms. Algorithm 2 shows the steps required in the training phase as well as Algorithm 3 shows the classification phase. The general testing/evaluation scheme for supervised and unsupervised algorithms is sketched in Figure 6 based on the use of a testing external set as reported in [31].

Training phase

Input: DS=(U, A ∪ {d}),A={Sr(xi, yj)},U={(xi, yj)}, ∀ xi∈G1, ∀ yj∈G2

Output: Model, Test_set

- Train_set, Test_set =

Selection_of_training_and_testing_sets (DS)

- Model = Model_building_with_imbalance_management %Learning/Training step

Algorithm 2. Training phase for big data supervised classification with imbalance management.

Classification phase

Input: Model, Test_set

Output: Ortholog_pairs_of_the_testing_set

1 Ortholog_pairs_of_the_testing_set = Pairwise_classification (Model, Test_set)

Algorithm 3. Classification phase for big data supervised ortholog detection with imbalance management.

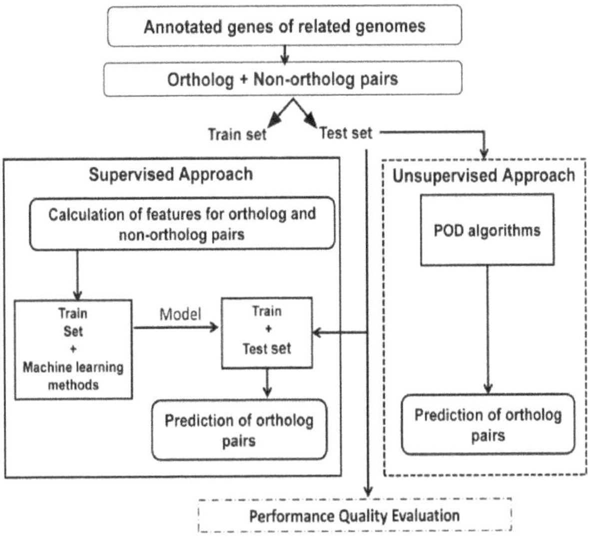

Figure 6. Workflow of the evaluation of supervised versus unsupervised pairwise ortholog detection algorithms.

The general scheme receives annotated genomes (proteomes) from related yeast species since some protein features as the membership to conserved regions require certain evolution closeness. For the evaluation of pairwise ortholog detection algorithms, we can compare the supervised solutions and the unsupervised reference algorithms such as RBH, RSD, and OMA. Firstly, protein pairs are separated into train and test sets and after that pairwise similarity measures for the pairs of both sets are calculated. Test sets are used for the assessment of the unsupervised algorithms while the training set is used to build the supervised models which will be further tested on the previously mentioned test set.

The classification performance will be assessed by evaluation metrics for imbalanced datasets. The Geometric Mean (G-Mean) is defined in [32] as:

$$G - Mean = \sqrt{sensitivity * specificity} \tag{6}$$

where sensitivity and specificity are traditionally calculated considering the true positives (TP), false negatives (FN), false positives (FP), and true negatives (TN).

The area under the curve (AUC) [33] is estimated from plotting of true positive rate (TPR) versus false positive rate (FPR) values. We approximate the AUC by averaging (TPR) and (FPR) values through the following equation:

$$AUC = \frac{1 + TPR - FPR}{2} \tag{7}$$

where $TPR = \frac{TP}{TP+FN}$ represents the percentage of orthologs (TP) correctly classified and $FPR = \frac{FP}{FP+TN}$, the percentage of non-orthologs (TN) misclassified. G-Mean measure maximizes the accuracy of the two classes (orthologs and non-orthologs) by considering the misclassification costs and AUC values in the estimation of the sensitivity and specificity, getting at a balance between them [34].

2.5. Experiments were conducted to find orthologs in closely related yeast species.

2.5.1. Settings

For the evaluation of pairwise ortholog detection algorithms in related yeast genomes we carried out (i) three exploratory experiments for an external validation of supervised classification models and (ii) an experiment to evaluate a model built with a pair of yeasts into two external pairs. In Experiment 1, we evaluated the algorithms inside a genome pair by partitioning the complete set of protein pairs at random (75% for training and 25% for testing). Specifically, we divided the S. cerevisiae – K. lactis dataset into 16,986,996 pairs for training and 5,662,332 pairs for testing. In Experiment 2, we evaluated the model in the first experiment into 8,095,907 pairs of S. cerevisiae and S. pombe genomes. In Experiment 3, we tested in S. cerevisiae and S. pombe as in the second experiment but with models obtained by training with the complete S. cerevisiae – K. lactis dataset (22,649,328 pairs). In Experiment 4, we used the model in the third experiment and tested it in two different pairs: 29,887,416 pairs of S. cerevisiae and C. glabrata and 20,318,472 pairs of C. glabrata – K. lactis.

The details for the data used in the experiments are specified in Table 3. For each genome pair we built four datasets (BLOSUM50, BLOSUM62_1, BLOSUM 62_2, and PAM250) from combinations of alignment parameter settings shown in Table 4. The gene pair features included in the datasets are the average of local and global alignment similarity measures, the length of sequences, the gene membership to conserved regions (synteny), and the physicochemical profiles within 3, 5, and 7 window sizes. We are aggregating global and local alignment similarities to combine structural and functional similarities of proteins.

The S. cerevisiae – S. pombe dataset represents the union of Inparanoid7.0 and GeneDB ortholog classifications as is described in [35]. On the other hand, the S. cerevisiae – K. lactis and S. cerevisiae – C. glabrata datasets contain all ortholog pairs in the gold groups reported in [10]. At the time of building the subset with all possible pairs, we just excluded 89 genes from S. cerevisiae, 37 from C. glabrata, and 1403 from K. lactis because the genome physical location data in the YGOB database [36], required for the LCB feature calculation, was not found.

Table 3. Genome data used in the experiments.

Genome pair	Number of features	Instances per class (majority; minority)	Imbalance ratio (*IR*)	Excluded genes
S. cerevisiae - *K. lactis*	6	(22,646,914; 2414)	9381.489	89 out of 5861 *S. cerevisiae* proteins 37 out of 5215 *C. glabrata* proteins 1403 out of 5327 *K. lactis* proteins
S. cerevisiae - *C. glabrata*	6	(29,884,575; 2841)	10519.034	
C. glabrata - *K. lactis*	6	(20,317,232; 1240)	16384.865	
S. cerevisiae - *S. pombe*	6	(8,090,950; 4957)	1632.227	

Table 4. Datasets built with different alignment parameter values.

Dataset	Substitution matrix	Gap open	Gap extended
BLOSUM50	BLOSUM50	15	8
BLOSUM62_1	BLOSUM62	8	7
BLOSUM62_2	BLOSUM62	12	6
PAM250	PAM250	10	8

The amount of pairs in each dataset prevented traditional machine learning methods from training and testing, so big data implementations were selected from scalable MapReduce Mahout [21] and Spark MLlib [37] libraries. The selected algorithms for the evaluation are listed in Table 5. Explicitly, we selected the random oversampling (ROS) and the cost-sensitive approaches for imbalance management. Besides, the big data framework details are specified in Table 6. On the other hand, the unsupervised algorithms parameter values used for the evaluation are specified in Table 7.

The infrastructure used to perform the big data experiments consists of 20 nodes connected via a 40 GB/s Infiniband network. Each node has two Intel Xeon E5-2620 microprocessors (at 2 GHz, 15 MB cache) and 64 GB of main memory working under Linux CentOS 6.5. The head node of the cluster has two Intel Xeon E5-2620 microprocessors (at 2.00 GHz, 15 MB cache) and 32 GB of main memory.

2.5.2. Results

Supervised classifiers that manage big data were compared from two different points of view:

- The efficacy and the execution time when the training process is performed with one partition of the genome pair dataset or with the complete dataset.

- The efficacy and the execution time in the classification process performed on datasets with potential fails for ortholog detection:
    - The one of related species (distant relatives), S. cerevisiae and S. pombe, false negatives can increase in the presence of divergent sequences.
    - Other two Saccharomycete yeast datasets which complexity is a possible increased number of false positives due to lots of paralog losses produced by the WGD.

Table 5. Supervised algorithms and parameter values selected for the experiments.

Algorithm	Parameter values
RF-BD1	**Number of trees:** 100 Random selected features per node: 32 Number of maps: 20
RF-BDCS	**Number of trees:** 100 Random selected features per node: 3 Number of maps: 20 $C(+\|-) = IR$ $C(-\|+) = 1$
ROS (100%) + RF-BD	RS3 = 100%
ROS (130%) + RF-BD	RS = 130%
SVM-BD	**Regulation parameter:** 1.0, 0.5, and 0.0 Number of iterations: 100 (by default) StepSize: 1.0 (by default) miniBatchFraction: 1.0 (percent of the dataset evaluated in each iteration 100%)
ROS (100%) + SVM-BD	RS = 100%
ROS (130%) + SVM-BD	RS = 130%

[1]BD means big data.

[2]int.($\log_2 N + 1$), where N is the number of features of the dataset.

[3]RS represents resampling size.

Table 6. Big data framework, applications and algorithms.

Big data framework	Application	Algorithms
Hadoop 2.0.0 (Cloudera CDH4.7.1) with the head node configured as name-node and job-tracker, and the rest, as data-nodes and task-trackers	– MapReduce ROS implementation – A cost-sensitive approach for Random Forest MapReduce algorithm (RF-BD) –MapReduce RF implementation (Mahout Library)	RF-BDCS ROS (100%) + RF-BD ROS (130%) + RF-BD
Apache Spark 1.0.0 with the head node configured as master and name-node, and the rest, as workers and data-nodes	– Apache Spark Support Vector Machines (MLLib)	ROS (100%) + SVM-BD ROS (130%) + SVM-BD

Supervised and unsupervised classifiers are compared in terms of the efficacy in complex Saccharomycete yeast datasets.

Best mean results of G-Mean, AUC, TPR, and TNR (true negative rate) in Experiments 1, 2, and 3 are shown in Tables 8–10, respectively, with the most outstanding values in bold face. The best performance of G-Mean, AUC and the balance between TPR and TNR mean values were obtained with the cost-sensitive Big Data Random Forest algorithm (RF-BDCS) in Experiment 1 (testing in a partition of S. cerevisiae – K. lactis) and with support vector machine-BigData (SVM-BD) (regParam: 0.5) in Experiments 2 and 3 (testing in S. cerevisiae – S. pombe). There are no significant changes in classifiers with the best results (those that manage data imbalance) when we use the four alignment parameter value combinations in Table 4. Similarly, good efficacy is obtained when we trained with a partition of S. cerevisiae – K. lactis and with the corresponding complete training set.

Table 7. Unsupervised algorithms and parameter values in the experiments.

Algorithm	Parameter values	Implementation
RBH	Soft filter and Smith Waterman alignment E-value = 1e-06	BLASTp program1 Matlab script
RSD	E-value thresholds: 1e-05, 1e-10 and 1e-20 Divergence thresholds α: 0.8, 0.5 and 0.2.	BLASTp program1 Python script2
OMA	Default parameter values	OMA stand-alone3
[1]Available in http://www.ncbi.nlm.nih.gov/BLAST/.		
[2]Available in https://pypi.python.org/pypi/reciprocal_smallest_distance/1.1.4/.		
[3]Available in http://omabrowser.org/standalone/OMA.0.99z.3.tgz.		

Table 8. Best mean *G-Mean, AUC, TPR* and *TNR* results in Experiment 1 in *S. cerevisiae – K. lactis*. Highlighted *TPR* and *TNR* with the best stability in these measures.

Algorithm	*G-Mean*	*AUC*	*TPR*	*TNR*
ROS (RS: 100%) + RF-BD	0.9807	0.9809	0.9621	1.0000
ROS (RS: 130%) + RF-BD	0.9812	0.9813	0.9629	0.9997
RF-BDCS	0.9889	0.9889	0.9788	0.9997
ROS (RS: 100%) + SVM-BD (regParam: 1.0)	0.9477	0.9477	0.9369	0.9586
ROS (RS: 100%) + SVM-BD (regParam: 0.5)	0.8791	0.8845	0.93038	0.96259
ROS (RS: 130%) + SVM-BD (regParam: 0.5)	0.8528	0.8629	0.93893	0.95476

Table 9. Best mean *G-Mean, AUC, TPR* and *TNR* results in Experiment 2 in *S. cerevisiae – S. pombe*. Highlighted *TPR* and *TNR* with the best stability in these measures.

Algorithm	*G-Mean*	*AUC*	*TPR*	*TNR*
RF-BDCS	0.7103	0.7542	0.5570	0.9997
ROS (RS: 100%) + SVM-BD (regParam: 1.0)	0.8543	0.8641	0.7432	0.9923
ROS (RS: 100%) + SVM-BD (regParam: 0.5)	0.8795	0.8848	0.7900	0.9796
ROS (RS: 100%) + SVM-BD (regParam: 0.0)	0.4312	0.5846	0.9781	0.1911

Table 10. Best mean *G-Mean, AUC, TPR* and *TNR* results in Experiment 3 in *S. cerevisiae – S. pombe*. Highlighted *TPR* and *TNR* with the best stability in these measures.

Algorithm	*G-Mean*	*AUC*	*TPR*	*TNR*
RF-BDCS	0.6745	0.7294	0.4590	0.9998
ROS (RS: 100%) + SVM-BD (regParam: 1.0)	0.8533	0.8632	0.7332	0.9933
ROS (RS: 100%) + SVM-BD (regParam: 0.5)	0.8791	0.8845	0.7895	0.9795
ROS (RS: 130%) + SVM-BD (regParam: 1.0)	0.7956	0.8164	0.6348	0.9979
ROS (RS: 130%) + SVM-BD (regParam: 0.5)	0.8528	0.8629	0.7331	0.9926

The SVM algorithm combined with ROS showed the best time results in Experiment 1 (SVM-BD (regParam: 0.5) executed in 14 minutes and 19 seconds) and (SVM-BD (regParam: 0.0) executed in 14 minutes and 21 seconds). These algorithms also showed good classification quality, mainly, in the balance between TPR and TNR. In Experiments 2 and 3 SVM variants combined with ROS (RS:100) run very rapidly, in particular, SVM-BD (regParam: 0.5) executed in 14 minutes and 29 seconds in Experiment 2 and in 16 minutes and 53 seconds in Experiment 3.

Analyzing measure values of Experiment 4 we noticed that the performance of supervised classifiers is almost the same as in the previous experiments in terms of the stability in the four different alignment parameter value datasets. Best mean results of G-Mean and AUC, TPR and TNR for supervised and unsupervised classifiers are shown in Table 11. The underlined values highlight the most effective method in the experiment while the bold face values identify the best performing supervised and unsupervised algorithms. Supervised Random Forest classifiers managing data imbalance overdid the unsupervised ones, specifically, when the cost-sensitive version is applied. Among the unsupervised classifiers, RSD reaches the highest G-Mean (0.9374) using the recommended parameter values (E-value = 1e-05 and α = 0.8) in [38]), higher values were also obtained for AUC and TPR. On the contrary, OMA was the best among the unsupervised algorithms in S. cerevisiae – S. pombe datasets.

The results of the Friedman test [39] for the AUC measure with the four datasets of S. cerevisiae – C. glabrata, C. glabrata – K. lactis, and S. cerevisiae – S. pombe and the supervised classifiers that manage imbalance are shown in Table 12 (first column). The test yielded significant differences among classifiers, being RF-BDCS the one with the high mean rank followed by the classifiers based on Random Forest with ROS preprocessing and the SVM classifier with ROS (RS:100) and 0.5 regulation. In the Friedman test for G-Mean (Table 12 second column) there are also significant differences with higher mean rank values for RF-BDCS and ROS (RS: 100%) + SVM-BD (regParam: 1.0). When we applied the same test to compare execution time (Table 12 third column) there are significant differences among the algorithms. In particular, SVM classifiers combined with ROS exhibited the best mean ranks.

Table 11. Best mean *G-Mean, AUC, TPR* and *TNR* results in Experiment 4 in *S. cerevisiae – C. glabrata* and *C. glabrata – K. lactis.* Highlighted *TPR* and *TNR* with the best stability in these measures.

Dataset/algorithm	*S. cerevisiae – C. glabrata*				*C. glabrata – K. lactis*			
	AUC	*G-Mean*	*TPR*	*TNR*	*AUC*	*G-Mean*	*TPR*	*TNR*
RBH	0.8196	0.7995	0.6392	0.9999	0.8052	0.8242	0.6484	0.9999
RSD 0.2 1e-20	0.9238	0.9206	0.8476	0.9999	0.9047	0.9092	0.8185	0.9999
RSD 0.5 1e-10	0.9340	0.9316	0.8680	0.9999	0.9267	0.9294	0.8589	0.9999
RSD 0.8 1e-05	0.9382	0.9362	0.8765	0.9999	0.9354	0.9374	0.8750	0.9999
OMA	0.9287	0.9259	0.8574	0.9999	0.9125	0.9163	0.8328	0.9999
ROS (RS: 100%) + RF-BD	0.9901	0.9900	0.9805	0.9997	0.9806	0.9811	0.9615	0.9997
ROS (RS: 130%) + RF-BD	0.9901	0.9901	0.9806	0.9997	0.9807	0.9807	0.9617	0.9997
RF-BDCS	0.9934	0.9934	0.9876	0.9993	0.9847	0.9862	0.9702	0.9992

Table 12. Results of the Friedman tests comparing supervised classifiers in terms of *AUC, G-Mean* and execution time in the four datasets of *S. cerevisiae – C. glabrata, C. glabrata – K. lactis* and *S. cerevisiae – S. pombe.*

Algorithm	*AUC* mean rank	*G-Mean* mean rank	Execution time mean rank
ROS (RS: 100%) + RF-BD	5.96	5.71	8.00
ROS (RS: 130%) + RF-BD	5.96	5.79	9.00
RF-BDCS	7.67	7.67	7.00
ROS (RS: 100%) + SVM-BD (regParam: 1.0)	2.62	6.12	2.33

Algorithm	*AUC* mean rank	*G-Mean* mean rank	Execution time mean rank
ROS (RS: 100%) + SVM-BD (regParam: 0.5)	5.50	5.79	2.25
ROS (RS: 100%) + SVM-BD (regParam: 0.0)	5.12	2.00	2.58
ROS (RS: 130%) + SVM-BD (regParam: 1.0)	2.92	4.46	4.33
ROS (RS: 130%) + SVM-BD (regParam: 0.5)	4.79	5.62	4.50
ROS (RS: 130%) + SVM-BD (regParam: 0.0)	4.46	1.83	5.00
Test statistics			
N	12.000	12.000	12.000
Chi-square	31.359	47.850	80.333
df	8.000	8.000	8.000
Asymp. Sig.	0.000	0.000	0.000

2.5.3. Discussion

As a general result, experiments showed that supervised classifiers changed only slightly with the selection of different alignment parameters, maybe because of the appropriate combination of scoring matrices and gap penalties in relation to the sequence diversity between the two yeast genomes. The four scoring matrices assessed have been previously recommended to detect homologs in a wide range of amino acid identities. Moreover, gap penalty settings were not low enough to affect the sensitivity of the alignment [40].

The ROS pre-processing method for big data made SVM effective for pairwise ortholog detection and improved the performance of Random Forest for big data even more with a higher value for the resampling size parameter of 130% [41]. Conversely, the experiments showed that the variation in this parameter value from 100 to 130% did not significantly influence on the performance of the SVM big data classifier with different regulation values.

The cost-sensitive classifier RF-BDCS showed the best performance in S. cerevisiae – C. glabrata, C. glabrata – K. lactis and S. cerevisiae – K. lactis because probably it improved the training from the minority class. The best tree split was selected following the misclassification costs assigned to the instances; such costs were also considered to associate certain class to a leaf [29]. This cost treatment does not imply changes in the sample distribution, and avoids possible overfitting that is commonly found in ROS solutions due to the presenceof duplicated instances. The setting of the cost values ((C(+|–) = IR) and C(–|+) = 1)) can also influence on the success of the algorithm.

In the case of SVM for big data classifier, the fixed regularization parameter defines the trade-off between the goal of minimizing the training error and minimizing the model complexity to avoid overfitting. The higher is its value, the simpler the model, however, a better performance in classification may be achieved by setting an intermediate regulation value, or one close to cero [37]. Specifically, the ROS (RS: 100%) + SVM-BD (regParam: 0.5) classifier exhibited the best AUC and G-Mean values in S. cerevisiae – S. pombe, and the best balance between TPR and TNR in the rest of the datasets.

All methodologies including the proposed supervised big data approach generally declined their performance for ortholog detection in S. cerevisiae – S. pombe datasets, probably because of S. pombe is a distant relative of S. cerevisiae [23]. The supervised classifiers performance was also negatively affected by differences in data distribution between the train and test sets [42]. On the contrary, ROS (RS: 100%) + SVM-BD (regParam: 0.5) remained stable in S. cerevisiae – C. glabrata, C. glabrata – K. lactis and S. cerevisiae – S. pombe datasets when considering the balance between TPRate and TNRate. Best results obtained in S. cerevisiae – C. glabrata are outstanding where algorithms are vulnerable to produce false positives since both genomes underwent a WGD and a subsequent differential loss of gene duplicates [10].

The initial assumption of RBH, RSD and OMA that the sequences of orthologous genes/proteins are more similar to each other than they are to any other genes from the compared organisms may produce classification errors [12]. The reduced quality shown by RBH, RSD and OMA, mainly in the case of RBH, could be caused by this assumption despite that BLAST parameters can be tuned as has been recommended in [43]. In particular, RBH infer orthology relationships simply based on reciprocal BLAST best hits. In contrast, the RSD procedure is less likely than RBH to be misled by existing close paralogs since it relies on both global sequence alignment and maximum likelihood estimation of evolutionary distances finding many putative orthologs missed by RBH. The OMA algorithm also displays advantages over RBH by using evolutionary distances instead of alignment scores. It allows the inclusion of one-to-many and many-to-many orthologs considering the uncertainty in distance estimations and detects potential differential gene losses.

However, the effectiveness of large data supervised classifiers in handling imbalance over RSD and OMA can be attributed to the mix of features and the knowledge gained from carefully selected classifications. By combining alignment measures, comparing sequence lengths, assessing gene membership to conserved regions (synteny), and analyzing the physicochemical profiles of amino acids, we were able to enhance the detection of homology and significantly improve the results of supervised classification on the test sets. This improvement was observed even when both species experienced whole genome duplication (WGD). More precisely, the combination of global and local alignment scores enables us to merge protein structural and functional connections between pairs of sequences, respectively. The periodicity of the physicochemical properties of amino acids can identify similarities between protein pairings that have functional similarities, even when their amino acid sequences have low identities (<35%). These sequences could impact the identification of orthologs in S. cerevisiae and S. pombe, which are somewhat similar and their orthologs may have undergone divergence. Synteny information allows us to infer that genes located within the same conserved section in the genomes of various species are likely to be orthologs. In addition, the length of sequences and the relative locations of amino acids within the same protein in various species, as well as in duplicated sections within the same species, can also contribute to improved results in supervised classification.

3. CONCISE SUMMARY

The integration of alignment measures with additional protein pair data, such as sequence lengths, gene membership to conserved areas, and physicochemical profiles, has enhanced the homology detection in the suggested supervised method for identifying pairwise orthologs. The combination of these attributes, coupled with curated ortholog pairs taken from a curated dataset, has resulted in an effective and efficient technique for classifying orthologs in a scenario involving huge data. This method addresses the challenge of the low ratio of orthologs to the total potential gene pairs between two genomes.

The supervised classifiers performed better than the prominent unsupervised techniques (RBH, RSD, and OMA) even when applied to yeast datasets that included "traps" designed to challenge ortholog discovery systems. Introducing novel gene pair features in future study could enhance the efficacy and efficiency of the supervised algorithms.

The scalability analysis of the suggested gene pair feature calculation demonstrated the potential improvements in scalability when parallelizing the computation and implementing a big data model for the same calculation.

AT A GLANCE

Orthologs are genes that exist in various species and have evolved from a shared ancestor. Ortholog identification is crucial for the analysis of phylogenies and the prediction of unknown gene functions. The scalability of comparing genes (or proteins) in pairs and the scalability of the classification process provide a difficulty because of the continuously growing number of sequenced genomes. Ortholog recognition techniques that rely solely on sequence similarity often struggle to accurately classify Saccharomycete yeasts due to the presence of several paralogies and gene deletions. This book chapter introduces a novel classification method that utilizes pairwise similarity measures in a decision system. The strategy takes into account the significant imbalance between ortholog and non-ortholog pairs. New gene pair similarity measurements are established using protein physicochemical characteristics, gene pair membership to conserved areas in related genomes, and protein lengths. An analysis is conducted to evaluate the efficiency and scalability of calculating these indicators, with the aim of recommending their application for big data. Overall, the assessment of supervised algorithms for handling large and imbalanced data shown significant efficacy in Saccharomycete yeast genomes.

REFERENCES

1. Fitch WM. Distinguishing homologous from analogous proteins. Systematic Zoology. 1970;19:99-113
2. Vashist A, Kulikowski C, Muchnik I. Screening for ortholog clusters using multipartite graph clustering by quasi-concave set function optimization. In: Slezak D et al., editors. RSFDGrC 2005. Vol. 3642. LNAI. Berlin Heidelberg: Springer-Verlag; 2005. pp. 409-419
3. Hirsh AE, Fraser HB. Protein dispensability and rate of evolution. Nature. 2001;411:1046-1049
4. Östlund G, Schmitt T, Forslund K, Köstler T, Messina DN, Roopra S, et al. InParanoid 7: New algorithms and tools for eukaryotic orthology analysis. Nucleic Acids Research. 2010;38:D196-D203
5. Chen F, Mackey AJ, Stoeckert CJ, Roos DS. OrthoMCL-DB: Querying a comprehensive multi-species collection of ortholog groups. Nucleic Acids Research. 2006;34(Database issue):D363-D3D8
6. Dessimoz C, Cannarozzi G, Gil M, Margadant D, Roth A, Schneider A, Gonnet GH. OMA, A Comprehensive, Automated Project for the Identification of Orthologs from Complete Genome Data: Introduction and First Achievements. In: Comparative Genomics: RECOMB 2005 International Workshop, Dublin, Ireland, September 18-20, 2005 Proceedings. Edited by McLysaght A, Huson DH. Berlin, Heidelberg, Germany; 2005:61-72
7. Wall DP, Fraser HB, Hirsh AE. Detecting putative orthologs. Bioinformatics. 2003;19(13):1710-1711
8. Kuzniar A, RCHJv H, Pongor S, JAM L. The quest for orthologs: Finding the corresponding gene across genomes. Trends in Genetics. 2008;30:1-13
9. Towfic F, VanderPlas S, Oliver CA, Couture O, Tuggle CK, Greenlee MHW, et al. Detection of gene orthology from gene co-expression and protein interaction networks. BMC Bioinformatics. 2010;11(Suppl 3):S7
10. Salichos L, Rokas A. Evaluating ortholog prediction algorithms in a yeast model clade. PLoS ONE. 2011;6(4):1-11
11. Koonin EV. Orthologs, paralogs, and evolutionary genomics. Annual Review of Genetics. 2005;39:309-338
12. Kristensen DM, Wolf YI, Mushegian AR, Koonin EV. Computational methods for gene orthology inference. Briefings in Bioinformatics. 2011;12(5):379-391
13. Kamvysselis MK. Computational Comparative Genomics: Genes, Regulation, Evolution. Massachusetts: Massachusetts Institute of Technology; 2003
14. Zheng XH, Lu F, Wang Z-Y, Zhong F, Hoover J, Mural R. Using shared genomic synteny and shared protein functions to enhance the identification of orthologous gene pairs. Bioinformatics. 2005;21(6):703-710

15. Lechner M, Hernandez-Rosales M, Doerr D, Wieseke N, Thévenin A, Stoye J, et al. Orthology detection combining clustering and synteny for very large datasets. PLoS ONE. 2014;9(8):e105015

16. Chen X, Zheng J, Fu Z, Nan P, Zhong Y, Lonardi S, et al. Assignment of orthologous genes via genome rearrangement. IEEE/ACM Transactions on Computational Biology and Bioinformatics. 2005;2(4):302-315

17. Fu Z, Chen X, Vacic V, Nan P, Zhong Y, Jiang T. MSOAR a high-throughput ortholog assignment system based on genome rearrangement. Journal of Computational Biology. 2007;14:16

18. Chen TW, Wu TH, Ng WV. DODO: An efficient orthologous genes assignment tool based on domain architectures. Domain based ortholog detection. BMC Bioinformatics. 2010;11(Suppl 7):S6

19. Sonnhammer ELL, Gabaldón T, Sousa da Silva AW, Martin M, Robinson-Rechavi M, Boeckmann B, Thomas PD, Dessimoz C. Big data and other challenges in the quest for orthologs. Bioinformatics. 2014, 30(21):1-6

20. Alexeyenko A, Tamas I, Liu G, Sonnhammer ELL. Automatic clustering of orthologs and inparalogs shared by multiple proteomes. Bioinformatics. 2006;22(14):e9-e15

21. Owen S, Anil R, Dunning T, Friedman E. Mahout in Action, 2011, Manning Publications Co., USA

22. Krishnan S, Smith V. MLlib. 2013 [enero 2015]. Available from: https://spark.apache.org/docs/

23. Wood V. Schizosaccharomyces pombe comparative genomics; from sequence to systems. In: Comparative genomics 2006 Jan 1 (pp. 233-285). Springer Berlin Heidelberg, Germany

24. Galpert D, Río Sd, Herrera F, Ancede-Gallardo E, Antunes A, Agüero-Chapin G. An effective big data supervised imbalanced classification approach for ortholog detection in related yeast species. BioMed Research International [serial on the Internet]. 2015;2015:748681

25. Grama A, Gupta A, Karypis G, Kumar V, editors. Introduction to Parallel Computing, Second Edition: Addison Wesley, Boston, USA; 2003

26. Bonvin N. Linear Scalability of Distributed Applications. Suisse: École polytechnique fédérale de lausanne; 2012

27. Dean J, Ghemawat S. MapReduce: Simplified data processing on large clusters. OSDI'04: Proceedings of the 6th Symposium on Operating System Design and Implementation. San Francisco, California, USA: USENIX Association; 2004. p. 137-150

28. Fernández A, Río Sd, López V, Bawakid A, Jesus MJd, Benítez JM, et al. Big Data with cloud computing: An insight on the computing environment, MapReduce, and programming frameworks. WIREs Data Mining and Knowledge Discovery [serial on the Internet]. 2014

29. Sd R, López V, Benítez JM, Herrera F. On the use of MapReduce for imbalanced Big Data using random forest. Information Sciences. 2014;285:112-137

30. Bishop CM, editor. Pattern Recognition and Machine Learning: Springer Science+Business Media, LLC, NY, USA; 2006.

31. Kuncheva LI, editor. Combining Pattern Classifiers Methods and Algorithms. Hoboken, New Jersey: John Wiley & Sons, Inc.; 2004

32. Barandela R, Sánchez JS, García V, Rangel E. Strategies for learning in class imbalance problems. Pattern Recognition. 2003;36(3):849-851

33. Bradley AP. The use of the area under the ROC curve in the evaluation of machine learning algorithms. Pattern Recognition. 1997;30(7):1145-1159

34. He H, Garcia EA. Learning from imbalanced data. IEEE Transactions on Knowledge and Data Engineering. 2009;21(9):1263-1284

35. Koch EN, Costanzo M, Bellay J, Deshpande R, Chatfield-Reed K, Chua G, et al. Conserved rules govern genetic interaction degree across species. Genome Biology. 2012;13(7)

36. Byrne KP, Wolfe KH. The yeast gene order browser: Combining curated homology and syntenic context reveals gene fate in polyploid species. Genome Research. 2005;15:1456-1461

37. Krishnan S, Smith V. Linear Support Vector Machines (SVMs). 2013 [enero 2015]. Available from: https://spark.apache.org/docs/latest/mllib-linear-methods.html#linear-support-vector-machines-svms

38. DeLuca TF, Wu I-H, Pu J, Monaghan T, Peshkin L, Singh S, et al. Roundup: A multi-genome repository of orthologs and evolutionary distance. Bioinformatics. 2006;22(16):2044-2046

39. Trawinski B, Smetek M, Telec Z, Lasota T. Nonparametric statistical analysis for multiple comparison of machine learning regression algorithms. International Journal of Applied Mathematics and Computer Science. 2012;22(4):867-881

40. Pearson WR. Selecting the right similarity-scoring matrix. Current Protocols in Bioinformatics. 2013;43:3.5.1-3.5.9

41. Triguero I, del Río S, López V, Bacardit J, Benítez JM, Herrera F. ROSEFW-RF: the winner algorithm for the ECBDL?14 big data competition: an extremely imbalanced big data bioinformatics problem. Knowledge-Based Systems. 2015;87:69-79

42. Moreno-Torres JG, Llorà X, Goldberg DE, Bhargava R. Repairing fractures between data using genetic programming-based feature extraction: A case study in cancer diagnosis. Information Sciences. 2013;222:805-823

43. Hagelsieb GM, Latimer K. Choosing BLAST options for better detection of orthologs as reciprocal best hits. Bioinformatics. 2008;24(3):319-324

SECTION 2

FOOD APPLICATIONS

CHAPTER 3

YEAST STRAIN SELECTION FOR FERMENTED BEVERAGES: A STUDY ON GROWTH KINETICS

1. INTRODUCTION: AN OVERVIEW

The preparation of alcoholic beverages is a really ancient culinary tradition. Their elaboration relies on a crucial process known as alcoholic fermentation (AF), which is a biological transformation of carbohydrates into ethanol facilitated by yeasts. The occurrence of AF can be classified into three methods [1, 2]: (i) spontaneous fermentation, which arises from the naturally existing microbiota in the musts or on equipment surfaces; (ii) fermentation induced by the addition of commercial yeasts; and (iii) fermentation initiated by the introduction of certain native strains. Currently, there is a growing global demand for autochthonous strains due to the recognition that this choice helps maintain the distinct and characteristic qualities that native microorganisms offer. The yeasts involved in alcoholic fermentation (AF) are classified as "fermentative yeasts" and can be generally separated into two groups: Saccharomyces and non-Saccharomyces (nS). Saccharomyces demonstratcs a remarkable ability to efficiently convert sugars and withstand high levels of ethanol and SO2. This makes it the ideal genus for fermentation [3]. The majority of nS, including Candida, Kloeckera, Hanseniaspora, Zygosaccharomyces, Schizosaccharomyces, Torulaspora, Brettanomyces, Saccharomycodes, Pichia, and Williopsis, have a limited tolerance to ethanol [4]. The significance of these yeasts in beverages such as wine lies in their metabolic characteristics, since the diverse range of enzymes they provide can increase the typical qualities and improve the sensory profile [4–7]. In contrast, in other beverages such as Mexican tequila and mescal, as well as Brazilian cachaça, non-saccharomyces yeasts (nS) are regarded as the primary contributors to ethanol synthesis [8–10]. Throughout the fermentation process, yeast cells experience various stressful situations, which ultimately lead to a drop in their growth and survival rate. This decrease in yeast cell performance results in a fall in fermentation efficiency. Yeasts that can withstand these conditions with minimal loss of viability are most suitable for these purposes [12].

2. STRESS TOLERANCE OF FERMENTATIVE YEAST

The primary goal of fermentative yeast is to rapidly and efficiently convert simple sugars into ethanol without developing unpleasant flavors [13]. Several factors affect the yeast ability to grow in the fermentative media related with the type and style of beverage produced; therefore, the ability to adapt and to cope with this hostile environment is considered the main feature to select fermentative yeast [12]. Some of the most relevant inhibiting conditions are summarized in this section.

2.1. Limiting conditions associated with musts

2.1.1. Carbohydrates

Carbohydrates are the most important nutrient since they are metabolized to form biomass, ethanol, and different by-products such as volatile compounds, glycerol, and others that will develop the sensorial characteristics [14]. However, they also produce the first stress due to osmotic pressure in the cells after their inoculation. Therefore, tolerance to high sugar concentration is one of the main criteria for yeast selection, especially for those designated for their use in the elaboration of liquorish beverages such as "Sauternes" wine. Yeast cells have developed mechanisms to adjust to high external osmolarity and maintain or reestablish an inside-directed driving force for water. Adaption to this stress usually takes several hours in which yeast cells accumulate glycerol and trehalose [15] and change their cell wall composition [16] to counter loss of water by the osmotic pressure. The stress level will depend on the type and concentration of sugars found. Concentrations range from diluted juices to the high gravity worst containing 16–18% of dissolved solids, rice mash for sake production with 20% of solids, and grape juice with 200 g L-1 of sugar content or even more.

Sugar composition of the media strongly impacts on yeast metabolic physiology [17]. Glucose and fructose are widely found in nature as free sugars or as polysaccharides, also in different proportions. Grape juice contains approximately 1:1 of glucose-fructose [18] apple juice 5:8 [19], and agave juice for the elaboration of tequila and mezcal, after thermal processing of inulins from the plant, contains 1:20 ratio [20]; in pulque production from raw agave juice, inulin is also predominant [21]. For beer fermentation maltotriose and maltose comprise 50% of sugars; in rum, cachaça, and tafia, produced by sugar cane juice or molasses, mainly sucrose is present [22].

2.1.2. Assimilable nitrogen

Deficiencies in the supply of assimilable nitrogen will lead to sluggish fermentation. The amounts of total nitrogen in musts for the production of fermentative beverages vary from 40 mg L-1 in agave juices [23], 50–150 mg L-1 in apple juice, 80 mg L-1 in sugar cane for cachaça [24], more than 150 mg L-1 in beer malt wort [25], and from 100 to 500 mg L-1 in grape must [26]. A minimum of 66 mg L-1 to sustain the growth of yeasts and to finish fermentation with a total consumption of sugars, and high ethanol yield is considered [26].

In the fermentative media, after yeasts are inoculated, a rapid uptake of nitrogen compounds used for the biosynthesis of macromolecules and storage in vacuoles is carried out if sufficient amount is present in the must. As yeasts have high preference for ammonium ions, it is used as exogenous nitrogen source in fermentations. Glutamate, aspartate, and glutamine are the first amino acids uptaken if they are present in the must. When a favorite nitrogen source is depleted, yeast will use a less preferred, resulting in reduced growth and fermentation rates [12].

A target for the improvement of fermentation is then the selection of yeasts with low nitrogen requirements, which will reduce the necessity of adding supplements to the must that will make the process expensive.

2.1.3. Sulfur dioxide (SO2)

During the beginning of wine fermentation, native microbiota mainly nS yeasts and bacteria dominate the media with low ethanol production that could negatively affect the quality of the final product. To inhibit them and to favor native or inoculated Saccharomyces strains to develop, sulfur dioxide (SO2) is added at levels ranging from 20 to 50 ppm [27]. Saccharomyces also produces SO2 during metabolism of sulfate ions [28]. If the amount of SO2 added and produced by Saccharomyces strain are high and remain until the end the process, then safety problems arise as this compound causes certain level of toxicity in human population that consume it. The selection of yeast strains resistant to SO2 and low SO2 producers is then desirable.

2.2. Fermentative-derived conditions

2.2.1. Interaction with other microorganisms

Different microorganisms interact during the elaboration of fermented beverages. The variety of these interactions and their impacts on efficiency and product quality should be individually determined, as they will depend on the fermentative strains, the native associated microbiota, and the type of beverage. One of the main metabolites exerting a clear effect on yeast growth and performance is ethanol, mainly produced by Saccharomyces cerevisiae. Other metabolites as medium-chain fatty acids and high amounts of acetic acid can negatively affect the growth of a co-fermenting yeast species [29]. Cell-to-cell contact as well as oxygen availability appears to be also involved in the interactions between S. cerevisiae and other nS species [30].

One special aspect is the "killer phenotype," which refers to those yeasts able to secrete polypeptide toxins which kill sensitive cells and which is believed to be a potential mechanism to prevent a competitor from gaining access to a resource [31]. Killer toxins differ between species and strains, thus varying the modes of action, from changing membrane permeability in sensitive cells to inhibiting DNA replication or stopping cell division at G1 phase. All killer toxins are usually active and stable at pH 4–5 and 20–25°C; nevertheless each toxin has an optimum pH and temperature at which it manifests its killer character more effectively [32, 33].

2.2.2. Ethanol

Ethanol produced during fermentation is known to inhibit yeast growth, resulting in a primary factor on yeast efficiency; in turn, the viability of yeast cells in the presence of ethanol constitutes a key feature on strain selection for fermentative purposes [34]. Ethanol affects many aspects of yeast survival, as the fluidity of the plasmatic membrane [35], the vacuole morphology [36], the activity of crucial glycolytic enzymes [37], and the mitochondrial DNA [38]. Ethanol also causes the denaturation of hydrophilic and hydrophobic proteins, affecting various transport systems such as the general amino acid permease and glucose uptake processes [18]. Regarding molecular response of S. cerevisiae in the presence of ethanol, it increases the expression of genes associated with glycolysis and mitochondrial function and decreases gene expression in energy-demanding growth-related processes; it also induces the production of heat shock-like proteins, lowering the rate of RNA and protein accumulation, enhancing the frequency of small mutations, altering metabolism, denaturing intracellular proteins and glycolytic enzymes, and reducing their activity [39].

Moreover, yeasts have developed diverse strategies to counteract the damages produced by ethanol [40], as the generation of fatty acid unsaturation of membrane lipids in S. cerevisiae [35]. Genes involved in intracellular pH homeostasis are also crucial for the resistance to ethanol and other alcohols [41]. This entails in better adapted strains that show a better capacity to activate these mechanisms and endure in the hostile environment formed through alcoholic fermentation.

2.2.3. Organic acids

As previously stated, microorganisms interacting with yeasts during fermentation produce organic acids, and some of them can affect the growth and fermentative efficiency. Lipophilic weak acids, such as acetic, may accumulate inside yeast cells in their undissociated form diffusing into the yeast cells where it dissociates, inducing an acidification of the cytosol [42]. Fatty acids of medium-chain length, as hexanoic, octanoic, and decanoic acids, and also their respective ethyl esters have shown to negatively affect the survival and growth capacity of yeast, being partly responsible for the premature stoppage of fermentations carried by S. cerevisiae [43]. These metabolites can be absorbed by the cell membrane; its toxic effect increases in the pH range of 3.0–5.4. This apparent disadvantage of permeability to fatty acids has been exploited, using the denominated "yeast ghosts," which are a commercially available product that can be added when a sluggish or stuck fermentation occurs in order to absorb this kind of inhibitors [44].

2.2.4. pH

During alcoholic fermentation, pH in the media tends to reduce [45], which is known to be a limiting factor on growth of microorganisms, including yeasts. The optimal pH value for yeast growth is around 4.5, and fermented beverages range in pH from 2.5 to 5.5, which by itself does not imply a restrictive condition, but a low pH (<3.5) combined with ethanol, as it usually occurs on fermented beverages, can prematurely inhibit yeast growth and/or fermentation rate [42].

3. YEAST SELECTION WORKFLOW

The aspects previously described define the medium in which yeasts will be developed, and they must be kept in mind onward. On the other hand, the selection of fermentative yeasts involves several sequential steps in which the final objective is the identification of strains capable to efficiently ferment the must and obtain a product with optimum sensory qualities [3]. The framework for selection usually starts determining the diversity of yeasts present during the spontaneous fermentation of the beverage. The isolation of yeasts usually takes place along with that stage; afterward, the aspects of interest ranging from tolerance conditions to sensory impact are evaluated in each strain in order to identify the best suited to be selected.

3.1. Determination of yeast diversity

As recently reviewed, in every fermented beverage produced around the world, a particular microbiota will develop as a cause of the raw material from which it starts (fruits, grains, dairy products, parts of plants, etc.), its characteristics (nutrients, pH, type of sugar, etc.), the geographic region, and the modifiable options in each elaboration process, including pre-fermentative manipulations, temperatures, and added substances, among others [45]. Furthermore, yeast diversity that develops will largely determine the sensory profile of the final product; therefore it is first necessary to become acquainted with the species present, their dynamics, and the possible role each one plays during the fermentation. For these studies, advances in molecular techniques have been exploited, being nowadays the area with most scientific contributions on fermented beverages worldwide. Culture-independent methods are the preferred, since they are neither affected by the viability of the microorganism nor by low populations of less abundant species [46]. Some of the preferred techniques are high-throughput or next generation sequencing (HTS or NGS, respectively), denaturing gradient gel electrophoresis (DGGE), and quantitative PCR (qPCR), as it is summarized in Table 1 [47–49].

Table 1. Recent studies on the diversity of yeasts present during the elaboration of fermented beverages.

Beverage	Country	Technique	Year	Main yeasts found	Reference
Caxiri (cassava, corn, and sweet potatoes)	Brazil	DGGE	2015	*S. cerevisiae, P. kluivery, C. tropicalis, D. fabryi*	[50]
Wine (cv. "Grenache")	Spain	NGS	2015	*Hanseniaspora, Saccharomyces, Candida, Issatchenkia*	[51]
Tarubá	Brazil	DGGE	2015	*T. delbrueckii, P. exigua, P. manshurica, P. kudriavzevii, C. tropicalis, C. ethanolica*	[52]
Grape must	Spain	DGGE, qPCR, NGS	2015	*H. uvarum, S. bacillaris, S. cerevisiae*	[53]
Fuzhou Hong Qui (Rice wine)	China	DGGE	2015	*Saccharomycopsis fibuligera, P. guilliermondii, S. cerevisiae, Wickerhamomyces anomalus, C. glabrata*	[54]
Cold soak (wine)	Spain	qPCR, DGGE	2016	*H. uvarum, S. bacillaris, S. cerevisiae*	[55]
Xaj-pitha (rice wine)	India	Whole genome shotgun sequencing	2016	*Meyerozyma guilliermondii, Wickerhamomyces ciferrii, S. cerevisiae, C. glabrata, D. hansenii, Ogataea Parapolymorpha, D. bruxellensis*	[56]
Taberna (palm wine)	Mexico	DGGE	2016	*H. guillermondii, S. cerevisiae, P. kudriavzevii, C. tropicalis, K. exigua*	[57]
Pitmud used in strong-flavor liquor	China	DGGE and NGS	2017	*Candida, Wickerhamomyces, Debaryomyces, Saccharomyces, Pichia*	[58]

In spite of the multiple advantages of the culture-independent techniques, it is necessary to become aware of possible gaps inherent to these approaches, as undetectability of minor populations, preferential amplifications, limited databases, and different effectiveness of lysis protocols on certain species are some of the principal. By applying both, culture-dependent and culture-independent techniques, these drawbacks can be surpassed, along with other benefits, as recovering the isolates needed to characterize individually and select possible starter cultures.

3.2. Isolation of strains

Although the sources of yeasts can be quite diverse, isolates are preferably obtained from the initial must and along the fermentation process, intending that this system itself directs to select the best adapted strains. Two possibilities arise at this stage: (1) to isolate random species using a general solid media (PDA or YEPDA) or with an agent leading to add visual differentiation in colony morphologies (WL nutrient agar) or (2) to focus on a specific group using selective media (lysine media for nS or nutrient medium supplemented with sodium metabisulfite and ethanol for Saccharomyces) [3, 81]. If one of the purposes is to complement the community study, the first would be the best choice, but if the selection is directed to certain species, the second would be best. After isolating several strains, characterization and selection will take place.

3.3. Characterization (qualitative approach)

The desirable aspects of fermentative yeast are quite variable; preference is given to one or the other, depending on each case [75]. Some of the most important and various studies in which they were applied are summarized in Table 2. Among this characteristics, it can be considered essential to evaluate: resistance to high concentrations of sugar and ethanol, high fermentation performance (this may not apply when selecting nS yeasts), low production of sulfurous compounds, and volatile acidity and implantation aptitude [79–82]. This last trait is not always performed, even though it is almost imperative, as it requires the design and implementations of molecular methods, such as pulsed field gel electrophoresis (PFGE), mini- or microsatellite markers, qPCR, or others [65, 66]. Once the survival capacity of the strains is determined, the next interesting feature is the impact they exert on sensory aspects. Olfactory qualities of yeast are usually assessed by determining their enzymatic activity, including glucosidases, lyases, proteases, or reductases, which can release or produce active odorant compounds [75]. Also, visual aspect can be affected by yeasts through their enzymatic activity (hydroxycinnamate decarboxylase) or their ability to excrete pyruvate and acetaldehyde, which can lead to the formation of highly stable colorant compounds as pyranoantocyanins that help to improve the visual quality of red wines [77, 83]. Besides, it is quite useful to analyze the nitrogen requirements of yeasts, if they possess the killer phenotype and flocculation ability (particularly important in beer and sparkling wines). Furthermore, some features can only be qualitatively studied such as the presence or absence of certain enzymes, as well as which killer phenotype they possess. Some other are by definition quantitative, as fermentation performance, and several characteristics can be determined both ways, qualitatively and quantitatively, which is the case of tolerance feature.

Table 2. Main aspects considered to select fermentative yeasts.

Characteristic of interest	References
Tolerance to inhibitors (ethanol, SO_2, sugar, pH)	Nikolau et al. [59]; Fiore et al. [60]; Arrizon et al. [61]; Capece et al. [62]; Tristezza et al. [63]
Implantation aptitude	Lopes et al. [64]; Capece et al. [62]; Perrone et al. [65]; Alonso del real [66]
Fermentation vigor (efficiency, ethanol yield, speed)	Tristezza et al. [63]; Ribeiro et al. [67]
Low nitrogen requirements	Arrizon et al. [61]; Julien et al. [68]; Gardner et al. [69]
Enzymatic activity	Fiore et al. [60]; Capece et al. [62]; Capece et al. [70], Csoma et al. [71]; Romano et al. [72]
Color and aroma enhancement	Belda et al. [75]; Steensels et al. [76]; Morata et al. [77]; Morata et al. [78].
Production of specific metabolites (sulfurous, glycerol, fatty acids, other alcohols)	Nikolau et al. [59]; Capece et al. [62, 70]; Tristezza et al. [63]
Killer phenotype	Zagorc et al. [73].
Flocculation	Silva et al. [74]

The methods to evaluate resistant traits of yeast strains are usually based in the qualitative measure of the growth of the microorganism in a synthetic medium containing the limiting compound after a specific time of incubation [82, 84–86]. In the case of ethanol tolerance, there are a broad number of tests that could define this characteristic [87]: (a) the ability to grow in the presence of ethanol, (b) the degree of survival after exposure to a certain concentration, or (c) the maximal ethanol production capacity. Some of these methods are qualitatively measured, and some other imply the set up and monitor of a fermentation, requiring large amounts of resources and hindering the possibility to evaluate a high number of strains. However, fermentation parameters such as ethanol yield, productivity, and maximum specific velocity of cell growth must be measured during the traditional process [22].

Additionally, the fermentative traits are often individually evaluated, while in fermentation several stress factors intervene together and increase in number and magnitude along the process. Testing more than one inhibitor in a qualitative evaluation complicates the selection based only in the presence or absence of growth; thus, a more objective method should be used to compare between strains.

4. QUANTITATIVE METHODS USED FOR THE SELECTION OF FERMENTATIVE YEAST

An alternative to assess the tolerance of yeast strains to limiting factors is to study the growth kinetics of the strains exposed to the inhibiting condition. This can be achieved by traditional methods (plate counts), implying more resources but being more reliable, or by means of optical density (OD) measurements of the yeast in the media to test. Microbial growth data obtained by absorbance measurements permits to obtain kinetic parameters, which can be transformed into quantitative and comparable variables such as (a) detection time, the time to reach the detection level of Bioscreen (Automated OD reader equipment) and its period includes the lag time; (b) maximum growth rate could be estimated with the slope of the tangent of exponential phase; and (c) maximum population density, asymptotic level of OD at the end of exponential phase (Figure 1).

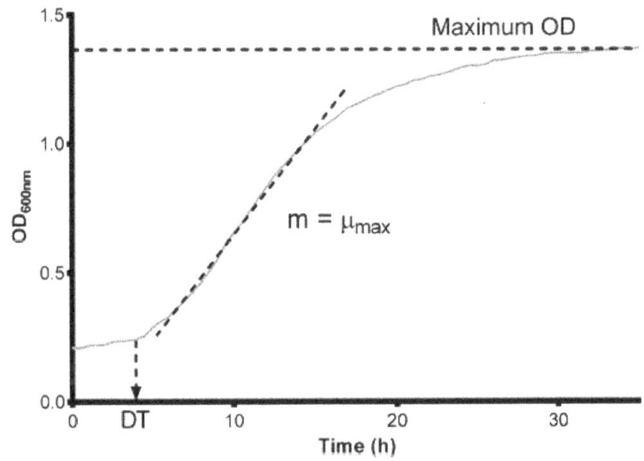

Figure 1. Kinetic parameters for microbiological growth using OD measurements. Detection time (DT), maximum growth rate (μmax), and maximum population density (maximum OD).

Measurement of microbial growth by using turbidity and cells' kinetic behavior description offer some benefits. A considerable amount of data could be generated faster and in real time. Nonetheless, this approach possesses some limitations such as the need to conduct the study in a high-transmittance liquid. Another factor to be considered is that cell concentration required to detect a shift in the measurement is at least 6 log CFU/mL.

4.1. Applications in food matrices

Analytical methods based on OD by assessing the final cell density have been used to evaluate the growth of foodborne pathogens in culture media, translucent liquid foods, and diluted food extracts [88–91]. Data from these studies have offered the possibility to obtain information to determine which food matrices and conditions could be inhibitory of pathogens growth. Also, this approach allows the comparison of strain variability of foodborne pathogens under several growth conditions, e.g., pH and salt concentration [92].

This method has allowed to rapidly evaluate the efficacy of antifungal compounds on the germination of fungal spores. The non-inhibitory concentration (NIC) and the minimum inhibitory concentration (MIC) for different environmental conditions can be calculated mathematically using the OD data [93]. MIC is considered the lowest concentration at which no growth is observed, while NIC is the lowest concentration at which any inhibitory effect is observed.

4.2. Potential applications for the selection of fermentative yeast

Quantitative measurement of the growth of microorganisms based in OD could be used to assess strain tolerance in fermentation-associated conditions. Some strategies and parameters have been proposed in base to OD data.

The MIC and NIC values are related to the susceptibility/tolerance of the microorganisms to stressful conditions. These parameters were proposed by Lambert and Pearson [94] to evaluate microorganisms not associated to fermentation processes, such as Escherichia coli, Pseudomonas aeruginosa, and Staphylococcus aureus. Arroyo-López et al. [18] used MIC and NIC to compare the growth of 29 strains of yeasts (S. cerevisiae, Saccharomyces paradoxus, Saccharomyces bayanus, Saccharomyces kudriavzevii, Saccharomyces mikatae, Saccharomyces arboriculus, Saccharomyces cariocanus, and some nS) in a medium containing ethanol. Álvarez-Pérez et al. [95] used the MIC among the criteria to select S. cerevisiae strains to produce Picudo Rosé wines with different aromatic profiles in Spain.

The area under the OD-time curve could be estimated, and it represents the population generated along the incubation period at a specific condition. The inhibitory effect could be estimated as a relative fraction of the area obtained under an optimal condition (control). Figure 2 shows an example of this parameter, in which the growth of S. cerevisiae in synthetic medium is contrasted with the yeast behavior at the presence of ethanol and SO2 [96].

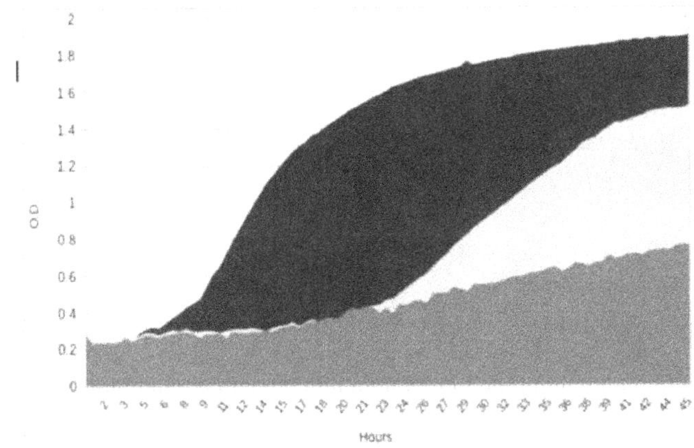

Inhibitor	Area (OD.h)	Fractional area
Ethanol	37.5	37.5/60.48=**0.62**
SO_2	23.5	23.5/60.48=**0.39**
Control	60.48	60.48/60.48=**1.00**

Figure 2. OD curves over time of S. cerevisiae growth in synthetic medium added with ethanol (12%), SO2 (200 mg/L) and without inhibitors (control).

When OD measurements are obtained with an automatized spectrophotometer, the detection time (DT) is reported for each tested condition. Low DT values are associated with the ability of microorganisms to adapt to a specific condition, and consequently, they could start their growth in short time.

Ortiz-Barrera et al. [97] reported DT for 90 nS strains grown in artificial medium containing 6% ethanol or 30 mg of total SO2 (Figure 3). This parameter was used to select strains with the best ability to grow under stressful conditions (ethanol and SO2) associated to wine (enclosed in the square in the figure).

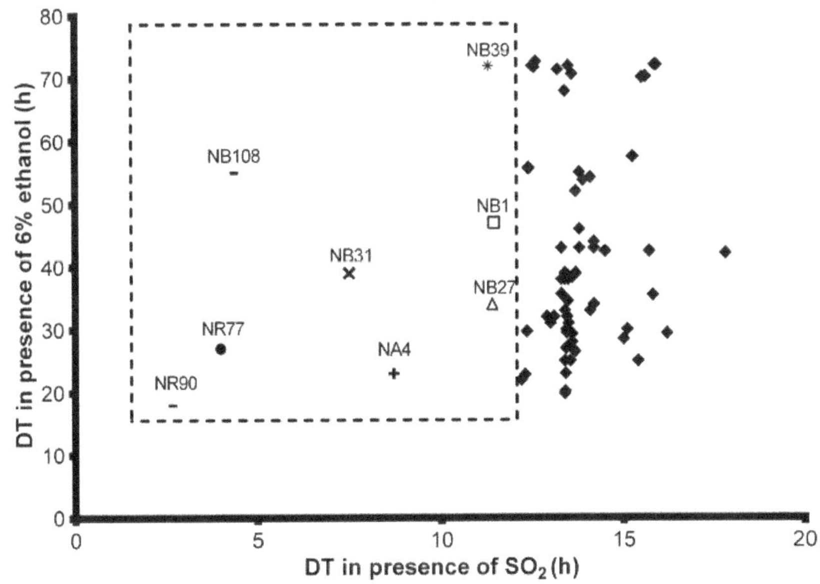

Figure 3. Detection times of 90 non-Saccharomyces strains in YPD (pH 3.5, 20°Bx) containing ethanol (6%) or SO2 (30 mg·L-1). Selected strains are shown in the inset.

In log growth phase, a close relationship between OD and cell concentration is apparent due to the fact that most of the cells are in an active multiplication process which in turn is used to estimate growth rate. In Figure 4, growth rate values of 12 Saccharomyces strains incubated in grape juice media with ethanol and SO2 are shown (unpublished data).

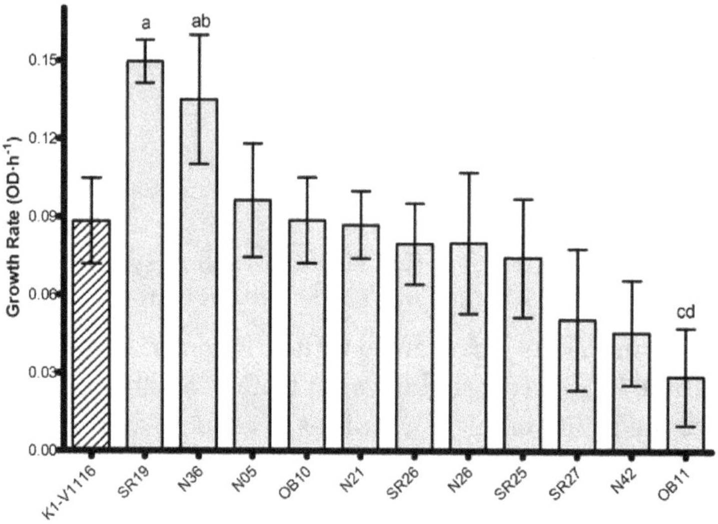

Figure 4. Growth rate of a commercial strain (stripped bar) and 12 Saccharomyces strains (solid bars) in grape juice media with 14% of ethanol, 50 mg/L of SO2, and pH 3.5. Different letters denote statistical difference. Bars without letters correspond to ABCD.

Growth rate and DT are complementary parameters to describe the ability of yeast to adapt and grow in adverse conditions, similar to those found in fermentative beverages. Thus, both parameters (growth rate and DT) could be simultaneously used to select fermentative yeast strains (Figure 5).

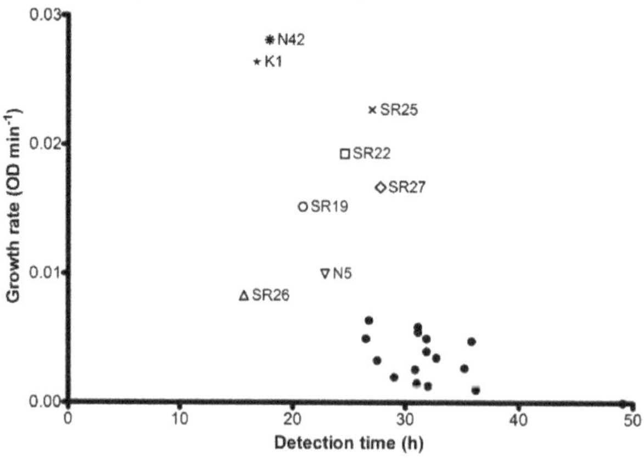

Figure 5. Detection times and growth rate of 29 strains of Saccharomyces grown in YPD medium with 12% of ethanol.

Based on OD data, Salvadó et al. [98] estimated kinetic parameters of yeast strains (S. cerevisiae and nS) grown in a medium with ethanol (0–194.45 g/l) or incubated at different temperatures (4–46°C) by using Gompertz equation modified by Zwietering et al. [99]

$$y = D \times \exp\{-\exp[((\mu_{max})/D) \times (\lambda - t))+1]\} \quad (1)$$

where y = ln(ODt/OD0), D = ln(ODmax/OD0) is the asymptotic maximum value reached, OD0 is the initial OD, ODt is the OD at time t, μmax is the maximum specific growth rate, and λ is the lag phase duration (h).

Using μmax as a basis, the authors developed a secondary model that relates ethanol concentration and fermentation temperature. The model prediction identified temperature during fermentation as the primary factor that promotes the dominance of S. cerevisiae over other yeast genera in the fermentation process. Only a limited number of scientists have utilized this modeling method in yeast comparison [66, 100].

5. CONCISE SUMMARY

The wide array of fermented drinks found globally, together with the diverse range of yeast strains present in each, has prompted the quest for excellent indigenous yeast strains. These strains are selected to enhance the quality of the goods and optimize the fermentation process. Molecular approaches have emerged as an effective approach to uncover the diversity and composition of yeast during any fermentation process. Nevertheless, the techniques used to assess the tolerance of yeast strains, which have been developed over the past few decades, mostly rely on qualitatively measuring the growth of the microorganism in a medium containing the limiting substance after a certain incubation period. By using optical density measurements, it is possible to study the growth of strains in media with limited resources. This method allows for the quick and cost-effective estimation of quantitative and comparable parameters. It provides valuable information about the fitness of each strain in a specific environment, the duration of the lag phase, the growth rate, and the maximum population. Additionally, it enables the rapid initial selection of strains.

AT A GLANCE

The selection of autochthonous yeast strains for use in fermented drinks is based on specific criteria. These criteria include the yeast's capacity to dominate the fermentation process and improve desired sensory characteristics. Additionally, it is important that the yeast does not produce undesirable substances such as biogenic amines or off-odors. One crucial aspect of yeast selection is its ability to withstand various pressures and perform well in fermentation. This necessitates the establishment and monitoring of the process, which can be resource-intensive. Yeast strain tolerance is often assessed by qualitatively measuring the growth of the microorganism in a medium containing the limiting chemical after a certain incubation period. Studying strain growth by optical density measurements allows for the estimation of quantitative and comparative parameters, which provide insight into the cell's fitness in a specific environment. These factors include lag phase duration, growth rate, maximum population, and others. Over the past few decades, culture-independent techniques have been employed to assess the changes in microbial populations during fermentation processes. This chapter presents a comprehensive overview of the latest developments in the selection of fermentative yeasts and the use of kinetic evaluation and molecular techniques in the context of choosing yeast strains for fermented beverages.

REFERENCES

1. Callejon RM, Clavijo A, Ortigueira P, Troncoso AM, Paneque P, Morales ML. Volatile and sensory profile of organic red wines produced by different selected autochthonous and commercial Saccharomyces cerevisiae strains. Analytica Chimica Acta. 2010;660(1-2):68-75. DOI: 10.1016/j.aca.2009.09.040
2. Fleet GH. Yeast interactions and wine flavour. International Journal of Food Microbiology. 2003;86:11-22. DOI: 10.1016/S0168-1605(03)00245-9
3. Rainieri S, Pretorius I. Selection and improvement of wine yeast. Annals of Microbiology. 2000;50:15-31. DOI: 10.1007/s13213-010-0098-0
4. Jolly NP, Varela C, Pretorius IS. Not your ordinary yeast: Non-Saccharomyces yeasts in wine production uncovered. FEMS Yeast Research. 2014;14(2):215-237. DOI: 10.1111/1567-1364.12111
5. Ciani M, Comitini F, Mannazzu I, Domizio P. Controlled mixed culture fermentation: A new perspective on the use of non-Saccharomyces yeasts in winemaking. FEMS Yeast Research. 2009;10(2):123-133. DOI: 10.1111/j.1567-1364.2009.00579.x-
6. Capozzi V, Garofalo C, Chiriatti MA, Grieco F, Spano G. Microbial terroir and food innovation: The case of yeast biodiversity in wine. Microbiological Research. 2015;181:75-83. DOI:10.1016/j.micres.2015.10.005
7. García M, Esteve-Zarzoso B, Arroyo T. Non-saccharomyces yeasts: Biotechnological role for wine production, grape and wine biotechnology. Morata A, editor. InTech, Rijeka, Croatia; 2016. DOI: 10.5772/64957
8. Lachance M. Yeast communities in a natural tequila fermentation. Antonie Van Leeuwenhoek. 1995;68:151-160. DOI: 10.1007/BF00873100
9. Escalante-Minakata P, Blaschek H, Barba de la Rosa A, Santos L, De Leon-Rodríguez A. Identification of yeast and bacteria involved in the mezcal fermentation of Agave Salmiana. Letters in Applied Microbiology. 2008;46:626-630. DOI: 10.1111/j.1472-765X. 2008.02359.x
10. Portugal C, Paron de Silva A, Bortoletto AM, Alcarde AR. How native yeasts may influence the chemical profile of the Brazilian spirit cachaça. Food Research International. 2016;91:18-25. DOI: 10.1016/j.foodres.2016.11.022
11. Zuzuarregui A, del Olmo M. Analyses of stress resistance under laboratory conditions constitute a suitable criterion for wine yeast selection. Antonie Van Leeuwenhoek. 2004;85:271-280. DOI: 10.1023/B:ANTO.0000020162.21248.53

12. Bauer FF, Pretorius IS. Yeast stress response and fermentation efficiency: How to survive the making of wine—A review. South African Journal of Enology and Viticulture. 2000;21:27-51

13. Pretorius IS. Tailoring wine yeast for the new millennium: Novel approaches to the ancient art of winemaking. Yeast. 2000;16:675-729. DOI: 10.1002/1097-0061(20000615)16:8<675::AID-YEA585>3.0.CO;2-B

14. Sanz ML, Martínez-Castro I. Carbohydrates. In: Moreno-Arribas V, Polo C, editors. Wine Chemistry and Biochemistry. New York: Springer Science +Business Media; 2009. pp. 231-248. DOI: 10.1007/978-0-387-74118-5

15. Hohmann S. Osmotic stress signaling and osmoadaption in yeasts. Microbiology and Molecular Biology Reviews. 2002;66(2):300-372. DOI: 10.1128/MMBR.66.2.300-372.2002

16. Slaninova I, Šesták S, Svoboda A, Farkaš V. Cell wall and cytoskeleton reorganization as the response to hyperosmotic shock in Saccharomyces cerevisiae. Archives of Microbiology. 2000;173:245-252. DOI: 10.1007/s002030000136

17. Rodrigues F, Ludovico P, Leao C. Sugar Metabolism in yeast: An overview of aerobic and anaerobic glucose catabolism. In: Péter G, Rosa C, editors. The Yeast Handbook: Biodiversity and Ecophysiology of Yeasts. Heidelberg, Berlin: Springer Berlin; 2006. pp. 101-121

18. Arroyo-López FN, Salvadó Z, Tronchoni J, Guillamón JM, Barrio E, Querol A. Susceptibility and resistance to ethanol in Saccharomyces strains isolated from wild and fermentative environments. Yeast. 2010;27:1005-1015. DOI: 10.1002/yea.1809

19. Wang D, Xu Y, Hu J, Zhao G. Fermentation kinetics of different sugars by Apple yeast Saccharomyces cerevisiae. Journal of the Institute of Brewing. 2004;110(4):340-346. DOI: 10.1002/j.2050-0416.2004.tb00630.x

20. Waleckx E, Gschaedler A, Colonna-Ceccaldi B, Monsan P. Hydrolysis of fructans from Agave tequilana Weber var. azul during the cooking step in a traditional tequila elaboration process. Food Chemistry. 2008;108:40-48. DOI: 10.1016/j.foodchem.2007.10.028

21. Lappe-Oliveras P, Moreno-Terrazas R, Arrizón-Gaviño J, Herrera-Suaréz T, García-Mendoza A, Gschaedler-Mathis A. Yeasts associated with the production of Mexican alcoholic nondistilled and distilled Agave beverages. FEMS Yeast Research. 2008;8(7):1037-1052. DOI: 10.1111/j.1567-1364.2008.00430.x

22. Souza E, Rosa C, Morgano M, Serra G. Fermentation characteristics as criteria for selection of cachaça. World Journal of Microbiology and Biotechnology. 2004;20:19-24. DOI: 10.5713/ajas.14.0508

23. Arrizon J, Gschaedler A. Increasing fermentation efficiency at high sugar concentrations by supplementing an additional source of nitrogen during the exponential phase of the tequila fermentation proccss. Canadian Journal of Microbiology. 2002;48:965-970. DOI: 10.1139/w02-093

24. Espinosa E, de Billerbec G, Ardaillon D, Schuler A, François J, de Morais M. Influence of nitrogen supply on the production of higher alcohols/esters and expression of flavour-related genes in cachaça fermentation. Food Chemistry. 2013;138:701-708. DOI: 10.1016/j.foodchem.2012.10.147

25. Lekkas C, Stewart G, Hill A, Taidi B, Hodgson J. Elucidation of the role of nitrogenous wort components in yeast fermentation. Journal of the Institute of Brewing. 2007;113(1): 3-8. DOI: 10.1002/j.2050-0416.2007.tb00249.x

26. Bell S, Henschke P. Implications of nitrogen nutrition for grapes, fermentation and wine. Australian Journal of Grape and Wine Research. 2005;11:242-295. DOI: 10-1111/j.1755-0238.2005.tb00028.x

27. Rankine B, editor. Making Good Wine. Melbourne, Australia: Sun Books ed; 1989

28. Wells A, Osborne J. Production of SO2 binding compounds and SO2 by Saccharomyces during alcoholic fermentation and the impact on malolactic fermentation. South African Journal of Enology and Viticulture. 2011;32(2):267-279. DOI: 10.21548/32-2-1387

29. Ciani M, Capece A, Comitini F, Canonico L, Siesto G, Romano P. Yeast interactions in inoculated wine fermentation. Frontiers in Microbiology. 2016;7:555. DOI: 10.3389/fmicb.2016.00555

30. Arneborg N, Siegumfeldt H, Andersen GH, Nissen P, Daria VR, Rodrigo PG, et al. Interactive optical trapping shows that confinement is a determinant of growth in a mixed yeast culture. FEMS Microbiology Letters. 2005;245:155-159. DOI: 10.1016/j.femsle.2005.03.008

31. Starmer WT, Ganter PF, Aberdeen V, Lachance MA, Phaff HJ. The ecological role of killer yeast in natural communities of yeasts. Canadian Journal of Microbiology. 1987;33:783-796. DOI: 10.1139/m87-134

32. Kagan BL. Mode of action of yeast killer toxins: Channel formation in lipid bilayers. Nature. 1983;302:709-711. DOI: 10.1038/302709a0

33. Schmitt MJ, Breinig F. The viral killer system in yeast: From molecular biology to application. FEMS Microbiology Reviews. 2002;26:257-276. DOI: 10.1111/j.1574-6976.2002.tb00614.x

34. Lam FH, Ghaderi A, Fink GR, Stephanopoulos G. Engineering alcohol tolerance in yeast. Science. 2014;346(6205):71-75. DOI: 10.1126/science.1257859

35. Alexandre H, Rousseaux I, Charpentier C. Relationship between ethanol tolerance, lipid composition and plasma membrane fluidity in Saccharomyces cerevisiae and Kloeckera apiculata. FEMS Microbiology Letters. 1994;124:17-22. DOI: 10.1111/j.1574-6968.1994.tb07255.x

36. Meaden PG, Arneborg N, Guldfeldt LU, Siegumfeldt H, Jakobsen M. Endocytosis and vacuolar morphology in Saccharomyces cerevisiae are altered in response to ethanol stress or heat shock. Yeast. 1999;15:1211-1222. DOI: 10.1002/(SICI)1097-0061(19990915)15:12<1211::AID-YEA448>3.0.CO;2-H

37. Salmon JM, Vincent O, Mauricio JC, Bely M, Barre P. Sugar transport inhibition and apparent loss of activity in Saccharomyces cerevisiae as a major limiting factor of enological fermentation. American Journal of Enology and Viticulture. 1993;44:56-64. DOI: 10.1139/w02-093

38. Ibeas JL, Jiménez J. Mitochondrial DNA loss caused by ethanol in Saccharomyces flor yeast. Applied and Environmental Microbiology. 1997;63:7-12

39. Hu XH, Wang MH, Tan T, Li JR, Yang H, Leach L, et al. Genetic dissection of ethanol tolerance in the budding yeast Saccharomyces cerevisiae. Genetics. 2007;175:1479-1487. DOI: 10.1534/genetics.106.065292

40. Stanley D, Bandara A, Fraser S, Chambers PJ, Stanley GA. The ethanol stress response and ethanol tolerance of Saccharomyces cerevisiae. Journal of Applied Microbiology. 2010;109(1):13-24. DOI: 10.1111/j.1365-2672.2009.04657.x

41. Alexandre H, Ansanay-Galeote V, Dequin S, Blondin B. Global gene expression during short-term ethanol stress in Saccharomyces cerevisiae. FEBS Letters. 2001;498:98-103. DOI: 10.1016/S0014-5793(01)02503-0

42. Pampulha ME, Loureiro-Dias MC. Combined effect of acetic acid, pH and ethanol on intracellular pH of fermenting yeast. Applied Microbiology and Biotechnology. 1989;31:547-550. DOI: 10.1007/BF00270792

43. Viegas CA, Rosa MF, Sa-Correia I, Novais JM. Inhibition of yeast growth by octanoic and decanoic acids produced during ethanolic fermentation. Applied and Environmental Microbiology. 1989;55(1):21-28

44. Lafon-Lafourcade S, Geneix C, Ribéreau-Gayon P. Inhibition of alcoholic fermentation of grape must by fatty acids produced by yeasts and their elimination by yeast ghost. Applied and Environmental Microbiology. 1984;47(6):1246-1249

45. Tamang JP, Watanabe K, Holzapfel WH. Review: Diversity of microorganisms in global fermented foods and beverages. Frontiers in Microbiology. 2016;7:377. DOI: 10.3389/fmicb.2016.00377

46. Cocolin L, Ercolini D, editors. Molecular Techniques in the Microbial Ecology of Fermented Foods. New York: Springer; 2008. DOI: 10.1007/978-0-387-74520-6

47. Ercolini D. High-throughput sequencing and metagenomics: Moving forward in the culture-independent analysis of food microbial ecology. Applied and Environmental Microbiology. 2013;79(10):3148-3155. DOI: 10.1128/AEM.00256-13

48. Cocolin L, Alessandria V, Dolci P, Gorra R, Rantsio K. Culture independent methods to assess the diversity and dynamics of microbiota during food fermentation. International Journal of Food Microbiology. 2013;167:29-43. DOI: 10.1016/j.ijfoodmicro.2013.05.008

49. De Filippis F, Parente E, Ercolini D. Metagenomics insights into food fermentations. Microbial Biotechnology. 2017;10(1):91-102. DOI: 10.1111/1751-7915.12421

50. Miguel G, Collela C, De Almeida E, Dias D, Schwan R. Physicochemical and microbiological description of Caxiri- a cassava and corn alcoholic beverage. International Journal of Food Science and Technology. 2015;50:2537-2544. DOI: 10.1111/ijfs.12921

51. Portillo M, Mas A. Analysis of microbial diversity and dynamics during wine fermentation of Grenache grape variety by high-throughput barcoding sequencing. LWT Food Science and Technology. 2015;72:317-321. DOI: 10.1016/j.lwt.2016.05.009

52. Ramos C, De Sousa E, Ribeiro J, Almeida T, Santos C, Abegg M, Schwan R. Microbiological and chemical characteristics of tarubá, an indigenous beverage produced from solid cassava fermentation. Food Microbiology. 2015;49:182-188. DOI: 10.1016/j.fm.2015.02.005.

53. Wang C, Mas A, Esteve-Zarzoso B. Interactions between Hanseniaspora uvarum and Saccharomyces cerevisiae during alcoholic fermentation. International Journal of Food Microbiology. 2015;206:67-74. DOI: 10.1016/j.ijfoodmicro.2015.04.022

54. Xu-Cong L, Zhi-Chao C, Rui-Bo J, Zhi-Bin L, Wen Z, Shao-Jun C, et al. Microbial community structure and dynamics during the traditional brewing of Fuzhou Hong Qu glutinous rice wine as determined by culture-dependent and culture-independent techniques. Food Control. 2015;57:216-224. DOI: 10.1016/j.foodcont.2015.03.054

55. Maturano YP, Mestre VM, Combina M, Toro ME, Vázquez F, Esteve-Zarzoso B. Culture-dependent and independent techniques to monitor yeast species during cold soak carried out at different temperatures in winemaking. International Journal of Food Microbiology. 2016;237:142-149. DOI: 10.1016/j.ijfoodmicro.2016.08.013

56. Sankar S, Keot J, Das S, Sarma K, Barooah M. Metagenomics analysis of microbial communities associated with a traditional rice wine starter culture (Xaj-pitha) of Assam, India. Biotechnology. 2016;6:153. DOI: 10.1007/s13205-016-0471-1

57. Santiago-Urbina J, Peña-Montes C, Nolasco-Cancino H, Ruiz-Terán F. PCR-DGGE analysis of the yeast population associated with natural fermentation of taberna. Journal of Microbiology, Biotechnology and Food Sciences. 2016;6:758-763. DOI: 10.15414/jmbfs.2016.6.2.758-763

58. Liu M, Tang Y, Zhao K, Liu Y, Guo X, Ren D, et al. Determination of the fungal community of pit mud in fermentation cellars for Chinese strong-flavor liquor, using DGGE and Illumina MiSeq sequencing. Food Research International. 2017;91:80-87. DOI: 10.1016/j.foodres.2016.11.037

59. Nikolaou E, Soufleros EH, Bouloumpasi E, Tzanetakis N. Selection of indigenous Saccharomyces cerevisiae strains according to their oenological characteristics and vinification results. Food Microbiology. 2006;23(2):205-211. DOI: 10.1016/j.fm.2005.03.004

60. Fiore C, Arrizon J, Gschaedler A, Flores J, Romano P. Comparison between yeast from grape and agave must for trades of technological interest. World Journal of Microbiology and Biotechnology. 2005;21(3):1141-1147

61. Arrizon J, Fiore C, Acosta G, Romano P, Gschaedler A. Fermentation behavior and volatile compound production by agave and grape must yeasts in high sugar Agave tequilana and grape must fermentations. Antonie Van Leeuwenhoek. 2006;89:181-189. DOI: 10.1007/s10482-005-9022-1

62. Capece A, Romaniello R, Siesto G, Pietrafesa R, Massari C, Poeta C. et al. Selection of indigenous Saccharomyces cerevisiae strains for Nero d'Avola wine and evaluation of selected starter implantation in pilot fermentation. International Journal of Food Microbiology. 2010;144:187-192. DOI: 10.1016/j.ijfoodmicro.2010.09.009

63. Tristezza M, Fantastico L, Vetrano C, Bleve G, Corallo D, Grieco F, et al. Molecular and technological characterization of Saccharomyces cerevisiae strains isolated from natural fermentation of Susumaniello grape must in Apulia, Southern Italy. International Journal of Microbiology. 2014;2014:897427. DOI: 10.1155/2014/897428

64. Lopes CA, Rodríguez ME, Sangorrín M, Querol A, Caballero A. Patagonian wines: Implantation of an indigenous strain of Saccharomyces cerevisiae in fermentations conducted in traditional and modern cellars. Journal of Industrial Microbiology & Biotechnology. 2007;34(2):139. DOI: 10.1007/s10295-006-0178-0

65. Perrone B, Giacosa S, Rolle L, Cocolin L, Rantsiou K. Investigation of the dominance behavior of Saccharomyces cerevisiae strains during wine fermentation. International Journal of Food Microbiology. 2013;165:156-162. DOI: 10.1016/j.ijfoodmicro.2013.04.023

66. Alonso-del-Real J, Lairón-Peris M, Barrio E, Querol A. Effect of temperature on the prevalence of Saccharomyces non cerevisiae species against a S. cerevisiae wine strain in wine fermentation: Competition, physiological fitness, and influence in final wine composition. Frontiers in Microbiology. 2017;8:150. DOI: 10.3389/fmicb.2017.00150

67. Ribeiro LS, Duarte WF, Dias DR, Schwan RF. Fermented sugarcane and pineapple beverage produced using Saccharomyces cerevisiae and non-Saccharomyces yeast. Journal of the Institute of Brewing. 2015;121:262-272. DOI: 10.1002/jib.218

68. Julien A, Roustan JL, Dulau L, Sablayrolles JM. Comparison of nitrogen and oxygen demands of enological yeasts: Technological consequences. American Journal of Enology and Viticulture. 2000;51:215-222

69. Gardner JM, Poole K, Jiranek V. Practical significance of relative assimilable nitrogen requirements of yeast: A preliminary study of fermentation performance and liberation of H2S. Australian Journal of Grape and Wine Research. 2002;8:175-179. DOI: 10.1111/j.1755-0238.2002.tb00253.x

70. Capece A, Romaniello R, Pietrafesa R, Romano P. Indigenous Saccharomyces cerevisiae yeasts as a source of biodiversity for the selection of starters for specific fermentations. BIO Web of Conferences.. 2014;3:02003. DOI: 10.1051/bioconf/20140302003

71. Csoma H, Zakany N, Capece A, Romano P, Sipiczki M. Biological diversity of Saccharomyces yeasts of spontaneously fermenting wines in four wine regions: Comparative genotypic and phenotypic analysis. International Journal of Food Microbiology. 2010;140:239-248. DOI: 10.1016/j.ijfoodmicro.2010.03.024

72. Romano P, Fiore C, Paraggio M, Caruso M, Capece A. Function of yeast species and strains in wine flavour. International Journal of Food Microbiology. 2003;86:169-180. DOI: 10.1016/S0168-1605(03)00290-3

73. Zagorc T, Maraz A, Cadez N, Jemec KP, Peter G. Indigenous wine killer yeasts and their application as a starter culture in wine fermentation. Food Microbiology. 2001;18:441-445. DOI: 10.1006/fmic.2001.0422

74. Silva CLC, Rosa CA, Oliveira ES. Studies on the kinetic parameters for alcoholic fermentation by flocculent Saccharomyces cerevisiae strains and non-hydrogen sulfide-producing strains. World Journal of Microbiology and Biotechnology. 2006;22(8):857-863. DOI: 10.1007/s11274-005-9115-z

75. Belda I, Ruiz J, Alastruey-Izquierdo A, Navascués E, Marquina D, Santos A. Unraveling the enzymatic basis of wine "Flavorome": A phylo-functional study of wine related yeast species. Frontiers in Microbiology. 2016;7:12. DOI: 10.3389/fmicb.2016.00012

76. Steensels J, Meersman E, Snoek T, Saels V, Verstrepen J. Large-scale selection and breeding to generate industrial yeasts with superior aroma production. Applied and Environmental Microbiology. 2014;80(22):6965-6975. DOI: 10.1128/AEM.02235-14

77. Morata A, González C, Suárez LA. Formation of vinylphenolic pyranoanthocyanins by selected yeasts fermenting red grape musts supplemented with hydroxycinnamic acids. International Journal of Food Microbiology. 2007;116(1):144-152. DOI: 10.1016/j.ijfoodmicro.2006.12.032

78. Morata A, Gómez-Cordovés MC, Suberviola J, Bartolomé B, Colomo B, Suárez LA. Adsorption of Anthocyanins by yeast cell walls during the fermentation of red wines. Journal of Agricultural and Food Chemistry. 2003;51(14):4084-4088. DOI: 10.1021/jf021134u

79. Borrull A, Poblet M, Rozès N. New insights into the capacity of commercial wine yeasts to grow on sparkling wine media. Factor screening for improving wine yeast selection. Food Microbiology. 2015;48:41-48. DOI: 10.1016/j.fm.2014.12.006

80. Walker G, Van Dijck P. Physiological and molecular responses of yeast to the environment. In: Querol A, Fleet G, editors. The Yeast Handbook: Yeast in Food and Beverages. Berlin: Springer-Verlag; 2006. ISBN 3-540-28388-9

81. Manzanares P, Vallés S, Viana F. Non-saccharomyces yeasts in the winemaking process. In: Carrascosa A, Muñoz R, González R, editors. Molecular Wine Microbiology. Academic Press. London, United Kingdom. 2011. ISBN: 978-0-12-375021-1

82. Regodón J, Perez F, Valdes M, De Miguel C, Ramírez M. A simple and effective procedure for selection of wine yeast strains. Food Microbiology. 1997;14:247-254. DOI: 10.1006/fmic.1996.0091

83. Morata A, Gómez-Cordovés MC, Colomo B, Suárez JA. Pyruvic acid and acetaldehyde production by different strains of Saccharomyces cerevisiae: Relationship with Vitisin A and B formation in red wines. Journal of Agricultural and Food Chemistry. 2003;51(25):7402-7409. DOI: 10.1021/jf0304167

84. Mauriello G, Capece A, D'Auria M, Garde-Cerdán T, Romano P. SPME-GC method as a tool to differentiate VOC profiles in Saccharomyces cerevisiae wine yeasts. Food Microbiology. 2009;26(3):246-252. DOI: 10.1016/j.fm.2009.01.003

85. Lee YJ, Choi YR, Lee SY, Park JT, Shim JH, Park KH, et al. Screening wild yeast strains for alcohol fermentation from various fruits. Mycobiology. 2011;39(1):33-39. DOI: 10.4489/MYCO.2011.39.1.033

86. de Souza AP, Vicente MA, Klein RC, Fietto LG, Coutrim MX, de Cássia Franco Afonso RJ, et al. Strategies to select yeast starters cultures for production of flavor compounds in cachaça fermentations. Antonie Van Leeuwenhoek. 2012;101(2):379-392. DOI: 10.1007/s10482-011-9643-5

87. Snoek T, Verstrepen K, Voordeckers K. How do yeast cells become tolerant to high ethanol concentrations? Current Genetics. 2016;62:475-480. DOI: 10.1007/s00294-015-0561-3

88. Nilsson RE, Ross T, Bowman JP. Variability in biofilm production by Listeria monocytogenes correlated to strain origin and growth conditions. International Journal of Food Microbiology. 2011;150:14-24. DOI: 10.1016/j.ijfoodmicro.2011.07.012

89. Ferreira V, Wiedmann M, Teixeira P, Stasiewicz MJ. Listeria monocytogenes persistence in food-associated environments: Epidemiology, strain characteristics, and implications for public health. Journal of Food Protection. 2014;77:150-170. DOI: 10.4315/0362-028X.JFP-13-150

90. Posada-Izquierdo G, Del Rosal S, Valero A, Zurera G, Sant'Ana AS, Alvarenga V, Pérez-Rodríguez F. Assessing the growth of Escherichia coli O157:H7 and Salmonella in spinach, lettuce, parsley and chard extracts at different storage temperatures. Journal of Applied Microbiology. 2006;120(6):1701-1710. DOI: 10.1111/jam.13122

91. Ortiz S, López V, Martínez-Suárez JV. The influence of subminimal inhibitory concentrations of benzalkonium chloride on biofilm formation by Listeria monocytogenes. International Journal of Food Microbiology. 2014;189:106-112

92. Lianou A, Koutsoumanis K. Effect on the growth environment on the strain variability of Salmonella enterica kinetic behavior. Food Microbiology. 2011;28(4):828-837. DOI: 10.1016/j.fm.2010.04.006

93. Medina A, Lambert RJ, Magan N. Rapid throughput analysis of filamentous fungal growth using turbidimetric measurements with the Bioscreen C: A tool for screening antifungal compounds. Fungal Biology. 2006;116(1):161-169. DOI: 10.1016/j.funbio.2011.11.001

94. Lambert RJW, Pearson J. Susceptibility testing: Accurate and reproducible minimum inhibitory concentration (MIC) and non-inhibitory concentration (NIC) values. Journal of Applied Microbiology. 2000;88:784-790. DOI: 10.1046/j.1365-2672.2000.01017.x

95. Álvarez-Pérez JM, Álvarez-Rodríguez M, Campo E, Sáenz de Miera LE, Ferreira V, Hernández-Orte P, et al. Selection of Saccharomyces cerevisiae strains applied to the production of Prieto Picudo Rosé wines with a different aromatic profile. South African Journal of Enology and Viticulture. 2014;35:242-256. DOI: 10.21548/35-2-1013

96. Miranda-Castilleja DE, Ortiz-Barrera E, Arvizu-Medrano SM, Pacheco R, Aldrete-Tapia JA, Martínez-Peniche RA. Aislamiento, selección e identificación de levaduras Saccharomyces spp. Nativas de viñedos en Querétaro, México. Agrociencia. 2015;49: 759-773

97. Ortiz-Barrera E, Miranda-Castilleja DE, Arvizu-Medrano SM, Pacheco-Aguilar JR, Aldrete-Tapia JA, Hernández-Iturriaga M, Martínez-Peniche RA. Enological potential of native non-Saccharomyces yeasts from vineyards established in Querétaro, Mexico. Revista Chapingo Serie Horticultura. 2015;21(2):169-183. DOI: 10.5154/r.rchsh.2015.01.001

98. Salvadó Z, Arroyo-López FN, Barrio E, Querol A, Guillamón JM. Quantifying the individual effects of ethanol and temperature on the fitness advantage of Saccharomyces cerevisiae. Food Microbiology. 2011;28:1155-1161. DOI: 10.1016/j.fm.2011.03.008

99. Zwietering MH, Jongenburger I, Rombouts FM, van't Riet K. Modeling of the bacterial growth curve. Applied and Environmental Microbiology. 1990;56(6):1875-1881

100. García-Ríos E, Gutiérrez A, Salvadó Z, Arroyo-López FN, Guillamon JL. The fitness advantage of commercial wine yeasts in relation to the nitrogen concentration, temperature, and ethanol content under microvinification conditions. Applied and Environmental Microbiology. 2014;80: 704-713. DOI: 10.1128/AEM.03405-13

CHAPTER 4

CONTROLLING WINE SPOILAGE YEASTS: A STRATEGY

1. INTRODUCTION: AN OVERVIEW

Since the wild yeast colonies in grape musts have been managed, the fermentation has resulted in wine with distinct organoleptic characteristics. The winemaker's ability to regulate yeast colonies throughout the entire winemaking process allows for the manipulation of these features in order to produce wine with improved quality parameters. While yeasts play a crucial role in converting grape must into wine through fermentation, certain yeasts have the ability to ruin it (Figure 1). Most of the time, spoilage yeasts are able to withstand extreme circumstances such as high levels of ethanol, low pH, and deadly amounts of sulphites (SO2) or dimethyl dicarbonate (DMDC), which are utilized as antimicrobial agents.

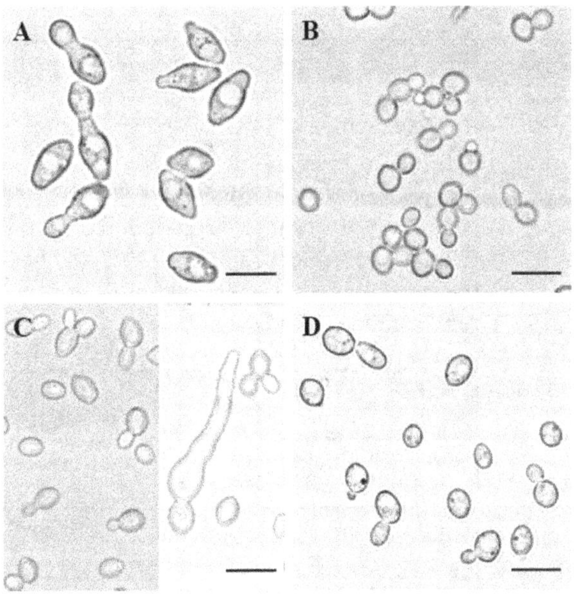

Figure 1. Optical microscopic images of some spoilage yeasts found in wine. (A) S.

ludwigii; (B) Z. rouxii; (C) Dekkera spp.; (D) P. anomala. Bar scale: 10 μm.

Certain yeast genera can be classified as spoiling germs because they have the ability to undesirably establish themselves in food, resulting in a decline in nutritional and sensory quality. This can ultimately lead to significant financial losses. Their association with public health is currently being questioned [1]. Table 1 provides a summary of yeast genera that are known to ruin wine and some food goods. It specifically highlights the key components that alter the quality of these products, along with the resulting effects and potential health hazards.

Table 1. Some of the most common yeasts often found in grape, musts and wines that can be considered spoilage yeast species in a wide range of food products.

Yeast species	Food product	Effect observed	Health hazard
Zygosaccharomyces rouxii	Products with 50% sugar [2, 3]	Gas production: bubbling and package expansion [3]	
	Sweet wines [4]	Refermentation and CO_2 production [4]	
	Mould-ripened soft cheeses [5]		
	Fruit juices, sauces, carbonated soft drinks, salad dressings, ketchup [3, 6]	Gas production: bubbling and package expansion [3]	
Brettanomyces bruxellensis	Bulk, barrel matured and bottled wines [7, 8]	Off aromas, cloudiness formation in sparkling wine, mousy aroma [7] Unpleasant mousy and medicinal taints [8]	
Pichia anomala	Dairy and baking products, beer, high salt environments and silage [9]	No restriction on handling and no risk to healthy persons [9]	
	Lactic acid-rich products [10]		
	Wine, winemaking [7, 8]	Oxidation of ethanol [8]	
Pichia guilliermondii	Grape juice [7]	Off-odours barnyard-like or horsey [7]	
	Soft drinks [6]		

Yeast species	Food product	Effect observed	Health hazard
Pichia membranaefaciens	Mould-ripened soft cheeses [5]	Chalky film layer [11]	
Saccharomycodes ludwigii	Food products with SO_2 as antiseptic [2]		
	Bottled wines [7]	Spoilage by sediment or cloudiness formation [7]	
	Wine [12]	Flocculent sediment [11]	
Candida tropicalis	Fresh fruits: orange (*Citrus sinensis*) and pineapple (*Ananas comosus*) [13]		Candidiasis has not been transmitted by food products [5]
	Grated raw carrots [14]	Increase in exudate and softening [14]	
Hanseniaspora/Kloeckera	Fresh must [15]		
	Various storage products [2]	Undesirable fermentation products [2]	
	Must under fermentation [7]	Aroma modification at early fermentation stage in winemaking [7]	

The indiscriminate utilization and improper administration of antibiotics has led to a growing resistance among a wide range of disease-causing microbes, including those found in food. These germs not only withstand the effects of antibiotics but also manage to survive the food preservation procedures [16].

This chapter provides a comprehensive review of various methods and treatments used to manage spoilage yeasts. These techniques were categorized into early and emerging technologies based on their uniqueness in the application to the winemaking sector. Additionally, this text includes a concise evaluation of monitoring strategies as an effective means of enhancing quality control in the winery.

2. METHODS FOR MANAGING SPOILAGE YEAST IN TECHNOLOGY

2.1. Early technologies

Traditionally, in the food business, thermal treatments have been used to restrict the growth of spoilage and pathogenic bacteria. These processes aim to partially or completely eliminate the microflora that is present [17, 18]. By utilizing aseptic and hermetic packaging, the shelf life of the food goods was successfully prolonged while also guaranteeing their microbiological safety [19, 20]. The primary disadvantage of these types of inactivation methods is the impairment of organoleptic quality caused by elevated processing temperatures.

An alternative method to combat undesired bacteria is by including natural or chemically produced preservatives, such as organic acids (ascorbic, citric, benzoic, sorbic, etc.) and salts (potassium sorbate, sodium benzoate, sodium metabisulphite, etc.) [21, 22]. Undesirable microbial growth can be controlled by adjusting specific environmental factors such as temperature, pH, water activity, nutrition availability, and the presence of hazardous substances during the production process. Within the realm of oenology, ethanol tolerance is considered a primary component that restricts the proliferation of yeast [23]. Furthermore, sulphur dioxide (SO2), dimethyl dicarbonate, and sorbic acid are among the antimicrobials frequently employed in the process of winemaking.

Yeast species that are resistant to one preservative also exhibit resistance to other preservatives that have similar chemical composition. These three compounds, benzoic acid, sorbic acid, and sulfur dioxide, are examples of such cases [24]. Furthermore, it has been shown that there is a strong correlation between resistance to sorbic acid and resistance to ethanol [25]. Typically, the ability of yeast to resist preservatives varies depending on the strain and the physiological features of the cells [6, 26].

Sulphur dioxide (SO2) is a chemical additive commonly employed in wineries as an antioxidant and preservative to regulate bacteria, moulds, and spoilage yeasts [11, 27]. Its ability to prevent microbial growth is influenced by the pH of the environment [28]. Nevertheless, in recent decades, there has been a reevaluation of its usage as a result of growing concerns regarding allergies. Scientists are currently seeking alternate approaches to decrease the quantities of additives typically used in grape juice and wine [29, 30]. Significant concentrations of SO2 are required to effectively inhibit the proliferation of Dekkera bruxellensis in red wine. According to Barata et al. [23], a concentration of 1 mg/L of molecular sulphur dioxide is recommended to inhibit the growth of D. bruxellensis during the aging of red wine in oak barrels. Furthermore, Saccharomycodes ludwigii and Zygosaccharomyces bailii exhibit significant resistance to SO2, and Z. bailii also demonstrates tolerance to organic acids [24]. As a result of prolonged exposure to SO2, many wine yeasts have acquired specific defense mechanisms to combat this antimicrobial agent [28]. It has been demonstrated that Brettanomyces bruxellensis can enter a viable but not culturable (VBNC) condition as a survival strategy in response to the presence of SO2. VBNC cells of several B. bruxellensis strains (tested in a synthetic wine medium) can be triggered by molecular SO2 levels as low as 0.2–0.4 mg/L [32]. Nevertheless, the metabolic activity of this yeast that causes spoiling remains active, resulting in the ongoing production and release of volatile phenols, even when the yeast is not reproducing [32]. According to Benito et al. [33], a minimum concentration of 20 mg/L of free SO2 is necessary to suppress the enzymatic activity of B. bruxellensis at a pH of 3.5.

Dimethyl dicarbonate (DMDC) is a compound derived from dicarbonic acid that is utilized as a cold sterilant in the food industry. It is officially sanctioned in the USA [34], Australia [35], and Europe [36] as a chemical preservative to ensure the microbiological stability of wines. The antibacterial efficiency of DMDC is highest when the pH is low, the ethanol level is high, and the microbial population is low [37]. DMDC exhibits greater efficacy against yeast compared to bacteria or molds at authorized usage levels [38]. The primary mode of operation of DMDC is to render cellular enzymes inactive by reacting with nucleophilic groups such as imidazoles, amines, and thiols. This reaction specifically targets enzymes like glyceraldehyde-3-phosphate dehydrogenase and alcohol dehydrogenase [39]. DMDC can be utilized to inhibit the growth of spoilage yeasts in wines [40], halt alcoholic fermentation in the creation of sweet wines [41], and disinfect musts by eliminating natural flora [42].

Regarding their impact on spoilage yeasts, Z. bailii, S. ludwigii, and Lachancea thermotolerans were discovered to be highly susceptible to DMDC. By adding 100-200 mg/L of DMDC to red wine with a 12% volume/volume and pH of 3.5, complete cell death (starting with an initial population of 106 CFU/mL) can be achieved [43]. Subsequent trials conducted by Zuehlke et al. [44] verified the harmful effects of DMDC on Z. bailii. In contrast, Schizosaccharomyces pombe and Saccharomyces cerevisiae are capable of surviving in the presence of elevated levels of DMDC, with a minimum inhibitory concentration (MIC) of ≥300 mg/L. Furthermore, it has been demonstrated that the inclusion of DMDC in wines at the highest concentration allowed by European regulations (200 mg/L; [36]) is not efficient in controlling lactic acid and acetic acid bacteria [43]. Resistance to DMDC is contingent upon the composition of the media. In the case of grape must, a larger dose of DMDC (about 400 mg/L) is required to limit the growth of Z. bailii and S. ludwigii, with a cell concentration of 106 cells/mL [42]. Considering the varying sensitivity of yeast to DMDC, it is recommended to utilize this molecule as a means to promote the growth of specific yeast species in the early stages of alcoholic fermentation. To ensure the dominance of the chosen yeast, it is recommended to apply the inoculum 12 hours after treating the grape must with DMDC [42].

Using DMDC before the completion of fermentations is not advisable due to its antibacterial properties, which can also affect the fermenting species such as S. cerevisiae and Oenococcus oeni [40]. According to these authors, the addition of DMDC throughout the process of making wine has a temporary effect, meaning that its ability to preserve the wine diminishes over time. Therefore, it may be necessary to add smaller amounts of DMDC at different stages, particularly if the aim is to safeguard the wine during the aging process in barrels [45].

There is a synergistic impact between DMDC and sulphur dioxide in both potassium and sodium metabisulphite salts, which enhances the inactivation of wine yeast and bacteria [40, 43]. The utilization of DMDC enables a substantial decrease in the sulphur concentration seen in grape juice and semi-sweet wines [46].

DMDC poses no health risks as it undergoes spontaneous, quick, and full hydrolysis into methanol and carbon dioxide when in contact with water-based solutions [46]. However, one drawback of using DMDC in winemaking is its possible toxicity when being handled [39]. Regarding the utilization of weak acids, benzoic and sorbic acids can effectively regulate the growth of spoilage yeasts in wines when applied at specific quantities. The lipophilic nature of benzoic and sorbic acids enables them to diffuse through cell membranes in their undissociated state [24]. Currently, benzoic acid is not approved for usage in wine [47].

Z. bailii is a yeast species that causes wine spoiling and has a high level of resistance to commonly used preservatives like benzoic and sorbic acids [48]. Zygosaccharomyces rouxii is a type of yeast that thrives in environments with high sugar concentration, and it is known to cause spoiling in beverages. Potassium sorbate, sodium benzoate, dimethyl dicarbonate, and vanillin are chemical compounds that have antibacterial capabilities. These chemicals have been found to effectively limit the development of Z. rouxii in concentrated grape juice, as demonstrated in a study [26]. The mathematical model determined that each of the four preservatives, when used individually, had a maximum inhibitory impact resulting in an average growth reduction of 40%. The predicted doses required for this effect were: potassium sorbate (>4.7 mM), sodium benzoate (>10.4 mM), dimethyl dicarbonate (>1 mM), and vanillin (>30 mM).

B. bruxellensis exhibits a high level of resistance to benzoic and sorbic acids. The necessary amount of sorbic acid needed to stop its development and enzymatic activity in wine surpasses the legal limit of 250 mg/L [33]. As previously stated, the use of benzoic acid as a wine ingredient remains prohibited. Loureiro and Malfeito-Ferreira [49] found that B. bruxellensis is capable of tolerating sorbic acid concentrations as high as 950 mg/L at a pH of 3.5.

The storage temperature and ethanol level in wine can synergistically restrict the growth of the wine spoilage yeast B. bruxellensis among the physicochemical parameters that can be controlled in the cellar. The most effective combination of these factors to stop the growth of these bacteria is when the ethanol level is above 14% v/v and the storage temperature is below 12°C [50]. Furthermore, experiments conducted with synthetic medium have demonstrated that the growth of D. bruxellensis can be impeded by ethanol concentrations exceeding

14.5–15.0% (v/v) [23]. Couto et al. [51] found that the temperature and time required to destroy Dekkera/Brettanomyces yeast cells through heat mostly relies on the levels of ethanol and phenolic compounds, specifically ferulic acid, present in the wine. Temperatures above 35°C can lead to substantial decreases in the population of contaminated wine samples, which initially contained roughly 1×107 CFU mL-1. It should be noted that B. bruxellensis has distinct behavior in white and red wines, with less progression observed in white wines [23]. Furthermore, lowering the storage temperatures can potentially decrease the necessary amount of SO2, as the ideal temperature range for the growth of Brettanomyces is 25–28°C [52]. The authors suggested that a temperature below 15°C and a molecular SO2 concentration above 0.4 mg/L can effectively reduce spoiling by Brettanomyces in stored red wines. However, it is important to note that total eradication of Brettanomyces cannot be guaranteed under these conditions.

2.2. Emerging technologies

2.2.1. High hydrostatic pressure

The high hydrostatic pressure (HHP) treatment is a non-thermal process used to inhibit pathogenic microorganisms including spoilage microbes as well as enzymes [53]. According to these authors, the food products that are treated do not change nutritional or modify their sensory quality. The HHP uses pressure-transmitting liquids (water, ethanol solutions, sodium benzoate solutions, etc.) to homogeneously transfer pressure to the food sample [54]; the process may be batch, semi-continuous or continuous. The use of HHP is also expected to increase the shelf life [55] of a wide range of food products able to be treated including meat, eggs, vegetables, seafood [53], fruits such as sweet cherries [56] or mango pulp [57], juices [55, 58] and Serrano ham [59], among others.

In the winemaking industry, the HHPs have been used to reduce or inhibit the presence of spoilage yeasts such as B. bruxellensis. In this matter, Van Wyk and Silva [60] have used 200 MPa for 3 min achieving a reduction of up to 5.8 log but higher than 6 log when increasing the pressure to 400 MPa during a much shorter period of time (5 seg). These parameters were also evaluated by González-Arenzana et al. [61] but also considering pH and amount of ethanol in wines; high ethanol and high pH needed less pressure (100 MPa) to inactivate B. bruxellensis cells, while at lower ethanol and lower pH the pressure used was higher (200 MPa). The HHP could also be used as a treatment for grapes in order to reduce wild yeast populations in order to be able to use yeast starters [62]; HHP has shown to increase pigment extraction and ethanol and methanol increase in wines after treatment. In the case of beer pasteurization, the use of 600 MPa allows beer with different ethanol content to have more than 7 log reductions of S. cerevisiae within 5 s [63].

2.2.2. Pulsed light irradiation

The pulsed light irradiation is a technique used to inactivate microorganisms in food products. The pulses are stored electricity in UV lamps that do not contain mercury but xenon [64], released in very short periods of time (fractions of millionths or thousandths of a second) that are rich in UV-C light (from 200 to 280 nm in the spectrum) [65]. This sterilization technique can cause DNA damage in the same way that continuous UV light does, nonetheless irradiation with pulsed light may also induce distortion in cell membranes and vacuoles [66]; therefore, the pulsed light irradiation may be considered an improved version of the UV treatment [64].

Pulsed light has been evaluated in the last decade as an efficient technique against foodborne spoilage microorganisms in commercial and fresh fruit juices using pulsed light against S. cerevisiae and pathogenic bacteria [67], or apple juices [68] where the combination of pulsed light with ultrasound has been evaluated as an alternative to inactivate S. cerevisiae: fresh strawberries [69] with reductions of 1 log colonies of yeast and microfungi lengthening shelf life 2 days; cured meat products [70] against pathogenic bacteria. Nevertheless, there are studies on the use of pulsed light on several food products; in wine or in the winemaking industry the evaluation of this technique appears to be scarce. The use of pulsed light might not be limited to just inactivating spoilage yeast in grapes and musts but also as a technique to potentially improve polyphenols extraction during maceration, thus increasing pigment compounds in wine.

2.2.3. Electric pulses

The electric pulses (EPs) are a method used for food preservation that avoids the use of chemical compounds as well as the use of thermal treatments [71]. Thus, it can be considered a cold temperature treatment that may preserve foodborne properties.

Hülsheger et al. [72] demonstrated that EPs have different effect on microorganisms. Yeasts and Gram-positive bacteria are more resistant than Gram-negative bacteria to the disruption caused by cellular structures by low-energy pulses while most of the microorganisms (>99%) die with high-energy pulses. The colonies were also more susceptible to die at logarithmic growth phase than when in steady state. Besides the growth phase, Gáskovà et al. [73] have shown that there are other factors involved in the efficiency of electric pulses for killing yeasts; such factors include the amplitude and the duration of the pulses, the size of the yeast cells and the temperature and conductivity of the media where the yeasts grow.

Electric pulses have been used lately in the production of fruit juices for its efficacy in microbial reduction and for keeping their sensorial and nutritional properties [74]. The use of EP to eliminate spoilage yeasts such as S. cerevisiae in orange juice requires less intensity pulses than those to inactivate other pathogens microorganisms such as the bacteria Escherichia coli or Listeria innocua [75].

The effect of EP could be combined with the use of other antimicrobial technology like the lyticase digestion; the combined effect of both techniques may be a biotechnological application for spoilage yeast control since the use of mild EP (2–4.5 kV/cm) affects the cell-wall porosity and therefore lower doses of lyticase are needed to effectively inactive S. carlsbergensis, Kluyveromyces lactis and K. marxianus [76].

Biotechnology applications of the electric pulses or the electroporation, other than spoilage yeasts control, may include genetic transformation of cells by incorporation of foreign DNA, extraction of intracellular metabolites and biomass drying [71].

2.2.4. Natural extracts

The use of antimicrobials extracted from nature such as chitosan, essential oils (active ingredients: eugenol, allicin, carvacrol, thymol and limonene), spices or nisin is very widespread in food preservation [21], but the limits of its antifungal activity and their potential applications as winemaking additives remain to be further explored.

Chitosan (β-1,4-d-glucosamine) is a linear heteropolysaccharide obtained by deacetylation of chitin. The high density of amino groups present in its structure makes chitosan a bioactive polymer [77]. This polysaccharide with antimicrobial activity is found in nature being part of the exoskeletons of arthropods, diatoms and algae, and the cell wall of some fungi (particularly zygomycetes) [78]. In a study carried out by Rojo et al. [26], where a dose of 300 mg/L of chitosan was added to a high sugar culture media in order to assess its inhibition properties to control the wine spoilage yeast Z. rouxii, it was found that with such a high dose of chitosan, tripling the allowable dose in wines (100 mg/L), no inhibition effect exists. Previously, Z. rouxii had been already reported as sensitive yeast to chitosan (completely inactivated by 100–400 mg/L), but this result was only based on the study of one strain isolated from a spoiled carbonated beverage [79]. Similarly, S. ludwigii has been identified as yeast very resistant to chitosan (5 g/L of chitosan was required to inactivate S. ludwigii). The minimum inhibitory concentration for Brettanomyces/Dekkera ranges from 0.2 to 0.5 mg/mL of chitosan, depending on the molecular weight [80]. Low-molecular-weight chitosan has higher antifungal properties. With respect to impact on sensory quality, the addition of chitosan may adversely affect the colour of the wine, mainly through a reduction of the colour intensity [80]. Moreover, Gómez-Rivas et al. [77] observed that in mixed fermentations B. bruxellensis and B. intermedius are more sensitive to chitosan than S. cerevisiae. Taillandier et al. [81] have also tested the use of chitosan of fungal origin to control B. bruxellensis in winemaking and concluded that its mechanism of action is a complex combination of both physical adsorption on cell wall and the biological interaction by changing permeability of cell membrane, together with an observed effect on the growing rate and the physiological state of B. bruxellensis. Petrova et al. [82] also reported on the efficacy of chitosan as strategy to control B. bruxellensis development in red wines, with an average reduction of 3 logs in the culturability. It has been proved that when applied in the vineyard chitosan can act as an inducer of the plant defence system, mainly by enhancing the polyphenolic phytoalexins synthesis, and therefore it represents a useful protection treatment against powdery mildew infection [83]. Chitosan can be also used as natamycin carrier to protect the cheese surface from fungal contamination and thus extend its shelf life [84]. It is also possible to use microencapsulation techniques with chitosan to dose oily compounds, such as essential oils, with the additional advantage of protecting the active ingredient (e.g. limonene) and thus preserving its antimicrobial properties [85].

S. cerevisiae has demonstrated strong resistant to chitosan, doses above 2 g/L are needed to fully inhibit it. Conversely, B. bruxellensis seems to be very susceptible to this natural extract, 0.2 g/L can effectively inactivate it. Hanseniaspora uvarum and Z. bailii, other spoilage wine yeasts, can be also inhibited by chitosan with 0.4 g/L [86]. However, all these results correspond to growth inhibition assays in laboratory media. As expected, these authors observed that the antimicrobial effect of chitosan is less under winemaking conditions. Regarding the performance of alcoholic fermentation, a slight delay in the beginning of the fermentation was observed, but without major repercussions [86].

Among the essential oils extracted from plants, those from cinnamon, clove, garlic, onion, oregano and thyme are the inhibitoriest against food spoilage yeasts [87]. Antimicrobial activity of these compounds is well documented [88].

Eugenol is the main component of clove oil (≈85%). So far, eugenol has proven to be effective as antibacterial and antifungal in several food products [89]. Both eugenol and thymol act on the membrane and cell wall, causing cell lysis [90]. These same authors suggested the use of eugenol and thymol as lysing agents to extract the genomic DNA from yeast cell instead of using zymolyase and sodium dodecyl sulphate (SDS). However, according to Kubo et al. [91] eugenol alone may not respond as effective antimicrobial against Z. bailii and S. cerevisiae, since the concentration required for ensuring the fungicidal effect is quite high (800 μg/mL).

Carvacrol and thymol are phenolic compounds being part of the essential oil from oregano and thyme [92]. According to Chavan and Tupe [93], low concentrations of carvacrol and thymol (≤64 mg/L; used independently) are effective in limiting the growth of several wine spoilage yeasts, including Metschnikowia pulcherrima, Z. rouxii, S. pombe, Debaryomyces hansenii and D. bruxellensis. Results for minimum inhibitory concentration values were comparable or even better than those of potassium metabisulphite. The antifungal mechanism of these two essential oils' active ingredients is based on the membrane damage and leakage of cytoplasmic content, thus increasing permeability.

As for allicin (diallyl thiosulphinate), it is the main bioactive component of garlic extract [94]. It is well known for its antioxidant, antibacterial and antifungal activities [95] and also for its anticancer activity [96, 97]. The main mechanism involved in the antimicrobial effect of allicin is based on its rapid reaction with thiol-containing proteins [94]. Thus, allicin can regulate the activity of enzymes containing very reactive or unshielded SH groups. With relation to its potent antifungal properties, allicin is effective against a wide range of yeasts, among them S. pombe, S. cerevisiae [98], Aspergillus spp. [99] and Candida spp. [100]. It also affects the synthesis of mycotoxins and inhibits the germination of spores as well as the growth of the hyphae [100]. Yoshida et al. [98] set the minimum inhibitory concentration (MIC) of allicin in 5 mg/mL for S. pombe and 10 mg/mL for S. cerevisiae. Moreover, chitosan and garlic extract, used independently, have also shown promising results as treatments for the control of fungal diseases of the vine due to pruning [101]. In the medical field, allicin and its derivatives have been proposed as antifungal prophylactic due to their strong efficacy in inhibiting microbial development [102]. However, further studies are still required to confirm the potential therapeutic use of allicin [103].

Despite their interesting antimicrobial properties, there still exist some limitations to the use of some of these natural extracts such as their solubility in water, their susceptibility to oxidation (part of its effectiveness is lost) and the impact on sensory attributes [21]. A possible solution to the low water solubility of some natural extracts is their transformation into nanoemulsions [104, 105] or the use of nanoencapsulation techniques to protect their antimicrobial properties [106].

The inhibition activity on the growth of some wine spoilage yeasts using different antimicrobial compounds was measured by agar diffusion tests (Figure 2). The dark areas around the disks suggest the inhibitory effect in growing colonies, thus demonstrating their effectiveness. This assay can be used to assess the susceptibility of the yeast species to each of the antimicrobials, taking into account the limitations due to the diffusion properties of the antifungal compounds in solid media.

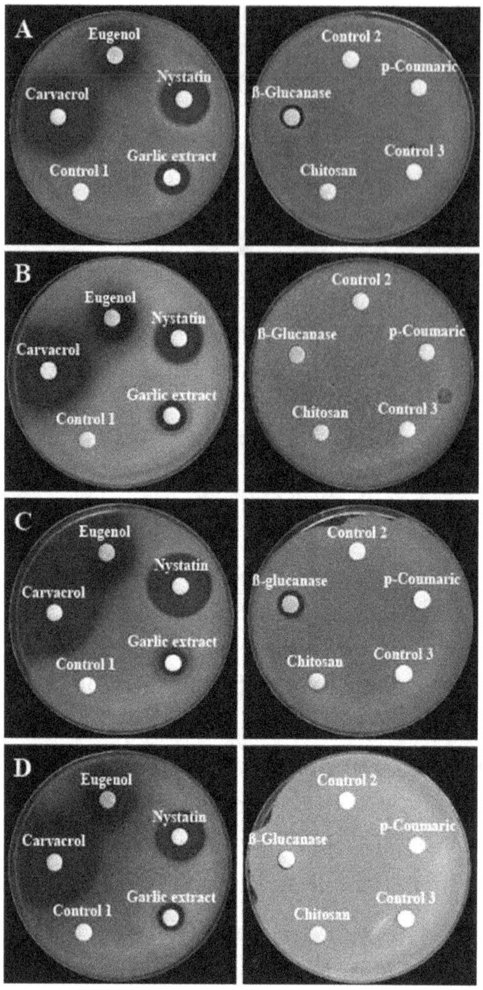

Figure 2. Agar diffusion tests for antimicrobial screening. Antimicrobials tested: eugenol, carvacrol, nystatin (antibiotic), garlic extract (allicin), chitosan, β-glucanase and p-coumaric acid. Controls used for the preparation of the antimicrobials: control 1-water, control 2-ethanol and control 3-acetic acid. Yeast species tested: (A) S. ludwigii, (B) D. bruxellensis, (C) P. anomala and (D) Z. rouxii.

Despite their interesting antimicrobial properties, there still exist some limitations to the use of some of these natural extracts such as their solubility in water and the impact on sensory attributes [21].

2.2.5. Antimicrobial peptides

Antimicrobial peptides are important molecules naturally occurring in the immune system [107] and may have different amino acids conformation. Their importance as antimicrobial agents increased when bacteria and other pathogens became more resistant to drugs [108]. Among the most studied peptides are tritrpticin and indolicidin, related cathelicidins [109] rich in Arg and Trp residues [110] as well as the lactoferricin. Other peptides having microbicidal activity are cecropins, defensins, magainins, melittin and alamethicin [111].

Tritrpticin is a positively charged peptide with tryptophan (Trp) residues [112] that has antifungal activity besides having antibacterial activity against Gram-positive and Gram-negative bacteria [113, 114]; it is naturally found in granules of bovine and porcine neutrophils [115] as well as in insects [107, 116, 117]. The electrostatic interaction between the cationic residue and the negative charge of phospholipids in the membrane has been recognized as an important factor in the microbicidal action found in these biomolecules [118].

Indolicidin is another peptide with antimicrobial activity against both fungi and bacteria (Gram-positive and Gram-negative) [119]. Similar to tritrpticin, indolicidin is a tryptophan (Trp)-rich peptide [120] that also contains proline (Pro) residues in its configuration [121]; its structure resembles that of detergent micelles and of phospholipid vesicles [122]. It has been isolated from cytoplasmic granules of bovine neutrophils [123] but it is also present in humans and other mammals as well as in some primitive vertebrates [124]. It acts directly on the lipidic bilayer [125] causing disruption in the cell membrane.

Lactoferricin B (Lfcin B) is a peptide obtained during the gastric digestion of the bovine protein lactoferrin [126]. Despite the fact that other mammals including humans also produce lactoferricins, the Lfcin B has higher antimicrobial potency [127]. Lfcin B shows different properties including antibacterial, antifungal, antiviral, antitumour, anti-inflammatory and immunoregulatory [127].

Antimicrobial peptides are important for the control of spoilage yeasts in food products. Lfcin B derivatives reduce spoilage yeasts such as D. bruxellensis [128] and Z. bailii and Z. bisporus [129] in wine production. Peptides are used to control the most common spoilage yeasts in dairy products such as K. marxianus, D. hansenii, Candida spp. and other spoilage yeasts such as Zygosaccharomyces microellipsoides, Rhodotorula mucilaginosa, Yarrowia lipolytica, Torulospora and Pichia [130]. Mozzarella cheese packaged with films of chitosan with up to 60% lysozyme has antimicrobial activity against yeast Pseudomonas fluorescens and other bacteria and moulds [131].

In medical applications, indolicidin has antimicrobial activity against different pathogens such as yeasts, viruses, bacteria, fungi and protozoa [132]. Indolicidin has been tested against candidiasis produced by fungi Candida albicans [133], and nosocomial pathogen yeasts Candida krusei and Candida parapsilosis [134] responsible for bloodstream infections [109]. It is also known that indolicidin directly binds DNA forming a complex with the catalytic HIV-1 integrase [124] producing inhibitory activity in it with an implication on the HIV virus replication cycle [135]. Lfcin B has fungicidal activity against the pathogenic yeast Candida tropicalis and fungicidal properties against the fungi Candida neoformans and C. albicans [136].

A drawback in the use of some antimicrobial peptides like tritrpticin and indolicidin for treatment of infectious diseases produced by pathogen agents is the hemolytic activity observed against blood hematocytes [113, 115, 132]; therefore, it is important to use alternative antimicrobial peptides with less toxicity in immunocompromised patients [133].

2.2.6. Killer toxins and β-glucanases

Killer toxins are known as pore proteins produced by yeast metabolism and are able to kill sensitive yeast cells by forming cell wall or cell membrane disruptions [137]. There are different killer toxins produced by either Saccharomyces and non-Saccharomyces yeasts used as antimicrobials.

Among the toxins produced by non-Saccharomyces, Tdkt is produced by the species Torulaspora delbrueckii with β-glucanase and chitinase activity against yeasts B. bruxellensis, Pichia guilliermondii, Peronospora manshurica and Pichia membranaefaciens [138]; toxin Kwkt is obtained from the species Kluyveromyces wickerhamii and toxin Pikt is produced by the species Wickerhamomyces anomalus (formerly Pichia anomala), both are used against spoilage yeasts Brettanomyces/Dekkera [139]; toxin Kpkt is produced by the species Tetrapisispora phaffii and it could be used to control spoilage yeasts in winemaking [140] since it has proven fungicidal activity against yeast Hanseniaspora/Kloeckera [141]. This last yeast is also used as killer yeast against some fungal pathogens in plants and also to fight infections produced by fungi in animals and humans due to its high β-1,3-glucanase activity producing disruptions in the cell wall [142].

S. cerevisiae has also shown the ability to produce killer toxins; four different killer yeasts from this species produce killer toxins K1, K2, K28 and Klus. These yeasts are able to produce the toxins during spontaneous wine fermentation as the population of other yeasts decreases [143]. The killer toxin K1, for example, having the glycosylated subunits α and β, is released out of the cytoplasm [144]. The killer toxin forms a channel in the membrane of target cells through which ions are conducted to the exterior producing cellular death. The activity observed in some killer toxins is due to β-glucanases, which has been used to produce a synthetic β-glucanases preparation as antibacterial material against yeasts D. bruxellensis and Z. bailii [145]. The effect of β-glucanases on the cell integrity can be observed in Figure 3 where cells of the species S. ludwigii have been disrupted after the addition of two different β-glucanases.

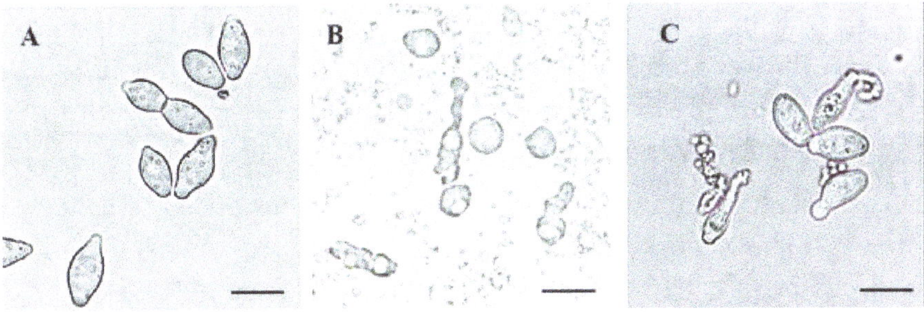

Figure 3. Effect of β-glucanases on the yeast's cell wall after the addition of β-glucanase I (B) and β-glucanase II (C) to S. ludwigii (A). Bar scale 10 μm.

Even though both compounds have same disruptive inhibition mechanism, the activity of the β-glucanases is not the same. Such an activity is compared in Figure 4 where YPD liquid growing media have been inoculated with four different wine spoilage yeasts: S. ludwigii, D. bruxellensis, P. anomala and Z. rouxii. The media where β-glucanase II has been used did not show any growth indicating total effectiveness of the treatment, while the media where β-glucanase I has been added showed CO2 production and turbidity formation in function of the sensibility of each yeast species. Wine spoilage yeast S. ludwigii appears sensitive to both β-glucanases, as no foam has been formed after the addition of either compound.

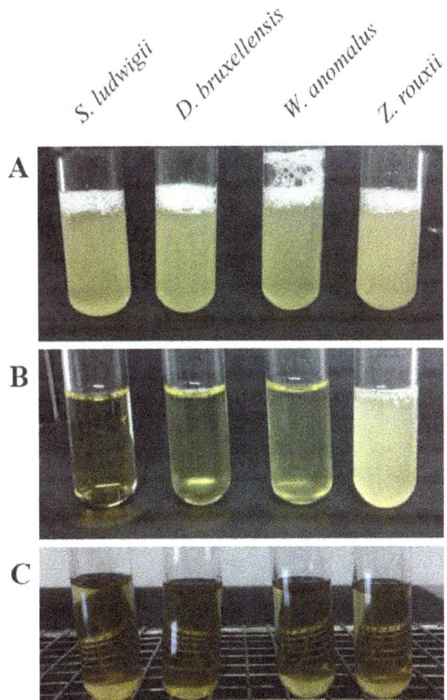

Figure 4. Comparison of the effect of two β-glucanases in YPD liquid growing media with same optical density (OD) of S. ludwigii, D. bruxellensis, P. anomala and Z. rouxii. The trials are grouped in (A) control, (B) β-glucanase I and (C) β-glucanase II.

In the case of β-glucanase, the inhibition zone observed in the agar plates suggests a low diffusion ability and thus limited antimicrobial activity; however, when grown in liquid media in the presence of same β-glucanase all the spoilage yeasts showed sensitivity (see Figure 2).

The yeasts could be neutral or sensitive to toxins produced by killer yeasts. In this matter, [146] have shown that the yeast species B. bruxellensis, B. anomalus, D. anomala, B. custerisanus, H. uvarum, S. cerevisiae and Z. bailii have either sensitive or neutral nature towards killer toxins produced by the species K. wickerhamii and C. pyralidae. Santos et al. [147] have also described the resistance of the species S. cerevisiae to the toxin CYC 1410 produced by the species Ustilago maydis during wine production; therefore, this procedure could inhibit the implantation of B. bruxellensis spoilage yeast. Cytoplasmatic β-glucanases have also shown, on the other hand, interesting contribution to wine-sensory properties as polysaccharides are released from the autolysis of non-Saccharomyces yeast cells during wine production [148]. Polysaccharides can improve mouth-feel properties of wines by changing their viscosity.

A summary of certain emerging antifungal compounds is presented in Figure 5.

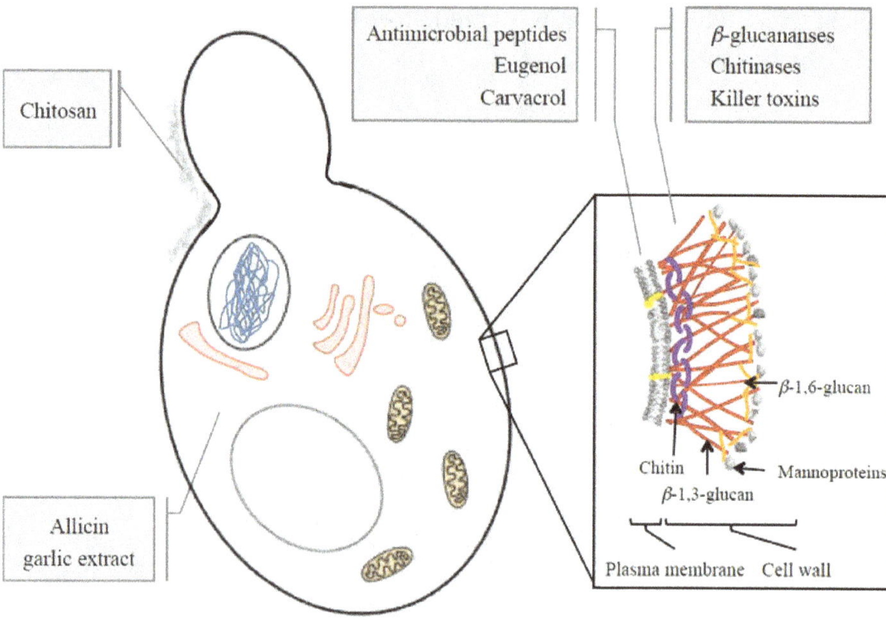

Figure 5. Schematic representation of the main mechanisms involved in the antimicrobial effect of different compounds on a yeast cell. The arrows indicate the target of each compound in the yeast structure: Plasma membrane (antimicrobial peptides, eugenol and carvacrol), cell wall (β-glucanases, chitinases and killer toxins), endoplasmatic enzymes (allicin) and physical adsorption on cell wall (chitosan).

3. YEAST POPULATIONS MONITORING TECHNIQUES

Parallel to the use of technologies that reduce the colonization of spoilage yeasts in food products, there are technologies able to monitor the yeasts and bacteria populations and their nature in the alcoholic and malolactic fermentations during wine production. One of these techniques is the flow cytometry (FCM) [149]; other techniques are fluorescence-activated cell sorting (FACS), quantum dots (QDs) and emerging monitoring technologies such as mass cytometry (Cy-TOF), imaging flow cytometry and, recently, spectral cytometry.

FCM, used for counts of microorganism populations in wine, allowed the simultaneous determination of yeasts for as low as 103 cells/mL and, despite its size, populations of malolactic bacteria higher than 104 cells/mL [150]. FCM is able to show real-time situation of microorganisms in different matrices although it is considered complicated implementation due to the cost of reagents and the need of recruiting trained staff. FCM that relies on the use of complementary fluorescence dying to selectively determine specific type of microorganisms [151] through, for example, the metabolic enzyme activity or antigen expression is known as fluorescence-activated cell sorting (FACS). This technology is useful in monitoring changes in yeast cellular organelles' biogenesis such as the mitochondria [152].

The QD is a technology based on the production of semiconductor nanoparticles made from cadmium salts crystals with chloride (Cl), tellurium (Te) and sulphide (S) by target microbial cells such as spoilage yeast [153]. These nanoparticles are used as biomarkers of nucleic acids and proteins and can be detected visually when they are excited by a light source. QDs are photostable and they have a wide range of absorption while they have a narrow emission peak [154].

Imaging flow cytometry has been employed to compare the mechanism by which several killer toxins impact cell architecture. Comitini et al. [140] observed that the toxin Kpkt from T. phaffii and the toxin K1 from S. cerevisiae have distinct modes of action. When yeast cells were exposed to toxin K1, which is known to disrupt the cell membrane and form potassium channels, all cells died and had their nucleic acids stained with propidium iodide (PI). However, when cells were exposed to Kpkt, not all of them were stained with PI despite being dead. These findings indicate a disparity in the mechanisms by which both poisons induce cell death in yeast. Flow cytometry investigations demonstrated that Kpkt induces cell-wall rupture through highly selective β-glucanase activity [15].

Spectral cytometry was developed to enhance the accuracy and precision of flow cytometry results by utilizing spectral analysis to obtain higher resolution from distinct bands of emission in multiple stained samples. This is particularly useful for microbial counts in foodborne matrices stained with PI. By utilizing the flow cytometry approach, spectral measurements can be conducted to analyze the fluorescence and RAMAN spectra of sizable particles involved in chemical and biological processes [156].

4. FUTURE TRENDS

Over the years, several antimicrobial strategies have been developed to effectively manage the prevalence of spoilage yeasts in food products. These approaches are designed to mitigate the adverse economic consequences of having tainted or spoiled items, as well as the potential health risks they may pose. Some of these approaches are now employed by the winemaking industry, while others may be investigated across the various phases of the process, from vine to bottled wine.

The current trend in the winemaking industry is to employ harmless control techniques to prevent the growth of spoilage yeasts throughout the entire process. This is done to preserve the distinct aromas of different grape varieties, safeguard and enhance the extraction of anthocyanins, and ultimately improve the overall quality of wines.

AT A GLANCE

Historically, sulphur dioxide (SO2) has been the most commonly used chemical for controlling microorganisms and preserving wine as an antibacterial agent. Additional chemicals that have been examined for targeted yeast management include sorbic and benzoic acids. In this discussion, we explore the efficacy and utilization of conventional and innovative therapies and biotechnologies for the management of wine spoilage yeasts by chemical and biological means. We shall examine the adaptability of lethal poisons and the antimicrobial characteristics of organic substances including carvacrol, essential oils, and bioactive peptides. The wine spoilage yeasts that are targeted for control include species from the genera Zygosaccharomyces, Saccharomycodes, and Dekkera/Brettanomyces. Additionally, non-Saccharomyces yeast species such as Hanseniaspora uvarum, Hansenula anomala, Metschnikowia pulcherrima, and Wickerhamomyces anomalus dominate the initial phase of fermentation. Depending on the type of wine being produced, other yeasts like Schizosaccharomyces pombe may also be present.

REFERENCES

1. Fleet G. The commercial and community significance of yeasts in food and beverage production. Yeasts in Food and Beverages. Springer-Verlag, Berlin Heidelberg; 2006. pp. 1-12
2. Palpacelli V, Ciani M, Rosini G. Activity of different 'killer' yeasts on strains of yeast species undesirable in the food industry. FEMS Microbiology Letters. 1991;84(1):75-78
3. Rojo MC, López FA, Lerena M, Mercado L, Torres A, Combina M. Effects of pH and sugar concentration in Zygosaccharomyces rouxii growth and time for spoilage in concentrated grape juice at isothermal and non-isothermal conditions. Food Microbiology. 2014;38:143-150

4. Alonso A, Belda I, Santos A, Navascués E, Marquina D. Advances in the control of the spoilage caused by Zygosaccharomyces species on sweet wines and concentrated grape musts. Food Control. 2015;51:129-134

5. Fleet G. Spoilage yeasts. Critical Reviews in Biotechnology. 1992;12(1-2):1-44

6. Martorell P, Stratford M, Steels H, Fernández-Espinar MT, Querol A. Physiological characterization of spoilage strains of Zygosaccharomyces bailii and Zygosaccharomyces rouxii isolated from high sugar environments. International Journal of Food Microbiology. 2007;114(2):234-242

7. Loureiro V, Malfeito-Ferreira M. Spoilage yeasts in the wine industry. International Journal of Food Microbiology. 2003 Sep 1;86(1-2):23-50

8. Fleet GH. Yeast interactions and wine flavour. International Journal of Food Microbiology. 2003;86(1):11-22

9. Passoth V, Fredlund E, Druvefors UA, Schnurer J. Biotechnology, physiology and genetics of the yeast Pichia anomala. FEMS Yeast Research. 2006 Jan;6(1):3-13

10. Schaffrath R, Breunig KD. Genetics and molecular physiology of the yeast Kluyveromyces lactis. Fungal Genetics and Biology. 2000;30(3):173-190

11. Du Toit M, Pretorius IS. Microbial spoilage and preservation of wine: Using weapons from nature's own arsenal—A review. South African Journal of Enology and Viticulture. 2000;21(Special Issue):74-96

12. Romano P, Suzzi G. Origin and production of acetoin during wine yeast fermentation. Applied and Environmental Microbiology. 1996 Feb;62(2):309-315

13. Akinmusire O. Fungal species associated with the spoilage of some edible fruits in Maiduguri Northern Eastern Nigeria. Advances in Environmental Biology. 2011:157-162

14. Babic I, Hilbert G, Nguyen-The C, Guiraud J. The yeast flora of stored ready-to-use carrots and their role in spoilage. International Journal of Food Science and Technology. 1992;27(4):473-484

15. Ciani M, Fatichenti F. Killer toxin of Kluyveromyces phaffii DBVPG 6076 as a biopreservative agent to control apiculate wine yeasts. Applied Environmental Microbiology. 2001 Jul;67(7):3058-3063

16. Gyawali R, Ibrahim SA. Natural products as antimicrobial agents. Food Control. 2014;46:412-429

17. Török T, King A. Thermal inactivation kinetics of food-borne yeasts. Journal of Food Science. 1991;56(1):6-9

18. Gaze J. Microbiological aspects of thermally processed foods. Journal of Applied Microbiology. 2005;98(6):1381-1386

19. Hotchkiss J. Safety Considerations in Active Packaging. Active Food Packaging. Springer, US; 1995. pp. 238-255

20. Brody AL, Bugusu B, Han JH, Sand CK, McHugh TH. Innovative food packaging solutions. Journal of Food Science. 2008;73(8):R107–R116

21. Raybaudi-Massilia RM, Mosqueda-Melgar J, Soliva-Fortuny R, Martín-Belloso O. Control of pathogenic and spoilage microorganisms in fresh-cut fruits and fruit juices by traditional and alternative natural antimicrobials. Comprehensive Reviews in Food Science and Food Safety. 2009;8(3):157-180

22. Davidson PM, Taylor TM, Schmidt SE. Chemical Preservatives and Natural Antimicrobial Compounds. In: Food microbiology. American Society of Microbiology, Washington, DC. 2013. pp. 765-801

23. Barata A, Caldeira J, Botelheiro R, Pagliara D, Malfeito-Ferreira M, Loureiro V. Survival patterns of Dekkera bruxellensis in wines and inhibitory effect of sulphur dioxide. International Journal of Food Microbiology. 2008;121(2):201-207

24. Warth AD. Resistance of yeast species to benzoic and sorbic acids and to sulfur dioxide. Journal of Food Protection. 1985;48(7):564-569

25. Stratford M, Anslow P. Evidence that sorbic acid does not inhibit yeast as a classic 'weak acid preservative'. Letters in Applied Microbiology. 1998;27(4):203-206

26. Rojo MC, López FA, Lerena M, Mercado L, Torres A, Combina M. Evaluation of different chemical preservatives to control Zygosaccharomyces rouxii growth in high sugar culture media. Food Control. 2015;50:349-355

27. Ough C, Crowell E. Use of sulfur dioxide in winemaking. Journal of Food Science. 1987;52(2):386-388

28. Divol B, du Toit M, Duckitt E. Surviving in the presence of sulphur dioxide: Strategies developed by wine yeasts. Applied Microbiology and Biotechnology. 2012;95(3):601-613

29. Santos MC, Nunes C, Saraiva JA, Coimbra MA. Chemical and physical methodologies for the replacement/reduction of sulfur dioxide use during winemaking: Review of their potentialities and limitations. European Food Research and Technology. 2012;234(1):1-12

30. Guerrero RF, Cantos-Villar E. Demonstrating the efficiency of sulphur dioxide replacements in wine: A parameter review. Trends in Food Science and Technology. 2015,3;42(1):27-43

31. Serpaggi V, Remize F, Recorbet G, Gaudot-Dumas E, Sequeira-Le Grand A, Alexandre H. Characterization of the "viable but nonculturable" (VBNC) state in the wine spoilage yeast Brettanomyces. Food Microbiology. 2012;30(2):438-447

32. Agnolucci M, Rea F, Sbrana C, Cristani C, Fracassetti D, Tirelli A, et al. Sulphur dioxide affects culturability and volatile phenol production by Brettanomyces/Dekkera bruxellensis. International Journal of Food Microbiology. 2010;143(1):76-80

33. Benito S, Palomero F, Morata A, Calderón F, Suárez-Lepe J. Factors affecting the hydroxycinnamate decarboxylase/vinylphenol reductase activity of Dekkera/Brettanomyces: Application for Dekkera/Brettanomyces control in red wine making. Journal of Food Science. 2009;74(1):M15–M22.

34. 27-CFR-24.246-2002 A. Code of Federal Regulations, Title 27, Chapter 1, Subchapter A, Part 24, Subpart L, Section 24.246. Materials Authorized for the Treatment of Wine and Juice. 2002. Available from: https://www.law.cornell.edu/cfr/text/27/24.246 [Accessed: 3 July 2017]

35. FAR A474-2004 A. Final Assessment Report, Application A474, Winemaking. Food Standards Australia New Zealand; 2004. Available from: http://www.foodstandards.gov.au/code/applications/pages/applicationa474winemaking/index.aspx [Accessed: 3 July 2017]

36. CR-EC 643-2006 A. Commission Regulation (EC) No 643/2006. Official Journal of the European Union, 28.4.2006, L115/6-L115/9. 27 April 2006. Available from: http://extwprlegs1.fao.org/docs/pdf/eur65203.pdf [Accessed: 3 July, 2017]

37. Ough C, Langbehn L, Stafford P. Influence of pH and ethanol on the effectiveness of dimethyl dicarbonate in controlling yeast growth in model wine systems. American Journal of Enology and Viticulture. 1978;29(1):60-62

38. Bizri JN, Wahem IA. Citric acid and antimicrobials affect microbiological stability and quality of tomato juice. Journal of Food Science. 1994;59(1):130-135

39. Bartowsky E. Bacterial spoilage of wine and approaches to minimize it. Letters in Applied Microbiology. 2009;48(2):149-156

40. Renouf V, Strehaiano P, Lonvaud-Funel A. Effectiveness of dimethyldicarbonate to prevent Brettanomyces bruxellensis growth in wine. Food Control. 2008;19(2):208-216

41. Divol B, Strehaiano P, Lonvaud-Funel A. Effectiveness of dimethyldicarbonate to stop alcoholic fermentation in wine. Food Microbiology. 2005;22(2):169-178

42. Delfini C, Gaia P, Schellino R, Strano M, Pagliara A, Ambrò S. Fermentability of grape must after inhibition with dimethyl dicarbonate (DMDC). Journal of Agriculture and Food Chemistry. 2002;50(20):5605-5611

43. Costa A, Barata A, Malfeito-Ferreira M, Loureiro V. Evaluation of the inhibitory effect of dimethyl dicarbonate (DMDC) against wine microorganisms. Food Microbiology. 2008;25(2):422-427

44. Zuehlke J, Glawe D, Edwards C. Efficacy of dimethyl dicarbonate against yeasts associated with Washington State grapes and wines. Journal of Food Processing and Preservation. 2014;39(6):1016-1026

45. Oelofse A, Pretorius I, Du Toit M. Significance of Brettanomyces and Dekkera during winemaking: A synoptic review. South African Journal of Enology and Viticulture. 2008;29(2): 128-144

46. Threlfall R, Morris J. Using dimethyldicarbonate to minimize sulfur dioxide for prevention of fermentation from excessive yeast contamination in juice and semi-sweet wine. Journal of Food Science. 2002;67(7):2758-2762

47. Resolution Oeno 6/2006. Method OIV-MA-AS313-20. Determination of sorbic, benzoic and salicylic acid content in wine by the use of high-performance liquid chromatography. Compendium of International Analysis of Methods – OIV. Available from: http://www.oiv.int/public/medias/2519/oiv-ma-as313-20.pdf [Accessed: 3 July 2017]

48. Steels H, James S, Roberts I, Stratford M. Sorbic acid resistance: The inoculum effect. Yeast. 2000;16(13):1173-1183

49. Loureiro V, Malfeito-Ferreira M. Spoilage activities of Dekkera/Brettanomyces spp. In: Blackburn C, editor. Food Spoilage Microorganisms. Cambridge, UK: Woodhead Publisher; 2006. pp. 354-398

50. Oswald TA, Edwards CG. Interactions between storage temperature and ethanol that affect growth of Brettanomyces bruxellensis in Merlot wine. American Journal of Enology and Viticulture. 2017;68:188-194. ajev. 2017.16102

51. Couto JA, Neves F, Campos F, Hogg T. Thermal inactivation of the wine spoilage yeasts Dekkera/Brettanomyces. International Journal of Food Microbiology. 2005,Oct 25;104(3):337-344

52. Zuehlke JM, Edwards CG. Impact of sulfur dioxide and temperature on culturability and viability of Brettanomyces bruxellensis in wine. Journal of Food Protection®. 2013;76(12):2024-2030

53. Wang C, Huang H, Hsu C, Yang BB. Recent advances in food processing using high hydrostatic pressure technology. Critical Reviews in Food Science and Nutrition. 2016;56(4):527-540

54. Yaldagard M, Mortazavi SA, Tabatabaie F. The principles of ultra high pressure technology and its application in food processing/preservation: A review of microbiological and quality aspects. African Journal of Biotechnology. 2008;7(16):2739-2767

55. Koutchma T, Popović V, Ros-Polski V, Popielarz A. Effects of ultraviolet light and high-pressure processing on quality and health-related constituents of fresh juice products. Comprehensive Reviews in Food Science and Food Safety. 2016;15(5):844-867

56. del Árbol JT, Pulido RP, La Storia A, Burgos MJG, Lucas R, Ercolini D, et al. Microbial diversity in pitted sweet cherries (Prunus avium L.) as affected by high-hydrostatic pressure treatment. Food Research International. 2016;89:790-796

57. Pulido RP, Burgos MJG, Gálvez A, Lucas R. Changes in bacterial diversity of refrigerated mango pulp before and after treatment by high hydrostatic pressure. LWT-Food Science and Technology. 2017;78:289-295

58. Shahbaz HM, Yoo S, Seo B, Ghafoor K, Kim JU, Lee D, et al. Combination of TiO2-UV photocatalysis and high hydrostatic pressure to inactivate bacterial pathogens and yeast in commercial apple juice. Food and Bioprocess Technology. 2016;9(1):182-190

59. Martínez-Onandi N, Castioni A, San Martín E, Rivas-Cañedo A, Nuñez M, Torriani S, et al. Microbiota of high-pressure-processed Serrano ham investigated by culture-dependent and culture-independent methods. International Journal of Food Microbiology. 2017;241:298-307

60. van Wyk S, Silva FV. High pressure inactivation of Brettanomyces bruxellensis in red wine. Food Microbiology. 2017;63:199-204

61. González-Arenzana L, Sevenich R, Rauh C, López R, Knorr D, López-Alfaro I. Inactivation of Brettanomyces bruxellensis by high hydrostatic pressure technology. Food Control. 2016;59:188-195

62. Morata A, Loira I, Vejarano R, Bañuelos MA, Sanz PD, Otero L, et al. Grape processing by high hydrostatic pressure: Effect on microbial populations, phenol extraction and wine quality. Food and Bioprocess Technology. 2015;8(2):277-286

63. Milani EA, Silva FV. Nonthermal pasteurization of beer by high pressure processing: Modelling the inactivation of Saccharomyces cerevisiae ascospores in different alcohol beers. High Pressure Research. 2016;36(4):595-609

64. Turtoi M. Inactivation of Saccharomyces cerevisiae using new non-thermal technologies. A review. Romanian Biotechnological Letters. 2014;19(1):8901-8909

65. Gomez-Lopez VM, Ragaert P, Debevere J, Devlieghere F. Pulsed light for food decontamination: A review. Trends in Food Science and Technology. 2007;18(9):464-473

66. Takeshita K, Shibato J, Sameshima T, Fukunaga S, Isobe S, Arihara K, et al. Damage of yeast cells induced by pulsed light irradiation. International Journal of Food Microbiology. 2003;85(1):151-158

67. Ferrario M, Alzamora SM, Guerrero S. Inactivation kinetics of some microorganisms in apple, melon, orange and strawberry juices by high intensity light pulses. Journal of Food Engineering. 2013;118(3):302-311

68. Ferrario M, Alzamora SM, Guerrero S. Study of the inactivation of spoilage microorganisms in apple juice by pulsed light and ultrasound. Food Microbiology. 2015;46:635-642

69. Luksiene Z, Buchovec I, Viskelis P. Impact of high-power pulsed light on microbial contamination, health promoting components and shelf life of strawberries. Food Technology and Biotechnology. 2013;51(2):284

70. Ganan M, Hierro E, Hospital XF, Barroso E, Fernández M. Use of pulsed light to increase the safety of ready-to-eat cured meat products. Food Control. 2013;32(2):512-517

71. Kotnik T, Frey W, Sack M, Meglič SH, Peterka M, Miklavčič D. Electroporation-based applications in biotechnology. Trends in Biotechnology. 2015;33(8):480-488

72. Hülsheger H, Potel J, Niemann E. Electric field effects on bacteria and yeast cells. Radiation and Environmental Biophysics. 1983;22(2):149-162

73. Gášková D, Sigler K, Janderova B, Plášek J. Effect of high-voltage electric pulses on yeast cells: Factors influencing the killing efficiency. Bioelectrochemistry and Bioenergetics. 1996;39(2):195-202

74. Buckow R, Ng S, Toepfl S. Pulsed electric field processing of orange juice: A review on microbial, enzymatic, nutritional, and sensory quality and stability. Comprehensive Reviews in Food Science and Food Safety. 2013;12(5):455-467

75. Aronsson K, Rönner U, Borch E. Inactivation of Escherichia coli, Listeria innocua and Saccharomyces cerevisiae in relation to membrane permeabilization and subsequent leakage of intracellular compounds due to pulsed electric field processing. International Journal of Food Microbiology. 2005;99(1):19-32

76. Ganeva V, Galutzov B, Teissie J. Evidence that pulsed electric field treatment enhances the cell wall porosity of yeast cells. Applied Biochemistry and Biotechnology. 2014;172(3):1540-1552

77. Gómez-Rivas L, Escudero-Abarca BI, Aguilar-Uscanga MG, Hayward-Jones PM, Mendoza P, Ramírez M. Selective antimicrobial action of chitosan against spoilage yeasts in mixed culture fermentations. Journal of Industrial Microbiology and Biotechnology. 2004;31(1):16-22

78. Raafat D, Sahl H. Chitosan and its antimicrobial potential—A critical literature survey. Microbial Biotechnology. 2009;2(2):186-201

79. Roller S, Covill N. The antifungal properties of chitosan in laboratory media and apple juice. International Journal of Food Microbiology. 1999;47(1):67-77

80. Ferreira D, Moreira D, Costa EM, Silva S, Pintado MM, Couto JA. The antimicrobial action of chitosan against the wine spoilage yeast Brettanomyces/Dekkera. Journal of Chitin and Chitosan Science. 2013;1(3):240-245

81. Taillandier P, Joannis-Cassan C, Jentzer J, Gautier S, Sieczkowski N, Granes D, et al. Effect of a fungal chitosan preparation on Brettanomyces bruxellensis, a wine contaminant. Journal of Applied Microbiology. 2015;118(1):123-131

82. Petrova B, Cartwright ZM, Edwards CG. Effectiveness of chitosan preparations against Brettanomyces bruxellensis grown in culture media and red wines. OENO One. 2016;50(1):49-56

83. Iriti M, Vitalini S, Di Tommaso G, D–AMICO S, Borgo M, Faoro F. New chitosan formulation prevents grapevine powdery mildew infection and improves polyphenol content and free radical scavenging activity of grape and wine. Australian Journal of Grape and Wine Research. 2011;17(2):263-269

84. Fajardo P, Martins J, Fuciños C, Pastrana L, Teixeira J, Vicente A. Evaluation of a chitosan-based edible film as carrier of natamycin to improve the storability of Saloio cheese. Journal of Food Engineering. 2010;101(4):349-356

85. Souza JM, Caldas AL, Tohidi SD, Molina J, Souto AP, Fangueiro R, et al. Properties and controlled release of chitosan microencapsulated limonene oil. Revista Brasileira de Farmacognosia. 2014;24(6):691-698

86. Elmacı SB, Gülgör G, Tokatlı M, Erten H, İşci A, Özçelik F. Effectiveness of chitosan against wine-related microorganisms. Antonie Van Leeuwenhoek. 2015;107(3):675-686

87. Conner D, Beuchat L. Effects of essential oils from plants on growth of food spoilage yeasts. Journal of Food Science. 1984;49(2):429-434

88. Curtis O, Shetty K, Cassagnol G, Peleg M. Comparison of the inhibitory and lethal effects of synthetic versions of plant metabolites (anethole, carvacrol, eugenol, and thymol) on a food spoilage yeast (Debaromyces hansenii). Food Biotechnology. 1996;10(1):55-73

89. Burt S. Essential oils: Their antibacterial properties and potential applications in foods—A review. International Journal of Food Microbiology. 2004;94(3):223-253

90. Bennis S, Chami F, Chami N, Bouchikhi T, Remmal A. Surface alteration of Saccharomyces cerevisiae induced by thymol and eugenol. Letters in Applied Microbiology. 2004;38(6):454-458

91. Kubo I, Cespedes CL. Antifungal activity of alkanols: Inhibition of growth of spoilage yeasts. Phytochemistry Reviews. 2013;12(4):961-977

92. Sivropoulou A, Papanikolaou E, Nikolaou C, Kokkini S, Lanaras T, Arsenakis M. Antimicrobial and cytotoxic activities of Origanum essential oils. Journal of Agricultural and Food Chemistry. 1996;44(5):1202-1205

93. Chavan PS, Tupe SG. Antifungal activity and mechanism of action of carvacrol and thymol against vineyard and wine spoilage yeasts. Food Control. 2014;46:115-120

94. Rabinkov A, Miron T, Konstantinovski L, Wilchek M, Mirelman D, Weiner L. The mode of action of allicin: Trapping of radicals and interaction with thiol containing proteins. Biochimica et Biophysica Acta (BBA)-General Subjects. 1998;1379(2):233-244

95. Chang T, Jang H, Lin W, Duan P. Antifungal activities of commercial rice wine extracts of Taiwanese Allium fistulosum. Advances in Microbiology. 2016;6(07):471

96. Oommen S, Anto RJ, Srinivas G, Karunagaran D. Allicin (from garlic) induces caspase-mediated apoptosis in cancer cells. European Journal of Pharmacology. 2004;485(1):97-103

97. Lee J, Gupta S, Huang J, Jayathilaka LP, Lee B. HPLC-MTT assay: Anticancer activity of aqueous garlic extract is from allicin. Analytical Biochemistry. 2013;436(2):187-189

98. Yoshida H, Katsuzaki H, Ishikawa K, Fukuda H, Fujino T, Suzuki A. Antimicrobial activity of the thiosulfinates isolated from oil-macerated garlic extract. Bioscience, Biotechnology, and Biochemistry. 1999;63(3):591-594

99. Shadkchan Y, Shemesh E, Mirelman D, Miron T, Rabinkov A, Wilchek M, et al. Efficacy of allicin, the reactive molecule of garlic, in inhibiting Aspergillus spp. in vitro, and in a murine model of disseminated aspergillosis. Journal of Antimicrobial Chemotherapy. 2004 May;53(5):832-836

100. Ankri S, Mirelman D. Antimicrobial properties of allicin from garlic. Microbial Infections. 1999;1(2):125-129

101. Cobos R, Mateos RM, Alvarez-Perez JM, Olego MA, Sevillano S, Gonzalez-Garcia S, et al. Effectiveness of natural antifungal compounds in controlling infection by grapevine trunk disease pathogens through pruning wounds. Applied Environmental Microbiology. 2015 Sep;81(18):6474-6483

102. Davis SR. An overview of the antifungal properties of allicin and its breakdown products—The possibility of a safe and effective antifungal prophylactic. Mycoses. 2005;48(2):95-100

103. Marchese A, Barbieri R, Sanches-Silva A, Daglia M, Nabavi SF, Jafari NJ, et al. Antifungal and antibacterial activities of allicin: A review. Trends in Food Science and Technology. 2016;52:49-56

104. Chang Y, McLandsborough L, McClements DJ. Physical properties and antimicrobial efficacy of thyme oil nanoemulsions: Influence of ripening inhibitors. Journal of Agriculture and Food Chemistry. 2012;60(48):12056-12063

105. Maté J, Periago PM, Palop A. When nanoemulsified, d-limonene reduces Listeria monocytogenes heat resistance about one hundred times. Food Control. 2016;59:824-828

106. Blanco-Padilla A, Soto KM, Hernandez Iturriaga M, Mendoza S. Food antimicrobials nanocarriers. Scientific World Journal. 2014;2014:837215

107. Fogaça AC, Lorenzini DM, Kaku LM, Esteves E, Bulet P, Daffre S. Cysteine-rich antimicrobial peptides of the cattle tick Boophilus microplus: Isolation, structural characterization and tissue expression profile. Developmental & Comparative Immunology. 2004;28(3):191-200

108. Schibli DJ, Epand RF, Vogel HJ, Epand RM. Tryptophan-rich antimicrobial peptides: Comparative properties and membrane interactions. Biochemistry and Cell Biology. 2002;80(5):667-677

109. Jung HJ, Park Y, Hahm K, Lee DG. Biological activity of Tat (47-58) peptide on human pathogenic fungi. Biochemical and Biophysical Research Communications. 2006;345(1):222-228

110. Chan DI, Prenner EJ, Vogel HJ. Tryptophan-and arginine-rich antimicrobial peptides: Structures and mechanisms of action. Biochimica et Biophysica Acta (BBA)-Biomembranes. 2006;1758(9):1184-1202

111. Zhao H, Mattila J, Holopainen JM, Kinnunen PK. Comparison of the membrane association of two antimicrobial peptides, magainin 2 and indolicidin. Biophysics Journal. 2001;81(5):2979-2991

112. Schibli DJ, Hwang PM, Vogel HJ. Structure of the antimicrobial peptide tritrpticin bound to micelles: A distinct membrane-bound peptide fold. Biochemistry. 1999;38(51):16749-16755

113. Salay LC, Procopio J, Oliveira E, Nakaie CR, Schreier S. Ion channel-like activity of the antimicrobial peptide tritrpticin in planar lipid bilayers. FEBS Letters. 2004;565(1-3):171-175

114. Yang S, Shin SY, Kim Y, Kim Y, Hahm K, Kim JI. Conformation-dependent antibiotic activity of tritrpticin, a cathelicidin-derived antimicrobial peptide. Biochemical and Biophysical Research Communications. 2002;296(5):1044-1050

115. Nguyen LT, de Boer L, Zaat SA, Vogel HJ. Investigating the cationic side chains of the antimicrobial peptide tritrpticin: Hydrogen bonding properties govern its membrane-disruptive activities. Biochimica et Biophysica Acta (BBA)-Biomembranes. 2011;1808(9):2297-2303

116. Tang Y, Shi Y, Zhao W, Hao G, Le G. Discovery of a novel antimicrobial peptide using membrane binding-based approach. Food Control. 2009;20(2):149-156

117. Tomie T, Ishibashi J, Furukawa S, Kobayashi S, Sawahata R, Asaoka A, et al. Scarabaecin, a novel cysteine-containing antifungal peptide from the rhinoceros beetle, Oryctes rhinoceros. Biochemical and Biophysical Research Communications.2003;307(2):261-266

118 .Yang S, Shin SY, Lee CW, Kim Y, Hahm K, Kim JI. Selective cytotoxicity following Arg-to-Lys substitution in tritrpticin adopting a unique amphipathic turn structure. FEBS Letters. 2003;540(1-3):229-233

119. Subbalakshmi C, Sitaram N. Mechanism of antimicrobial action of indolicidin. FEMS Microbiology Letters. 1998 Mar 1;160(1):91-96

120. Falla TJ, Karunaratne DN, Hancock RE. Mode of action of the antimicrobial peptide indolicidin. Journal of Biological Chemistry. 1996;271(32):19298-19303

121. Ladokhin AS, Selsted ME, White SH. Bilayer interactions of indolicidin, a small antimicrobial peptide rich in tryptophan, proline, and basic amino acids. Biophysics Journal. 1997;72(2):794-805

122. Rozek A, Friedrich CL, Hancock RE. Structure of the bovine antimicrobial peptide indolicidin bound to dodecylphosphocholine and sodium dodecyl sulfate micelles. Biochemistry. 2000;39(51):15765-15774

123. Friedrich CL, Rozek A, Patrzykat A, Hancock RE. Structure and mechanism of action of an indolicidin peptide derivative with improved activity against gram-positive bacteria. Journal of Biological Chemistry. 2001 Jun 29;276(26):24015-24022

124. Marchand C, Krajewski K, Lee HF, Antony S, Johnson AA, Amin R, et al. Covalent binding of the natural antimicrobial peptide indolicidin to DNA abasic sites. Nucleic Acids Research. 2006;34(18):5157-5165

125 .Lee DG, Kim HK, Am Kim S, Park Y, Park S, Jang S, et al. Fungicidal effect of indolicidin and its interaction with phospholipid membranes. Biochemical and Biophysical Research Communications. 2003;305(2):305-310

126. Bellamy W, Takase M, Wakabayashi H, Kawase K, Tomita M. Antibacterial spectrum of lactoferricin B, a potent bactericidal peptide derived from the N-terminal region of bovine lactoferrin. Journal of Applied Bacteriology. 1992;73(6):472-479

127. Vogel HJ, Schibli DJ, Jing W, Lohmeier-Vogel EM, Epand RF, Epand RM. Towards a structure-function analysis of bovine lactoferricin and related tryptophan-and arginine-containing peptides. Biochemistry and Cell Biology. 2002;80(1):49-63

128. Enrique M, Marcos JF, Yuste M, Martínez M, Vallés S, Manzanares P. Inhibition of the wine spoilage yeast Dekkera bruxellensis by bovine lactoferrin-derived peptides. International Journal of Food Microbiology. 2008;127(3):229-234

129. Enrique M, Marcos JF, Yuste M, Martínez M, Vallés S, Manzanares P. Antimicrobial action of synthetic peptides towards wine spoilage yeasts. International Journal of Food Microbiology. 2007;118(3):318-325

130. Frank JF. Milk and Dairy Products. Food Microbiology: Fundamentals and Frontiers. 3rd ed. American Society of Microbiology, Washington, DC. 2007. pp. 141-155

131. Duan J, Park S, Daeschel M, Zhao Y. Antimicrobial chitosan-lysozyme (CL) films and coatings for enhancing microbial safety of mozzarella cheese. Journal of Food Science. 2007;72(9):M355–M362

132. Ryge T, Doisy X, Ifrah D, Olsen J, Hansen P. New indolicidin analogues with potent antibacterial activity. The Journal of Peptide Research. 2004;64(5):171-185

133. Benincasa M, Scocchi M, Pacor S, Tossi A, Nobili D, Basaglia G, et al. Fungicidal activity of five cathelicidin peptides against clinically isolated yeasts. Journal of Antimicrobial Chemotherapy. 2006 Nov;58(5):950-959

134. Sader HS, Fedler KA, Rennie RP, Stevens S, Jones RN. Omiganan pentahydrochloride (MBI 226), a topical 12-amino-acid cationic peptide: Spectrum of antimicrobial activity and measurements of bactericidal activity. Antimicrobial Agents and Chemotherapy. 2004 Aug;48(8):3112-3118

135. Krajewski K, Marchand C, Long Y, Pommier Y, Roller PP. Synthesis and HIV-1 integrase inhibitory activity of dimeric and tetrameric analogs of indolicidin. Bioorganic & Medicinal Chemistry Letters. 2004;14(22):5595-5598

136. Vorland LH, Ulvatne H, Andersen J, Haukland HH, Rekdal Ø, Svendsen JS, et al. Lactoferricin of bovine origin is more active than lactoferricins of human, murine and caprine origin. Scandinavian Journal of Infectious Diseases. 1998;30(5):513-517

137. Bussey H. K1 killer toxin, a pore-forming protein from yeast. Molecular Microbiology. 1991;5(10):2339-2343

138. Villalba ML, Sáez JS, del Monaco S, Lopes CA, Sangorrín MP. TdKT, a new killer toxin produced by Torulaspora delbrueckii effective against wine spoilage yeasts. International Journal of Food Microbiology. 2016;217:94-100.

139. Oro L, Ciani M, Bizzaro D, Comitini F. Evaluation of damage induced by Kwkt and Pikt zymocins against Brettanomyces/Dekkera spoilage yeast, as compared to sulphur dioxide. Journal of Applied Microbiology. 2016;121:207-214.

140. Comitini F, Di Pietro N, Zacchi L, Mannazzu I, Ciani M. Kluyveromyces phaffii killer toxin active against wine spoilage yeasts: Purification and characterization. Microbiology. 2004;150(8):2535-2541

141. Oro L, Zara S, Fancellu F, Mannazzu I, Budroni M, Ciani M, et al. TpBGL2 codes for a Tetrapisispora phaffii killer toxin active against wine spoilage yeasts. FEMS Yeast Research. 2014 May;14(3):464-471

142. Anonymous N. Tetrapisispora phaffii. 2011. Available from: https://www.ncbi.nlm.nih.gov/genome/?term=Tetrapisispora+phaffii [Accessed: July 3, 2017]

143. Maqueda M, Zamora E, Alvarez ML, Ramirez M. Characterization, ecological distribution, and population dynamics of Saccharomyces sensu stricto killer yeasts in the spontaneous grape must fermentations of southwestern Spain. Applied and Environmental Microbiology. 2012 Feb;78(3):735-743

144. Marquina D, Santos A, Peinado J. Biology of killer yeasts. International Microbiology. 2002;5(2):65-71

145. Enrique M, Ibáñez A, Marcos J, Yuste M, Martinez M, Vallés S, et al. β-glucanases as a tool for the control of wine spoilage yeasts. Journal of Food Science. 2010;75(1):M41–M45

146. Mehlomakulu NN, Setati ME, Divol B. Characterization of novel killer toxins secreted by wine-related non-Saccharomyces yeasts and their action on Brettanomyces spp. International Journal of Food Microbiology. 2014;188:83-91

147. Santos A, Navascués E, Bravo E, Marquina D. Ustilago maydis killer toxin as a new tool for the biocontrol of the wine spoilage yeast Brettanomyces bruxellensis. International Journal of Food Microbiology. 2011;145(1):147-154

148. García M, Esteve-Zarzoso B, Arroyo T. Non-Saccharomyces yeasts: Biotechnological role for wine production. In Grape and Wine Biotechnology. InTech. 2016

149. Longin C, Petitgonnet C, Guilloux-Benatier M, Rousseaux S, Alexandre H. Application of flow cytometry to wine microorganisms. Food Microbiology. 2017April;62:221-231

150. Malacrinò P, Zapparoli G, Torriani S, Dellaglio F. Rapid detection of viable yeasts and bacteria in wine by flow cytometry. Journal of Microbiology Methods. 2001;45(2):127-134

151. Veal D, Deere D, Ferrari B, Piper J, Attfield P. Fluorescence staining and flow cytometry for monitoring microbial cells. Journal of Immunological Methods. 2000;243(1):191-210

152. Walker GM. Yeasts. Desk Encyclopedia of Microbiology. 2nd ed. London: Elsevier/Academic Press; 2009

153. Khiyami MA, Almoammar H, Awad YM, Alghuthaymi MA, Abd-Elsalam KA. Plant pathogen nanodiagnostic techniques: Forthcoming changes? Biotechnology & Biotechnological Equipment. 2014;28(5):775-785

154. Lim DV, Simpson JM, Kearns EA, Kramer MF. Current and developing technologies for monitoring agents of bioterrorism and biowarfare. Clinical Microbiology Reviews. 2005 Oct;18(4):583-607

155. Goddard G, Martin JC, Naivar M, Goodwin PM, Graves SW, Habbersett R, et al. Single particle high resolution spectral analysis flow cytometry. Cytometry Part A. 2006;69(8):842-851

156. Watson DA, Gaskill DF, Brown LO, Doorn SK, Nolan JP. Spectral measurements of large particles by flow cytometry. Cytometry Part A. 2009;75(5):460-464.

CHAPTER-5

THE ROLE OF YEAST IN ENHANCING THE AROMA AND FLAVOR OF WINE

1. INTRODUCTION: AN OVERVIEW

Wine is an alcoholic beverage that is typically produced by fermenting fresh grape must. Wine primarily contains 86.8% water and has a content, measured by weight, of the following substances: The composition of wine includes 11.2% ethanol, 0.5% volatile and non-volatile acids, 1% trace components such as sugars, anions, and cations, and a very small amount of 0.5% of volatiles that contribute to its aroma. These volatiles are typically referred to as odour-active compounds (OAC) [1, 2]. These chemicals contribute to the unique scent profile of each wine. The composition and proportion of different organic acid compounds (OAC) are distinct not just for each wine, but also for each terroir, vinification style, and aging process. Yeast plays a significant role in both the creation and adjustment of the scent in wine, in addition to its role in alcohol production. Fleet [3] outlines six processes via which yeast influences the final aroma of wine.

Involve in the biocontrol of moulds on grape, which influences quality before harvest. Perform alcoholic fermentation of must sugars and transform juice into wine; the de novo biosynthesis of the flavour and aroma compounds. Enzymatic conversion of flavour neutral, grape components into odour-active compounds. Alterations of OAC profile through the yeast autolysis products. Absorption of grape juice components. Spoilage of bulk wine throughout the storage period and even after packaging. Influence growth of other spoilage microorganism, for example, lactic acid bacteria, acetic acid bacteria.

The fermentation process is the most crucial mechanism among the listed ones, and it may be controlled by the winemaker. The highest concentration of the final organic acid (OAC) in wine is created during the fermentation process [3-5]. The winemaker has the ability to control and shape the final product by making decisions regarding various vinification practices. These practices include controlling the fermentation temperature within a range of 6-10 degrees, inoculating the must with specific yeast strains, adding yeast assimilable nitrogen (YAN), and determining the initial total concentration of suspended solids in the must before inoculation.

During the fermentation process, the yeast operates in an environment that undergoes constant modifications [22]. During fermentation, yeast generates a significant quantity of heat. However, this is well regulated by contemporary temperature control tanks. Temperature plays a crucial role in controlling yeast development. Furthermore, yeast must employ multiple methods in response to changes in the environment, known as 'stress', due to the conversion of sugar and the utilization of oxygen and YAN [22].

- Osmotic pressure alterations
    - Limitation of essential nutrients
    - Ethanol toxicity
    - Production of by-products toxic to the cell

The persistent requirement for adjustment to this type of environment necessitates equivalent adaptations in numerous responses to uphold intracellular metabolic activity [22]. The process of yeast fermenting sugars can be categorized into two distinct phases: primary metabolism and secondary metabolism. Primary metabolism refers to the metabolic processes that are crucial for yeast development and cell division, resulting in the production of substances such as ethanol, glycerol, acetaldehyde, and acetic acid. Secondary metabolism is not necessary for growth and results in the production of tiny molecules. Yeast adaptation processes primarily affect the ultimate aroma profile of the wine through secondary metabolism [23].

Namely, the FAC generating mechanisms in which the yeast is engaged are the following [3]:

- Utilising grape juice constituents
- Producing ethanol and other solvents that help to extract flavour components from grape solids
- Producing enzymes that transform neutral grape compounds into flavour active compounds
- Producing many hundreds of flavour active, secondary metabolites (e.g. acids, alcohols, esters, polyols, aldehydes, ketones and volatile sulphur compounds)
- Autolytic degradation of dead yeast cells

2. WINE AROMA: ORIGIN AND TYPE

In general, the scent of wine can be categorized into three main types: primary, which comes directly from the fruit; secondary, which is produced during the process of making the wine; and tertiary, which is a result of the wine's maturation and aging processes [24, 25].

The primary organic acid compounds (OAC) can be categorized into four main groups: esters, aldehydes, alcohols, and terpenes [25, 26]. Furthermore, there are two other groups that exhibit distinct characteristics for various grape types. These groups include pyrazines, which contribute to primary fragrance, and sulphur compounds such as polyfunctional sulphur compounds (e.g., 4-mercapto-4-methyl-pentan-2-one, 3-mercaptophenol) and dimethyl sulphide (DMS) [27]. Each of these groups has a distinct function in determining the perceived fragrance characteristics of the finished wine. Furthermore, all of these groups have a varied formation trajectory.

2.1. Flavour active groups contributing in wine aroma

2.1.1. Esters formation and contribution

Esters are the group with the highest importance in wine and are usually the most predominant in the formation of the flavour character of the final product [22]. Esters are formed with the combination of alcohols and organic acid with the elimination of water [25]. In wine, two types of this group can be met: first, the one that is modulated enzymatically by the yeast enzyme pool and second, the one that is formed during ageing [2, 28]. The enzymatic biosynthesis of ester is catalysed mainly by two types of enzymes: esterases and lipases. The final profile of esters in wine depends on various parameters, many authors pointed the variety and quantities of esters, in Table 1, are the main esters found in commercial wine [29]. The group of ethyl acetate, isoamyl acetate, isobutyl acetate, ethyl caproate and 2-phenyl acetate are descripted as the most important esters affecting wine flavour [28, 30, 31].

The net concentration of ester in wine at any given time varies due to the fact that wine is a fairly complex matrix substrate, with a number of different compounds involve in various procedures [2]. This depends on the enzymatic activities of synthesis and ester hydrolysis. Maximum concentrations of esters during fermentation observe around 9–12% of ethanol [8, 32].

Esters formed during alcoholic fermentation, by enzymes; fall into two main categories. The ethyl esters of organic acids and the acetates of higher alcohols [23]. The ethyl esters comprise of an alcohol group (ethanol) and an acid group (small, medium-chain fatty acid) (Figure 1). The acetate esters are comprised of an acid group (acetate) and an alcohol group which is either ethanol or a higher alcohol derived from amino acid metabolism (Figure 2). The latter are responsible for the pleasant fruity aroma of wines [2].

Table 1. Range of ester contents in commercial white wines [29].

Compound	Sweet wines (mg/l)	Dry wines (mg/l)
Ethyl acetate	96.56 ± 39.75	85.00 ± 12.54
Isobutyl acetate	0.07 ± 0.02	0.07 ± 0.04
Ethyl butyrate	0.31 ± 0.09	0.41 ± 0.05
Isoamyl acetate	1.81 ± 0.91	2.37 ± 0.62
Ethyl hexanoate	0.87 ± 0.41	1.06 ± 0.19
Hexyl acetate	0.06 ± 0.04	0.14 ± 0.14
Ethyl lactate	13.5 ± 6.6	23.00 ± 18.88
Ethyl octanoate	1.57 ± 0.73	2.11 ± 0.49
Ethyl decanoate	0.65 ± 0.26	0.56 ± 0.06
Benzyl acetate	0.004 ± 0.004	0.003 ± 0.001
2-Phenylethyl acetate	0.23 ± 0.17	0.21 ± 0.05
Ethyl dodecanoate	0.079 ± 0.053	0.021 ± 0.007

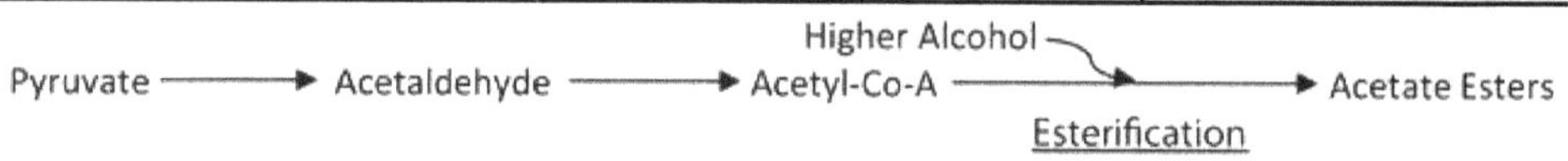

Figure 1. Formation of acetate esters by the esterification of acetyl-Co-A with a higher alcohol [23].

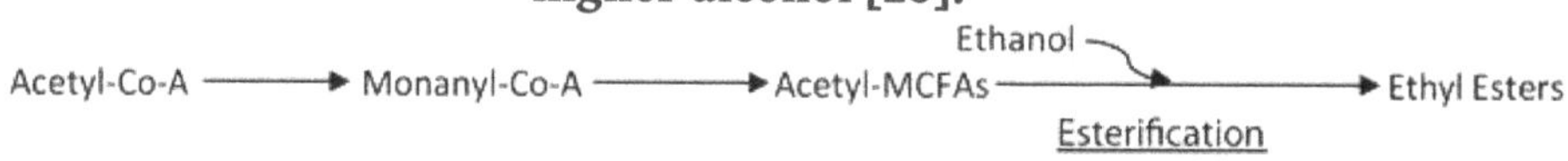

Figure 2. Formation of ethyl esters by the esterification of medium-chain fatty acid (MCFA) with a ethanol [23].

Formation of these two groups of esters during alcoholic fermentation involves a series of various proteins and genes. Today six genes have been identified, with their corresponding protein, to be involved in either the synthesis or hydrolysis of esters in yeast cells. Namely, these are ATF1, Lg-ATF1, ATF2, EHT1, EEB1 and IAH1 [2, 5]. The first three are involved in the mechanism of alcohol acetyltransferase with the first to be the most studied and important in the total quantity of esters formed. EHT1 is involved in the ethanol hexanol-transferase mechanism for the synthesis and hydrolysis of medium-chain fatty acids ethyl esters. EEB1 is involved in the ethanol acyltrasferase and ethyl hydrolase mechanism. Lastly, the IAH1 is a gene involved in the mechanism of esterase, for the hydrolysis of acetate esters [2, 5].

The acetate ester synthesis is an energy requiring mechanism [31], the reasons why esters are formed during the fermentation procedure are not quite clear. One approach to this is that the synthesis is associated with the detoxification effect by the removal of fatty acids [31]. Even though the transfer of esters through the membrane is directly associated with the length of their chain that is varying from 100% small medium-chain to 8–17% for a long chain (ethyl decanoate) [2]. Another systematic approach for the formation is that of the maintenance in the balance of the acetyl-CoA and CoA-SH pool. It is proposed that these are formed as overspill products from the fermentation through the sugar metabolism [31].

2.2. Higher alcohols or fusel alcohols origin and contribution

Higher alcohols refers to the group of alcohols that have more than two atoms of carbon on their molecule. This group of compounds with esters are the two biggest groups contributing to the aroma of wine. The vital step in the synthesis of these compounds is the formation of α-ketoacid [33]. Based on the origin of α-ketoacid fusel alcohols can be divided into two categories. The first has origin of α-ketoacid the amino acids and the second, the anabolic pathway of sugars [27]. From the first group, a list of higher alcohols is shown in Table 2, were as from the former 1-butanol and 1-pentanol are formed. Their contribution to wine aroma is consider positive when the concentration of these compounds is up to 300 mg/l, above this level the pungent odour is profound [25, 31]. The utilisation of nitrogen sources is strongly associated with the biosynthesis of these higher alcohols [17, 34, 35]. The nitrogen composition and nature (organic or mineral), of the must, are influencing the biosynthesis of these volatiles. It has been shown by many studies that the initial concentration and type of amino acids in the must in some cases is strongly associated with varietal aromas [5]. The observation that the increase in the concentration of certain amino acids led to the increase of the production of specific fusel alcohols let to the formulation of the Ehrlich pathway [33]. Also the well documented Ehrlich pathway intermediates can be found in bibliography, stating, the following Table 2 of the 'substrate' amino acid and their corresponding fusel alcohol [5].

Table 2. Flavour-producing amino acid catabolism via the Ehrlich pathway [5].

Amino acid	Higher alcohol
Leusine—Leu	3-Methylbutanol
Valine—Val	2-Methylpropanol
Isoleucine—Ile	2-Methylbutanol
Phenylalanine—Phe	2-Phenylethanol
Tyrosine—Tyr	2-(4-Hydroxylphenyl) ethanol
Tryptophan—Trp	2-(Indol-3-yl) ethanol
Methionine—Met	3-(methyl thio) propanol

The biosynthesis in the Ehrlich pathway starts with the transamination of the amino acid producing a α-ketoacid [33]. Followed by the decarboxylation α-ketoacid to a fusel aldehydes [33]. Finally is the decisive step were the fusel aldehyde is reduced to fusel alcohol or oxidised to the corresponding fusel acid. This step is highly dependable on the growing condition during the utilisation of amino acids [27, 33]. In aerobic conditions, amino acids are converted predominantly to fusel acids where as in anaerobic the product of the pathway in almost entirely fusel alcohol. However, the procedure is far from simple, since a vast number of genes and their corresponding proteins are involve in every stage.

Four S. cerevisiae proteins have been implicated in the initial transamination step of the Ehrlich pathway (Figure 3). Twt1p (also known as Bat1p or Eca39p) is the mitochondrial branched-chain amino acid aminotransferase, and Twt2p (Bat2p or Eca40p) is the cytosolic isozyme. The mitochondrial isozyme is highly expressed in batch cultures during exponential growth and is repressed during stationary phase, while the cytosolic isozyme has the opposite expression pattern [33].

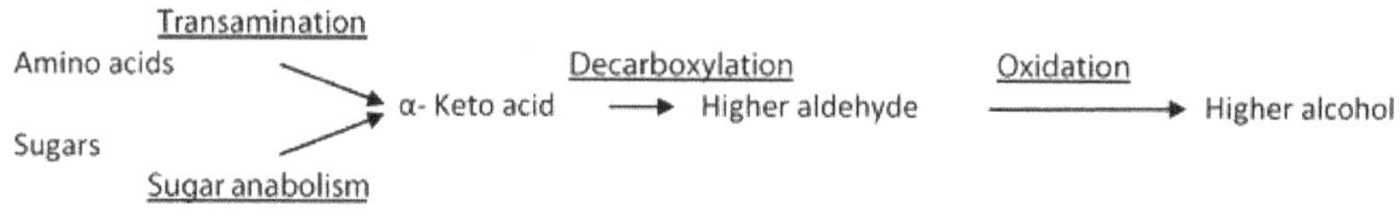

Figure 3. Formation of higher alcohols by the Ehrlich pathway and the anabolism of sugars.

The course of the formation of various higher alcohols was the study of different researchers. According to Fraile et al. [14, 36], the formation of different alcohols takes place at the end of the fermentation, when most of the amino acids have been consumed, whereas, according to Rapp and Versini [37], this synthesis occurs at the same time as ethanol production. A more recent study by Hernandez-Orte et al. [38] monitoring the course of formation of alcohols, among other volatiles, clearly shows that the formation of isoamyl, isobutanol and β-phenylethanol are generated throughout the entire alcoholic fermentation.

2.3. Carbonyl compounds origin and contribution

The two major compounds in this group are acetaldehyde and diacetyl. Acetaldehyde, which is one of the main metabolic intermediates in alcoholic fermentation, is the last precursor in the anaerobic pathway before ethanol. The pyruvate, end product of glycolysis, is converted to acetaldehyde by the pyruate decarboxylase enzymes, which is further converted to ethanol, by the dehydrogenase enzymes. Another source of acetaldehyde is the oxidation of ethanol during ageing or the activity of film forming yeast to the wine [25, 27] 'flor' effect.

Diacetyl is formed in small quantities in wine by the yeast. This can further metabolised to the corresponding end product of 2,3-butanediol or the intermediate, acetoin. Concentrations of up to 100 mg/L of acetaldehyde and 1–4 mg/L of diacetyl can be described as desirable and that these are contributing to the complexity of the aroma of wine [25, 27].

2.4. Thiols contribution and origin

Thiols are the group of alcohol compounds that oxygen was replaced by sulphur in the hydroxyl group. This compound group is very characteristic for wine aroma, especially for the Sauvignon blanc variety, and is mostly depended on yeast metabolism. The three thiols that have been identified in wine are 3-mercaptohexan-1-ol (3MH), 3-mercaptohexyl acetate (3MHA) and 4-mercapto-4-methylpentan-2-one (4MMP). These compounds apart from been characteristic are also very interesting for studying, exhibiting very low perception thresholds, that is, for 4MMP 3 ng/L, for 3MH 50–60 ng/L and for the acetylated form of 3MH the 3MHA is down to 2–4 ng/L [39]. In addition, the 3MH and 3MHA both have two enantiomers, as chiral molecules, R and S. The aroma the S-3MH and S-3MHA forms were described as passion fruit whereas the form R-3MH as grape fruit and R-3MHA as box tree [40].

Over the years, on Sauvignon blanc variety, two different molecule classes have been proposed as precursors of thiols, the amino acid-based compounds [41, 42] and the non-amino acid [43]. Most studies are focused on the first class of compounds and more specific to Cysteine-conjugates and Glutathione-conjugates. Interestingly, the amino compounds not only are the main precursors but also there was the evidence that glutathionylated precursors can be converted into cysteinylated precursors [44]. The non-amino acid precursor compounds were suggested to be the mesityl oxide and E-2-hexenal [43].

2.5. Terpenes: de novo synthesis and/or biotransformations by yeasts

Clearly terpenes are responsible for some of the most prominent, characteristic and important aromas in grapes and wines. It has been documented early on (1978) that beside grapes, yeasts are also capable of producing terpenes (citronellol, linalool and geraniol by Kluyveromyces lactis [45]. Enzymatic activity by yeasts is also possible in relation to liberation of terpenes from sugar molecules and β-glucosidase is well documented [46].

Saccharomyces cerevisiae shows some enzyme activity in different strains [47] but most studies demonstrate significantly higher enzyme production from non-Saccharomyces species [48–51]. Efforts to identify the most efficient non-Saccharomyces are showing the potential these yeasts have in modern wine making and mix cultures. Torulaspora delbrueckii and Metschnikowia pulcherrima are enhancing a very good aromatic profile if used in combination with S. cerevisiae [52–55]. Efforts to further our knowledge on the related pathways [56] as well as better exploiting the capacity of mixed cultures (Saccharomyces with non-Saccharomyces) are copious [57–62].

3. FERMENTATION CONDITIONS AND INFLUENCE TO WINE AROMA

The winemaker through the process of vinifications has various parameters that can use in his/her benefit, in order to manipulate the outcome of the process. These parameters are the temperature and molecular oxygen availability during fermentation, maturation and ageing, the nitrogen source, for the growth and propagation of yeast, inoculation size and yeast strain of the starting culture [3, 31], as well as the nature and quantity of the solids derived from the grapes. All these play a decisive role in the vinification strategy and style that the winemaker wants to follow. Although for most of the OAC, the formation pathway and production promoting parameters are clear in some cases the knowledge behind synergies between these parameters are not quite apparent. The winemaker also has to deal with the changing environment of the fermentation, and more specific to the metabolism of sugar into ethanol and carbon dioxide [9, 22]. These two metabolites, but more essentially, ethanol build-up consecration plays a significant role for the physiology of the yeast. First, high concentrations of ethanol are related to the reduced water activity; this has a triggering effect to the production of various compounds to counter-balance this, but most importantly is the functional alteration of cell membrane that is influencing the uptake of various essential nutrients, important for the yeast survival and growth, including nitrogen compounds, YAN [9, 22]. Although fermentation is a well-known anaerobic pathway, carbon dioxide concentration also is influencing indirectly, the availability of proline utilisation. Specifically, the saturation of must with CO2 is having as a consequence the elimination of dissolve molecular oxygen, which is needed by oxidase for the first step in proline degradation. This is precluding the utilisation of proline, which is the main amino acid in grape must.

3.1. Temperature effect on aroma formation

Temperature conditions are associated with all enzymatic reactions rate so forth the metabolism and growth of yeast among other microorganism. The temperature range between 15 and 25°C, during wine fermentation is considered favourable for yeast growth under winemaking conditions. The fact that aromatic profile can be modulated during fermentation was noticed very early, since temperature not only affects the volatile composition but also in the case of red wines the extraction of phenolic compounds from the skin and grape seeds [7, 10]. The most noticeable and well-known effect, to winemakers, of temperature is on the fermentation rate and completion which is defined by the total consumption of sugars. Fermentation at 28°C compared to one at 15°C was observed to be 2.5 times faster [7]. Temperature is influencing not only the production of FAC but also the concentration of primary metabolites like ethanol and glycerol, on which it seems temperature to have a reverse effect. In low temperature, the production of ethanol is counter to the glycerol production [7, 9]. From very early studies it was pointed out that the final concentration of esters, contributing to the fruity flavour of the wine, was favoured by low temperatures during fermentation [8]. Particularly esters associated with pleasant fruity aroma, like isoamyl acetate and n-hexyl-acetate, accumulated in higher consecrations at low temperatures. Whereas in high temperature fermentation higher accumulation of ester characterised as heavy odorants like ethyl-octanoate and ethyl decanoate, was observed. Higher final consecration of 2-phenylethyl-acetate was favoured in higher temperatures, by some authors is consider pleasant with rose like odour [6–8]. For thiols temperatures high as 20°C are more favour for their modulation, whereas low temperatures around 13°C show significant less modulation of the 3MH [63].

3.2. Nitrogen source (yeast assimilable nitrogen (YAN)) effect on aroma formation

Yeast assimilable nitrogen (YAN) concentration in grape must is a vital parameter not only completion of the fermentation but also for the production of volatile and nonvolatile metabolites [64, 65]. The depletion of YAN in grape juice during the early stages of fermentation is also triggering the entry to the stationary phase of yeast growth [35]. YAN, source in grape must is categorised into two types, the organic and the non-organic. The organic fraction, often referred to free amino (or amino acid) nitrogen FAN, is the total amount of the amino acids and some small peptides that can be utilised by the yeast. Ammonium nitrogen is the inorganic fraction. An initial concentration of 140 mg/L of YAN in the grape juice is considered to be the lowest threshold for the completion of an industrial fermentation, with low fermentation temperatures and low suspended solids [35]. Measurement of the initial YAN and supplementation of ammonium salts or mixtures of amino acids, to reach the lowest threshold of concentration, is a common practice in most of the wineries, as a prevention measurement to sluggish or incomplete fermentations. Supplementation of nitrogen during the early stages or even through the fermentation course not only results in high fermentation kinetics and yeast growth but also to the formation of various volatile and non-volatile compounds [66, 67]. Timing of the nutrition supplementation is also important since this is influencing the type of nitrogen intake by the yeast cell [35, 65]. Specifically ammonium ion has an inhibitory role in the uptake of amino acids, since in high concentration at the early stages of growth, the general amino acid permease (GAP) is not synthesised [15, 16]. This results in low uptake of amino acid during later stages of fermentation. Another parameter that has an inhibitory role in the amino acid uptake by the yeast is the CO2. High pressure of CO2 was observed to reduce the rate by which the amino acids are absorbed. Wines with high concentration levels more than 300 mg/L have showed high esters concentration and low concentrations of acids and higher alcohols [35]. The basic information regarding the initial concentration and ratio between organic and non-organic nitrogen can be obtained by a rather easy enzymatically or chemical method. Nature of the YAN, organic or not, plays also an important role in the outcome of the volatile profile [67]. Addition of amino acids in order to increase the YAN in low concentration grape juice, under the current regulation is forbidden. Use of amino acid enriched dry yeast preparation can provide the mean to serve this purpose; also these types of preparations are high in small peptides. Ratio between the two nitrogen sources is a good tool for the winemaker, to modify the aroma profile composition of the produced wine. It has been proven that the type of nitrogen supplementation resulted in quantitative differences for most of yeast metabolites related compounds, suggesting the importance of the supplementation decision-making process [35]. The concentration of acetates and medium-chain fatty acid esters, contributing to the fruity aroma, is favoured by the higher concentration of amino acids rather than ammonium concentrations. Also higher amino acid concentration is leading to higher concentration of fusel alcohols. High concentrations of ammonium as the sole nitrogen supplement, results in the increase of ethyl acetate and acetic acid [35, 64]. Other recent studies show that there is a close relation between the initial concentration and also most importantly profile of various amino acids for the production of certain aroma profile. Also the same study gives importance values to specific amino acids. Namely in the case of S. cerevisiae, they are leucine, isoleusine, valine, histidine, glutamine and proline under certain conditions. [35]. Whereas other researchers' show that, for the formation of volatile compounds: threonine, phenylalanine and aspartic acid are amino acids with the most important value. For thiols, high addition of assimilable nitrogen in the early stage of fermentation, in the form of ammonium (di-ammonium phosphate) seems to reduce the 3MH production [68]. In [56], researchers documented that the highest concentration of terpenes is obtained under conditions that stimulate glycolytic flux. Microaerobic and high assimilable nitrogen conditions, favour terpene accumulation.

3.3. Suspended solids and contribution to wine aroma

During the process of vinification and especially during the first stages of destemming and pressing it is inevitable the presence of grape solids in the must. These solids of various, origin, nature and size are generally referred to as 'sludge' [18]. The measurements by the winemakers assess the presence of these solids are the turbidity units (nephelometric turbidity units, NTU) and a wt% on suspended solids (total wet suspended solids, TWSS or dry TDSS% (w/w) [18]. The ease to measure NTU makes this measurement, the most widely used and accepted method of reference in wineries. A limitation of NTU measurement is that is not in a direct relation with the suspended solids quantity since this is a nephelometric measurement and is being influenced by the size and shape of the particles and refractive index of the medium [18]. Also NTU does not give us information regarding the composition of the suspended solids. Must suspended solids can influence white wines aroma profile in many different ways, directly and indirectly. Suspended solids are consider a good nutrient source, specially for amino acids, the role of which is consider to be the most crucial of all must substrates for the formation of volatile compounds [18–21, 69]. Another direct role of suspended solids is the high content of oxidative enzymes, which is enhanced by various grape moulds contaminations. Also some evidence suggests that grape tissues contain esterase, a limitating factor for the accumulation and final concentration of esters at the final product. Presence of SS on fermentation apart from the direct role on chemical composition, have also an important indirect role, that of the nucleation of the CO2 [18] and the further release it to atmosphere. High accumulation of CO2 produces higher concentrations of acetic acid by limiting the long chain fatty acid synthesis. Also over oversaturation of CO2, is affecting the transport and utilisation by the yeast of amino acids [18].

3.4. Inoculation rate and contribution to wine aroma

Yeast starting cultures are extensively used by many wineries as a mean to control the course of fermentation avoiding slow or sluggish fermentations [13]. Now available on market are a large number of dry yeast cultures ready to use. Primarily the need to use is the easy completion of fermentation, without any technological folds like reduction smell SO2 and high concentration of volatile acidity. The size though of the inoculation it seems to play a catalytic role in the overall behaviour and physiology of the yeast during fermentation [11–13]. In some cases [12], it was observed that the size of inoculation enhanced stress protectants like glycerol and proline production in high inoculation rate. Also in the same study, an observation of the reduction of citric acid cycle intermediate metabolites was made. On another study dealing with three inoculation concentrations 1×104, 1×105 and 1×106 [11] were studied. It was clear that the most favourable results, for the increase concentration desirable volatiles, like esters, and the simultaneous decrease of high concentrations of unfavourable volatiles, like higher alcohols was observed at inoculation rate of 105. It is obvious that through this process, of inoculation, the outcome of the fermentation can be altered but since there is not a lot of research done to this direction, is something that need to be investigated further.

4. YEAST AUTOLYSIS

4.1. Yeast autolysis and contribution to wine flavour

At the final stages of winemaking, the settlement of yeast cells at the bottom of fermentation tanks is inevitable, since there is not any CO_2 production. From that stage on an autolysis of dead yeast cells is observed. During this period, the hydrolysis of cell wall is taking place, releasing various compounds that up to that moment were either part of yeast cell wall or were capture inside the cell cytoplasm. Understanding the nature of the cell wall is vital in this stage. Yeast cell wall is compromising around 15–25% of total dry cell mass [70, 71] depending on the growth conditions of yeast. Yeast cell wall consists of polysaccharides, inner layer, mannoproteins and outer layer [70]. Mannans and glucans consist of about the 94–98% of the total structural cell wall mass with a small fraction of chitin. Mannoproteins play an important role in stabilising various fractions of wine, tartaric salts and also proteins [72]. Moreover, for the red wines, this fraction can make wine to feel less astringent due to mannoproteins/tannin condensation (46, 48) and also increase the colour stability [72]. Apart from these, the autolysis can also influence the aroma character of the wine [73, 74]. First, by the absorption of volatiles like 4-ethylphenol and 4-ethylguaiacol, which they have a negative contribution to the wine aroma [74]. OAC compound profile also is affected by the yeast autolysis. The overall reduction in ester concentration is observed due to release of esterase [75]. But in some case, volatiles like diethyl succinate, vitispirane and 1,1,6-trimethyl-1,2-dihydronaphthalene (TDN) levels are increased over the period of maturation of sparkling wine with the lees [76]. Also the reduction of oak character is observed due to the absorption of this group of volatiles by the yeast [75], in some case is considered to be positive and negative in some cases.

5. Prospects for the future

A comprehensive comprehension of the chemical composition of the volatile compounds inside the wine matrix is a pivotal factor in comprehending the unique qualities and characteristics of wine. In the majority of wine-producing countries, the existing legislation permits significant modifications to the taste of wine, such as acidification, deacidification, and fining. As a result, the winemaker has full authority to choose the ultimate mouthfeel of the wine. However, it is strictly prohibited to include any additional substances that could affect the characteristics of wine aroma. The fermentation process is the most crucial stage for winemakers to influence and alter the flavor of the wine. During this relatively brief phase in the life of the wine, the process of decision-making is of utmost importance. The selection of the fermentation temperature, incorporation of FAN (Free Amino Nitrogen), and duration of autolysis are techniques employed to achieve this objective. The subsequent action involves monitoring the timing of several procedures while establishing a link among various parameters. The correlation between the quality of wine and the quality of grapes is well acknowledged and accepted. Despite the potential of utilizing advanced omics approaches in commercial yeast to gather extensive information on physiology and, more crucially, predict the outcome of fermentation in terms of fragrance profile, these techniques have not yet been implemented in industrial applications. As these technologies become more accessible and cost-effective, they will offer winemakers additional tools to enhance the quality of wine. This includes adding nutrients, adjusting temperature, and selecting specific commercial yeast strains to fully showcase the grapes' unique characteristics.

Furthermore, there is ongoing interest in studying and developing the role and utilization of non-Saccharomyces yeasts that are naturally present in grapes. These yeasts contribute to the intricate aroma of wine and can also be used to determine the geographical origin of the wine. Given these recent advancements, it is crucial to have a comprehensive grasp of the physiological characteristics of various genera, species, and strains of yeasts, as well as the influence of the environment on them, in order to make any further progress. In order to understand the role of yeasts in aroma and flavor, it is essential to use advanced tools and instruments from molecular biology, biotechnology, and analytical chemistry that can process large amounts of data efficiently. These tools and devices are experiencing rapid technological advancement and a simultaneous decrease in pricing. They are now being used in several laboratories worldwide and will soon be accessible for industrial application on a large scale, particularly in wineries.

AT A GLANCE

The organoleptic properties of wine, specifically the range encompassing flavor and fragrance, are the crucial criterion for evaluating the wine's quality. These traits originate from four primary sources: grapes, vinification, maturity, and aging. The yeast during fermentation has a significant impact on the final quantities of several odour-active components (OAC). The primary organic acid compounds (OAC) that are produced during fermentation consist of volatile molecules such as esters, higher alcohols, and carbonyl compounds. By deciphering the source and impact of these organic acid compounds (OAC), the contemporary winemaker can actively control and regulate the yeast during the fermentation process to their advantage. These compounds are derived from the secondary metabolism of yeast. It is crucial to comprehend the impact of key factors such as temperature, yeast assimilable nitrogen (YAN), and suspended solids on OAC formation during fermentation in order to determine the final sensory characteristics of wine.

REFERENCES

1. Bisson LF, Karpel JE. Genetics of yeast impacting wine quality. Annual Review of Food Science and Technology. 2010;1:139-162
2. Sumby KM, Grbin PR, Jiranek V. Microbial modulation of aromatic esters in wine: Current knowledge and future prospects. Food Chemistry. 2010;121(1):1-16
3. Fleet GH. Yeast interactions and wine flavour. International Journal of Food Microbiology. 2003;86(1-2):11-22
4. Bartowsky EJ, Pretorius IS. Microbial formation and modification of flavor and off-flavor compounds in wine. In: König H, Unden G, Fröhlich J, editors. Biology of Microorganisms on Grapes, in Must and in Wine. Berlin, Heidelberg: Springer Berlin Heidelberg; 2009. p. 209-231
5. Styger G, Prior B, Bauer FF. Wine flavor and aroma. Journal of Industrial Microbiology and Biotechnology. 2011;38(9):1145-1159
6. Beltran G et al. Integration of transcriptomic and metabolic analyses for understanding the global responses of low-temperature winemaking fermentations. FEMS Yeast Research. 2006;6(8):1167-1183
7. Molina AM et al. Influence of wine fermentation temperature on the synthesis of yeast-derived volatile aroma compounds. Applied Microbiology and Biotechnology. 2007;77(3):675-687
8. Killian E, Ough CS. Fermentation esters—Formation and retention as affected by fermentation temperature. American Journal of Enology and Viticulture. 1979;30(4):301-305
9. Fairbairn S et al. Environmental stress and aroma production during wine fermentation. South African Journal of Enology and Viticulture. 2014;35(2):168-177
10. Torija MJ et al. Effects of fermentation temperature and Saccharomyces species on the cell fatty acid composition and presence of volatile compounds in wine. International Journal of Food Microbiology. 2003;85(1-2):127-136
11. Carrau F et al. Effect of Saccharomyces cerevisiae inoculum size on wine fermentation aroma compounds and its relation with assimilable nitrogen content. International Journal of Food Microbiology. 2010;143(1-2):81-85
12. Ding MZ et al. Inoculum size-dependent interactive regulation of metabolism and stress response of Saccharomyces cerevisiae revealed by comparative metabolomics. Journal of Biotechnology. 2009;144(4):279-286
13. Mateo JJ et al. Yeast starter cultures affecting wine fermentation and volatiles. Food Research International. 2001;34(4):307-314
14. Garde-Cerdán T, Ancín-Azpilicueta C. Effect of the addition of different quantities of amino acids to nitrogen deficient must on the formation of esters, alcohols, and acids during wine alcoholic fermentation. LWT: Food Science and Technology. 2008;41(3):501-510
15. Gutiérrez A, Chiva R, Guillamón JM. Arginine addition in the stationary phase influences the fermentation rate and synthesis of aroma compounds in a synthetic must fermented by three commercial wine strains. LWT: Food Science and Technology. 2015;60(2):1009-1016
16. Gutiérrez A et al. Nitrogen requirements of commercial wine yeast strains during fermentation of a synthetic grape must. Food Microbiology. 2012;31(1):25-32
17. Vilanova M et al. Effect of ammonium nitrogen supplementation of grape juice on wine volatiles and non-volatiles composition of the aromatic grape variety Albariño. Food Chemistry. 2012;133(1):124-131

18. Casalta E et al. Review: Characterization and role of grape solids during alcoholic fermentation under enological conditions. American Journal of Enology and Viticulture. 2016;67(2):133-138

19. Groat M, Ough CS. Effects of insoluble solids added to clarified musts on fermentation rate, wine composition, and wine quality. American Journal of Enology and Viticulture. 1978;29(2):112-119

20. Puig-Deu M et al. Influence of must racking and fining procedures on the composition of white wine. VITIS-GEILWEILERHOF. 1996;35:141-146

21. Klingshirn LM, Liu JR, Gallander JF. Higher alcohol formation in wines as related to the particle size profiles of juice insoluble solids. American Journal of Enology and Viticulture. 1987;38(3):207-210

22. Bauer F, Pretorius IS. Yeast stress response and fermentation efficiency: How to survive the making of wine—A review. South African Journal for Enology and Viticulture. 2000;21:27-51

23. Hirst MB, Richter CL. Review of aroma formation through metabolic pathways of Saccharomyces cerevisiae in beverage fermentations. American Journal of Enology and Viticulture. 2016;67(4):361-370

24. Ebeler SE, Thorngate JH. Wine chemistry and flavor: Looking into the crystal glass. Journal of Agricultural and Food Chemistry. 2009;57(18):8098-8108

25. Bakker J, Clarke RJ. Wine: Flavour Chemistry. John Wiley & Sons; The Atrium, Southern Gate, Chichester, West Sussex PO19 8SQ, UK 2011

26. Parker M et al. Aroma precursors in grapes and wine: Flavor release during wine production and consumption. Journal of Agricultural and Food Chemistry. 2017; DOI: 10.1021/acs.jafc.6b05255

27. Ugliano M, Henschke PA. Yeasts and wine flavour. In: Moreno-Arribas MV, Polo MC, editors. Wine Chemistry and Biochemistry. New York, NY: Springer New York; 2009. p. 313-392

28. Rojas V et al. Acetate ester formation in wine by mixed cultures in laboratory fermentations. International Journal of Food Microbiology. 2003;86(1-2):181-188

29. Rodrıguez-Bencomo J et al. Determination of esters in dry and sweet white wines by headspace solid-phase microextraction and gas chromatography. Journal of Chromatography A. 2002;963(1):213-223

30. Romano P et al. Function of yeast species and strains in wine flavour. International Journal of Food Microbiology. 2003;86(1-2):169-180

31. Swiegers JH, Pretorius IS. Yeast modulation of wine flavor. Advances in Applied Microbiology. 2005;57:131-175

32. Saerens SM et al. Production and biological function of volatile esters in Saccharomyces cerevisiae. Microbial Biotechnology. 2010;3(2):165-177

33. Hazelwood LA et al. The Ehrlich pathway for fusel alcohol production: A century of research on Saccharomyces cerevisiae metabolism. Applied and Environmental Microbiology. 2008;74(8):2259-2266

34. Torija MAJ et al. Effect of the nitrogen source on the fatty acid composition of Saccharomyces cerevisiae. Food Microbiology. 2003;20(2):255-258

35. Torrea D et al. Comparison of inorganic and organic nitrogen supplementation of grape juice—Effect on volatile composition and aroma profile of a Chardonnay wine fermented with Saccharomyces cerevisiae yeast. Food Chemistry. 2011;127(3):1072-1083

36. Fraile P, Garrido J, Ancín C. Influence of a Saccharomyces cerevisiae selected strain in the volatile composition of rose wines. Evolution during fermentation. Journal of Agricultural and Food Chemistry. 2000;48(5):1789-1798

37. Rapp A, Versini G. Influence of nitrogen compounds in grapes on aroma compounds of wines. Developments in Food Science. 1995;37:1659-1694

38. Hernández-Orte P et al. Addition of amino acids to grape juice of the Merlot variety: Effect on amino acid uptake and aroma generation during alcoholic fermentation. Food Chemistry. 2006;98(2):300-310

39. Tominaga T, Darriet P, Dubourdieu D. Identification of 3-mercaptohexyl acetate in Sauvignon wine, a powerful aromatic compound exhibiting box-tree odor. VITIS-GEILWEILERHOF. 1996;35(4):207-210

40. Tominaga T et al. Stereoisomeric distribution of 3-mercaptohexan-1-ol and 3-mercaptohexyl acetate in dry and sweet white wines made from Vitis vinifera (Var. Sauvignon Blanc and Semillon). Journal of Agricultural and Food Chemistry. 2006;54(19):7251-7255

41. Tominaga T, Peyrot des Gachons C, Dubourdieu D. A new type of flavor precursors in Vitis v inifera L. cv. Sauvignon blanc: S-cysteine conjugates. Journal of Agricultural and Food Chemistry. 1998;46(12):5215-5219

42. Peyrot d, Gachons C, Tominaga T, Dubourdieu D. Sulfur aroma precursor present in S-glutathione conjugate form: Identification of S-3-(hexan-1-ol)-glutathione in must from Vitis vinifera L. cv. Sauvignon blanc. Journal of Agricultural and Food Chemistry. 2002;50(14):4076-4079

43. Schneider R et al. Evidence for an alternative biogenetic pathway leading to 3-mercaptohexanol and 4-mercapto-4-methylpentan-2-one in wines. Analytica Chimica Acta. 2006;563(1):58-64

44. Baudouin-Cornu P et al. Glutathione degradation is a key determinant of glutathione homeostasis. Journal of Biological Chemistry. 2012;287(7):4552-4561

45. Drawert F, Barton H. Biosynthesis of flavor compounds by microorganisms. 3. Production of monoterpenes by the yeast Kluyveromyces lactis. Journal of Agricultural and Food Chemistry. 1978;26(3):765-766

46. Cordonnier R, Bayonove CL. Etude de la phase préfermentaire de la vinification: Extraction et formation de certains composés de l'arôme; cas des terpenols, des aldehydes et des alcools en C 6. OENO One. 1981;15(4):269-286

47. Mateo J, Di Stefano R. Description of the β-glucosidase activity of wine yeasts. Food Microbiology. 1997;14(6):583-591

48. Rosi I, Vinella M, Domizio P. Characterization of β-glucosidase activity in yeasts of oenological origin. Journal of Applied Microbiology. 1994;77(5):519-527

49. Miklosy E, Pölös V. Yeasts with β-D-glucosidase activity: Properties and possible application in winemaking processes. Acta Alimentaria. 1995;24(2):167-179

50. Mendes Ferreira A, Climaco MC, Mendes Faia A. The role of non-Saccharomyces species in releasing glycosidic bound fraction of grape aroma components—A preliminary study. Journal of Applied Microbiology. 2001;91(1):67-71

51. Fernández-González M, Di Stefano R, Briones A. Hydrolysis and transformation of terpene glycosides from muscat must by different yeast species. Food Microbiology. 2003;20(1):35-41

52. Fernández M, Úbeda JF, Briones AI. Typing of non-Saccharomyces yeasts with enzymatic activities of interest in wine-making. International Journal of Food Microbiology. 2000;59(1):29-36

53. Azzolini M et al. Effects of Torulaspora delbrueckii and Saccharomyces cerevisiae mixed cultures on fermentation and aroma of Amarone wine. European Food Research and Technology. 2012;235(2):303-313

54. Azzolini M et al. Contribution to the aroma of white wines by controlled Torulaspora delbrueckii cultures in association with Saccharomyces cerevisiae. World Journal of Microbiology and Biotechnology. 2015;31(2):277-293

55. Sadoudi M et al. Yeast-yeast interactions revealed by aromatic profile analysis of Sauvignon Blanc wine fermented by single or co-culture of non-Saccharomyces and Saccharomyces yeasts. Food Microbiology. 2012;32(2):243-253

56. Carrau FM et al. De novo synthesis of monoterpenes by Saccharomyces cerevisiae wine yeasts. FEMS Microbiology Letters. 2005;243(1):107-115

57. Bozoudi D, microbiome TDG. Potential and opportunities as a source of starter cultures. In: Morata A, Loira I, editors. Grape and Wine Biotechnology. Rijeka: InTech; 2016 Ch. 10

58. Sun SY et al. Selected non-Saccharomyces wine yeasts in controlled multistarter fermentations with Saccharomyces cerevisiae on alcoholic fermentation behaviour and wine aroma of cherry wines. Food Microbiology. 2014;44:15-23

59. Jolly NP, Varela C, Pretorius IS. Not your ordinary yeast: Non-Saccharomyces yeasts in wine production uncovered. FEMS Yeast Research. 2014;14(2):215-237

60. Padilla B, Gil JV, Manzanares P. Past and future of non-Saccharomyces yeasts: From spoilage microorganisms to biotechnological tools for improving wine aroma complexity. Frontiers in Microbiology. 2016;7:411

61. Lencioni L et al. Controlled mixed fermentation at winery scale using Zygotorulaspora florentina and Saccharomyces cerevisiae. International Journal of Food Microbiology. 2016;234:36-44

62. Wang XC et al. Evaluation of aroma enhancement for "Ecolly" dry white wines by mixed inoculation of selected Rhodotorula mucilaginosa and Saccharomyces cerevisiae. Food Chemistry. 2017;228:550-559

63. Masneuf-Pomarède I et al. Influence of fermentation temperature on volatile thiols concentrations in Sauvignon blanc wines. International Journal of Food Microbiology. 2006;108(3):385-390

64. Barbosa C et al. Nitrogen addition influences formation of aroma compounds, volatile acidity and ethanol in nitrogen deficient media fermented by Saccharomyces cerevisiae wine strains. Journal of Bioscience and Bioengineering. 2009;108(2):99-104

65. Barbosa C, Mendes-Faia A, Mendes-Ferreira A. The nitrogen source impacts major volatile compounds released by Saccharomyces cerevisiae during alcoholic fermentation. International Journal of Food Microbiology. 2012;160(2):87-93

66. Moreira N et al. Relationship between nitrogen content in grapes and volatiles, namely heavy sulphur compounds, in wines. Food Chemistry. 2011;126(4):1599-1607

67. Procopio S et al. Significant amino acids in aroma compound profiling during yeast fermentation analyzed by PLS regression. LWT: Food Science and Technology. 2013;51(2):423-432

68. Subileau M et al. Nitrogen catabolite repression modulates the production of aromatic thiols characteristic of Sauvignon Blanc at the level of precursor transport. FEMS Yeast Research. 2008;8(5):771-780

69. Ancin C, Ayestaran B, Garrido J. Clarification by vacuum filtration of Grenache must. Utilization of free amino acids during fermentation and bottle-aging of wine. American Journal of Enology and Viticulture. 1996;47(3):313-322

70. Walker GM. Yeast Physiology and Biotechnology. John Wiley & Sons; 1998

71. Klis FM, Boorsma A, De Groot PW. Cell wall construction in Saccharomyces cerevisiae. Yeast. 2006;23(3):185-202

72. Escot S et al. Release of polysaccharides by yeasts and the influence of released polysaccharides on colour stability and wine astringency. Australian Journal of Grape and Wine Research. 2001;7(3):153-159

73. Comuzzo P et al. Yeast derivatives (extracts and autolysates) in winemaking: Release of volatile compounds and effects on wine aroma volatility. Food Chemistry. 2006;99(2):217-230

74. Tao Y, Garcia JF, Sun DW. Advances in wine aging technologies for enhancing wine quality and accelerating wine aging process. Critical Reviews in Food Science and Nutrition. 2014;54(6):817-835

75. Gambetta JM et al. Factors influencing the aroma composition of Chardonnay wines. Journal of Agricultural and Food Chemistry. 2014;62(28):6512-6534

76. Kemp B et al. Effect of production phase on bottle-fermented sparkling wine quality. Journal of Agricultural and Food Chemistry. 2015;63(1):19-38

77. Pretorius IS. Tailoring wine yeast for the new millennium: Novel approaches to the ancient art of winemaking. Yeast. 2000;16(8):675-729

78. Ciani M et al. Controlled mixed culture fermentation: A new perspective on the use of non-Saccharomyces yeasts in winemaking. FEMS Yeast Research. 2010;10(2):123-133

79. García M, Esteve-Zarzoso B, Arroyo T. Non-Saccharomyces yeasts: Biotechnological role for wine production. In: Morata A, Loira I, editors. Grape and Wine Biotechnology. Rijeka: InTech; 2016 Ch. 11

80. Bokulich NA et al. Associations among wine grape microbiome, metabolome, and fermentation behavior suggest microbial contribution to regional wine characteristics. mBio. 2016;7(3):e00631-16. doi: 10.1128/mBio.00631-16

81. Bokulich NA et al. Microbial biogeography of wine grapes is conditioned by cultivar, vintage, and climate. Proceedings of the National Academy of Sciences. 2014;111(1):E139-E148

CHAPTER 6

THE YEAST ECOLOGY OF FERMENTED TABLE OLIVES AND ITS POTENTIAL APPLICATIONS IN BIOTECHNOLOGY

1. INTRODUCTION: AN OVERVIEW

Table olives are a significant and widely recognized type of fermented vegetables, with a current global production estimated at 2.5 million tons per year [1]. The culture and nutrition of several Mediterranean countries, including Greece, Spain, Italy, France, Portugal, Egypt, and Turkey, are closely tied to both the production and consumption of this food [2]. Additionally, countries such as the USA, Peru, Argentina, and Australia are emerging as formidable producers [3]. Yeasts are crucial in the fermentation process of table olives, primarily affecting the safety, quality, and flavor of the end product [4]. Yeasts play a crucial role in stimulating fermentation by being consistently present throughout the fermentative process. It is widely acknowledged that they have the ability to create compounds that have significant organoleptic characteristics, which play a crucial role in defining the quality and flavor of the end product [3]. Specifically, these microorganisms have the capability to generate beneficial biochemical activities such as lipase, esterase, β-glucosidase, catalase, and the generation of killer factors. These activities have various practical uses in the manufacturing of fermented table olives. However, yeasts can also contribute to the deterioration of olives during fermentation, packing, and storage. This is due to their production of unpleasant odors and flavors, the build-up of $CO2$ resulting in swollen containers and the formation of gas pockets, the clouding of brines, the softening of fruits, and the breakdown of lactic acid. The technical implications and functional features of yeasts have been thoroughly examined in previous studies [5–9]. According to reports, they are believed to play a crucial part in the fermentation process of both green-treated olives and naturally fermented black olives [6]. In addition, several yeast species commonly found in the processing of table olives, including Wickerhamomyces anomalus, Saccharomyces cerevisiae, Pichia membranifaciens, and Kluyveromyces lactis, have been observed to possess intriguing properties. These properties include improving the sensory qualities of the fruits, breaking down polyphenols, promoting the growth of lactic acid bacteria (LAB), and exhibiting antioxidant effects [5, 9]. Hence, the identification of the most suitable yeast strains for use as starters, whether in conjunction with LAB or not, is a potential objective for the future that could enhance the overall quality of the end product [5].

2. THE BIOLOGY AND PHYSIOLOGY OF YEASTS FOUND IN FERMENTED TABLE OLIVES

Yeasts are single-celled microbes, larger than bacteria, belonging to the kingdom of Fungi. Nowadays 1500 species are identified which corresponds approximately to 1% of all described fungal species [10]. Both their survival and growth are ensured by consuming starches and sugars. Plenty of yeasts are commercially available and can provide an inexpensive source for biochemical and biotechnological applications (classic and molecular) [11].

Generally, in the heterotrophic organisms, such as yeasts, the energy and carbon metabolism have an interconnection between them, i.e., anabolism is connected with catabolism. ATP is provided by the oxidation of organic molecules which also play an important role in the biosynthesis, because they can be used firstly as carbon sources and eventually as energetic currency for all kinds of cellular work. Although, for instance, table olive yeasts species contain a wide set of carbon sources (polyols, alcohols, organic acids and amino acids) that can help in their growth, they prefer to metabolize sugars. There are a lot of data related to the metabolism of different carbon sources, with the most widely studied sugars being hexoses (glucose, fructose, galactose or mannose) and disaccharides (maltose or sucrose), as well as compounds with two carbons (ethanol or acetate) [5, 12].

The metabolic pathway of the central carbon metabolism is mainly identical between different yeast species, suggesting that these microorganisms might constitute a metabolic homogenous group. Nevertheless, the mechanisms for nutrient uptake, the number of different isoenzymes and the regulation of fermentation and respiration differ substantially [13], making yeasts a highly heterogeneous and complex metabolic group.

Moreover, the presence of volatile compounds, such as ethanol, glycerol, higher alcohols, and, to a lesser extent, acetaldehyde in brines can be directly attributed to the metabolic activity of yeasts together with heterofermentative bacteria [12].

On the contrary, methanol, detected in brine [14], is not related with the yeast metabolism, because its activity is due to pectinolytic enzymes. The presence of these enzymes in the drupes is favored by an improper handling of raw material and by poor fermentation conditions [15].

2.1. Enzymatic activities

Yeasts ensure their survival in tissues, and are connected with the digestion of host's proteins and other organic molecules due to a large array of physiological traits and enzymatic activities [16, 17]. It is characteristic that the kind of medium used for their isolation and growth is an important adaptive factor which determines the technological activities of the strains selected [18, 19]. Although many surveys have underlined the potential of unconventional and extreme environments as a source of natural biodiversity for the isolation and selection of useful microorganisms [20], it is the secretion of extracellular proteases which has been studied extensively in yeasts isolated from these environments [21, 22]. Esterases from yeasts are also gaining industrial interest with applications in laundry detergents and in dairy industries [23, 24], while little attention has been paid to lipases from yeasts [7, 25, 26]. Although peptic enzymes for industrial uses have been so far produced by molds and bacteria [27], the pectinolytic activity of yeasts has, however, been studied with ambiguous results [28, 29].

Yeasts can also produce plenty of other interesting enzymes like invertase, zymase, hexose phosphatase, maltase, reductase, carboxylase, melibiase, and endo-tryptase. However, all kind of yeasts do not comprise the same range of enzymes. Hence, different yeasts are functioning differently toward the degradation of sugars. It is generally accepted that one yeast can ferment one sugar [30]. Since today, it is known that three of the aldo-hexose sugars, called dextrose, d-mannose, and d-galactose, are exclusively fermentable by yeasts. It has been reported that yeasts which ferment dextrose, can also use mannose and laevulose [30]. Due to the enzyme invertase, yeasts can firstly reverse and secondly ferment cane-sugar. However, they are unable to ferment milk sugar, because of the lack of the enzyme lactase.

2.2. Functional properties

Despite the fact that research has been focused almost exclusively on LAB, as being extremely useful in table olives microbiota [31], recent studies have been negotiating the significant contribution of yeasts to table olives fermentation [3, 5, 9], during the whole process and adding value to the final product. The selection of yeasts to be used as starter cultures is an unclear and complicated procedure, which includes three main steps. The first one is the selection of promising species, which should have some interesting properties. The second step is the validation of those microorganisms on laboratory scale and finally the third one, and the most important, is the demonstration at large scale [32].

Starter cultures must be selected for their functional traits, and for their ability to dominate at the fermentation process, as well as leading to a stable final product. Thus, the screening of those yeast strains is very important. Esterase and lipase enzymatic activities are desirable in yeasts because of their ability to improve the organoleptic characteristics of olives through the configuration of volatile compounds that can be produced by the catabolism of free fatty acids [9, 32–34].

Yeasts with β-glucosidase activity are also good candidates, because they can be used for the hydrolysis of oleuropein, removing the natural bitterness present in brined olives and thus, avoiding the placement of olives into large amounts of water [3, 35]. However, limited data are available in the literature about the contribution of these species on table olive fermentation process [7, 17, 34]. Furthermore, very few strains which produce nontoxic compounds, like biogenic amines (BA) (spermine, spermidine, agmatine) was proposed as a new criterion for the selection of yeasts as starter cultures for many fermented products [36, 37]. Many yeast species which have the ability to produce BA can be found in olives. As a coincidence, this trait should be investigated as another critical point for the selection of proper starter cultures in the near future.

It is noteworthy to mention the probiotic profile of some table olives yeasts. Generally, probiotics are microorganisms which could have beneficial impact on the human organism [38]. This positive impact would exist only if those microorganisms are safe for the host, metabolically active within the gastrointestinal tract and are being consumed, either as food components or as non-food preparations. For instance, Saccharomyces boulardii, a member of S. cerevisiae species, is the only yeast with clinical role and proven probiotic efficiency in double blind clinical studies [39]. Thus, finding other yeast strains with probiotic characteristics and especially those isolated from table olives is of great interest.

The probiotic potential of yeasts isolated from fermented table olives has been documented to some extent [8, 35]. In particular, Psani & Kotzekidou, [35] found Torulaspora delbrueckii and Debaryomyces hansenii strains to be quite tolerant in high bile salt concentrations and low pH values. Moreover, the fact that the inoculation of these strains were able to inhibit food borne pathogens such as Listeria monocytogenes, Bacillus cereus and Salmonella typhimurium is remarkable. Furthermore, data from Ref. [8] showed that P. membranifaciens and Candida oleophila strains have similar properties but with a different spectrum of inhibition zones (for Escherichia coli, Salmonella enteritidis and Staphylococcus aureus). Another significant number of yeast species, such as K. lactis, D. hansenii, T. delbrueckii and S. cerevisiae, have shown tolerance to traverse the gastrointestinal tract, inhibit the enteropathogens, adhere the intestinal CaCO2 cell line and present an immunostimulatory activity [40–42]. In particular, S. cerevisiae has shown its ability to prevent the survival of E. coli O157:H7 under simulated gastrointestinal conditions by the production of ethanol [43]. Although research of olive yeast strains with the properties mentioned above is a promising task, it is important for olive yeasts to adhere on to the olive skin and survive during storage and/or packaging, in order to be ingested by consumers at elevated numbers.

Except from their probiotic properties, yeasts could positively affect human health in other ways, as well. For instance, diverse strains of K. lactis, S. cerevisiae, and Issatchenkia orientalis show great ability to reduce cholesterol serum levels [40]. Secondly, phytate has a strong chelating ability to form insoluble complexes with divalent minerals of nutritional importance such as zinc, calcium, magnesium and iron. Due to a lack of some required enzymes, humans cannot degrade phytate complexes in the gastrointestinal tract. It is known that dephosphorylation of phytate is catalyzed by phytases, which are widespread in yeast species such as I. orientalis, W. anomalus, S. cerevisiae, T. delbrueckii, and K. lactis [41, 44]. Thus, yeasts could be included in humans' diet in order to help them to do so.

Folates (vitamin B9) are considered essential co-factors in the biosynthesis of nucleotides and play an important role in cellular replication and growth. It is common that mammals cannot synthesize folates and a potential solution is for yeasts to help them to do so because they contain a folate biosynthesis pathway and can produce natural folates. Some of those species which have a high folate biosynthesis pathway are the S. cerevisiae and Candida glabrata [41].

Other species of yeasts, such as the diverse yeast strains isolated from table olives, belonging to the P. membranifaciens and P. farinosa species have the ability to produce B-complex vitamins [8, 45]. This means that yeasts can synthesize a number of bioactive compounds which can serve as natural antioxidants.

Moreover, researchers are interested in screening of yeasts for free-radical-scavenging activity. For instance, W. anomalus produced the highest activity in a laboratory medium [46]. The production of bioactive antioxidants may retard the oxidative degeneration of fatty substances and improve human health. In any case, probiotic yeasts could be able to adhere onto the olive epidermis, and thus, be ingested by consumers.

Once researchers are very interested in the organoleptic features of olives, yeasts have some very essential and strain-specific metabolic properties, such as esterase and lipolytic activities. The former has frequently been detected, while the isolation of strains with lipoytic activities has been reported to a lesser extent [7]. Enterase positive yeasts are covetable because the fact that they are able to meliorate the flavor of olives from the formation of esters coming from free fatty acids. Strong lipase activity, as well as, weak activity have been detected both in vitro in some yeast species. The former has been reported in Candida boidinii, D.

hansenii and T. delbrueckii while the latter only in P. membranifaciens [34,35]. Those authors have emphasized the change of the free fatty acids composition of olives in the presence of yeast populations in contrast with sterile conditions, indicating that lipases produced by these microorganisms modify the characteristics of fruit lipids and therefore its organoleptic characteristics.

3. YEAST ECOLOGY OF TABLE OLIVES

According to Ref. [4] it has been reported, that yeasts are responsible for the fermentation process of natural black olives. However, to green olives, the fermentation is driven by LAB. Black olive fermentations have been studied in different countries, mainly in the Mediterranean (Italy, Spain, Portugal, Greece, and Morocco) and showed some yeast species biodiversity. Despite this fact, a few species were dominant, such as P. membranifaciens, Saccharomyces oleaginosus, Pichia anomala, C. boidinii and T. delbrueckii [47–50]. However, since LAB are partially inhibited in directly brined green and natural black olives due to the presence of phenolic compounds, yeasts became highly important for the fermentation process [3].

Nowadays, several studies have focused on yeasts microbiota situated on the surface of olive fruits. Some of those studies are summarized in Table 1. In the past, the characterization of yeasts associated with table olives was mainly made by morphological and biochemical methods comparing the obtained results with diverse taxonomic keys [10, 61]. However, molecular methods have started to be used recently, for the identification of yeasts coming from table olives fermentation. One of the most applicable methods is the Denaturating Gradient Gel Electrophoresis (DGGE-PCR), which is more precise than any classical method and is based on (i) restriction fragment length polymorphism (RFLP) analysis obtained after cutting the amplified 5.8S rRNA gene and the associated intergenic spacers ITS with endonucleases [62], and (ii) the direct sequencing of the D1/D2 domains of the 26S rRNA gene amplified with primers NL1 and NL4 [63] or the 5.8S-ITS region amplified with primers ITS1 and ITS4 [62].

Table 1. Scientific reports of microbial ecology studies reporting the isolation of yeasts from different cultivars of table olives and different fermentation processes and different Mediterranean countries.

Yeasts/stage	Country/variety	Method	Ref.
Saccharomyces cerevisiae, Issatchenkia occidentalis, Geotrichum candidum (during process)	Spanish green seasoned table olives	Sequencing of 26S rRNA, D1/D2 region	[12]
Candida boidinii, Hanseniaspora guilliermondii (final product)	Ripe black olives	Sequencing of 26S rRNA, D1/D2 region	[12]

Yeasts/stage	Country/variety	Method	Ref.
P. anomala, C. boidinii, Debaryomyces etchelsii (during process)	French black olives	RFLP, Sequencing of D1/D2 region	[51]
C. boidinii, Candida sorsoba, Candida diddensiae, P. membranifaciens, Kluyveromyces lactis, P. anomala, P. kluyveri, Rhodotorula glutinis (during process)	Spain, Arbequina table olives	RFLP	[52, 53]
Pichia caribbica, Lachancea fermentati, Nakazawaea holstii (during process)	Spain, Arbequina and Cornicabra	Sequencing of 5.8 rRNA, ITS1-ITS2 region	[54]
Metschnikowia pulcherrima, Debaryomyces hansenii, Aureobasidium pullulans (during process)	Greece, Conservolea black olives	Sequencing of 5.8 rRNA, ITS1-ITS2 region	[55]
P. membranifaciens, P. anomala (final product)	Greek black olives	Sequencing of 5.8 rRNA, ITS1-ITS2 region	[55]
Candida parapsilosis, P. guilliermondii, P. kluyveri (during process)	Sicilian green table olives.	RFLP, Sequencing of 26 s rRNA D1/D2 region	[56]
S. cerevisiae, Pichia galeiformis, C. boidinii (final product), and *P. membranifaciens* (during possess)	Hojiblanca	RFLP, Sequencing of 5.8 s rRNA, ITS1-ITS2 region	[34]
C. diddensiae, S. cerevisiae and *P. membranifaciens* (during process)	Spain/ Alorena	RFLP, Sequencing of 26 s rRNA D1/D2 region	[7]
Candida tropicalis, P. galeiformis and *P. anomala* (during process)	Spain/Gordal and Manzanilla	RFLP, Sequencing of 26 s rRNA D1/D2 region	[7]
S. cerevisiae and *Candida apicola* (during process)	Spain/ Alorena de Malaga	PCR-DGGE of 26SrRNA	[57]
S. cerevisiae, P. anomala, C. diddensiae, Issatchenkia orientalis (during process)	Sicilian green table olives	PCR-DGGE of 26 s rRNA	[58]
P. membranifaciens, Pichia fermentans, S. cerevisiae, Candida oleophila. (during process)	Portuguese brined olive	RFLP, Sequencing of 26 s rRNA D1/D2 region	[8]

Yeasts/stage	Country/variety	Method	Ref.
Citeromyces matritensis, Zygotorulaspora mrakii, S. cerevisiae (during process)	Portugal, Manzanilla	Sequencing of 26 s rRNA D1/D2 region	[59]
P. membranifaciens (during process)	Greece, Conservolea	Sequencing of 5.8 s rRNA, ITS1-ITS2 region	[60]

The information following below is a short description of microbial ecology studies at different cultivars and processing steps within the major table olives production countries of Mediterranean in an attempt to draw conclusions on their yeast colonization (Figure 1).

Figure 1. (A) Cyprus naturally black olives with salt, (B) Kalamata naturally black olives in brine, (C) Cyprus green cracked olives in brine, (D) Colonized surface of Cyprus green cracked olives (arrows show the formation of microbial colonies viewed under stereomicroscope).

3.1. Spain

According to Ref. [12], S. cerevisiae, Issatchenkia occidentalis and Geotrichum candidum were identified from Spanish naturally green seasoned table olives (cv. Alorena). In the same study the researchers identified C. boidinii and Hanseniaspora guilliermondii from the preservation stage of ripe black olives. During the fermentation of Arbequina naturally green table olives in Spain [52, 53], C. boidinii, C. sorsoba, Candida diddensiae, K. lactis, P. membranifaciens, W. anomalus, P. kluyveri, and Rhodotorula glutinis were found. In a study looking at the Spanish yeast biodiversity of oleic ecosystems, the yeast biodiversity in the fresh table olive, crushed olives and olive pomace from Arbequina and Cornicabra varieties, has been found to contain Pichia caribbica, Lachancea fermentati and Nakazawaea holstii, as the most important isolated species [54]. According to Ref. [7], researchers carried out the molecular identification by means of a RFLP analysis and sequencing of a total of 199 yeast isolates obtained from Spanish industrial green table olive fermentation. C. diddensiae, S. cerevisiae and P. membranifaciens were the most abundant yeast species isolated from directly brined Alorena olives, but for Gordal and Manzanilla cultivars, Candida tropicalis, P. galeiformis and W. anomalus were found. Recently, other scientists [57] used a culture-independent approach based on the PCR-DGGE analysis for the identification of yeasts associated with Alorena de Malaga olive fermentation, and found that in cold fermented olives the most essential yeasts were S. cerevisiae and Candida apicola. Finally, it has been shown that the most important species throughout the storage period of table olives were S. cerevisiae and Pichia galeiformis, although C. boidinii was present at the last stages of the process and P. membranifaciens was detected at an earlier stage of the Hojiblanca cultivar storage [34].

To summarize, the hierarchically most prominent yeast on Spanish table olives are P. membranifaciens, C. boidinii, C. diddensiae, S. cerevisiae and P. anomala.

3.2. Greece

In relevant scientific work of Greek scientists it has been discovered that Metschnikowia pulcherrima, D. hansenii and Aureobasidium pullulans were the dominant yeast species at the beginning of the fermentation process of Greek Conservolea black olives [55]. These researchers found a new yeast species associated with this type of fermentation, named Candida olivae. Species heterogeneity changed during fermentation and both P. membranifaciens and P. anomala became the only dominant yeasts at the end of the fermentation. A similar work focusing on microbial heterogeneity during aerobic and modified atmosphere packaging storage of Conservolea natural black olives found that P. membranifaciens was dominated in all pouches regardless of gas composition with a frequency more than 80% during storage [60]. Summarizing P. membranifaciens is the dominant yeast on Conservolea black olives followed by P. anomala.

3.3. Italy

In Ref. [56], it has been revealed that in Sicilian environment in four Italian olive cultivars (Bradofina, Castriciana, Nocellara del Belice and Passalunara) and one Spanish (Manzanilla) the presence of P. kluyveri followed by Candida parapsilosis and P. guilliermondii during the entire fermentation period is clear. In a similar work [58] also in Sicily with green table olives of cultivar Nocellara dell' Etna, researchers found S. cerevisiae, P. anomala, C. diddensiae, and I. orientalis during the process. Finally, when researchers [6] studied the technological and spoiling characteristics of yeast microflora isolated from cultivar Bella Di Cerignola table olives, most prominent yeasts isolated were Candida famata and C. guilliermondii. Summarizing the Italian yeast biodiversity in green olives is seems that P. guilliermondii and P. kluyveri are likely to play a crucial role in their fermentation.

3.4. France

In Ref. [51] it has been identified that P. anomala, C. boidinii and Debaryomyces etchelsii were dominant species in French black olives of Nyons area (South France).

3.5. Portugal

According to Ref. [8], researchers studied the yeast population associated with Portuguese brined green olives of cultivars Galega and Cordovil, fermentation was mainly driven by P. membranifaciens, P. fermentans, S. cerevisiae and C. oleophila. In a similar work [59], it has been found that during the initial phases of cracked green Manzanilla olive fermentation a great diversity of yeasts was observed; however, as the process was evolving, the biodiversity decreased with the fermentative yeasts Citeromyces matritensis, Zygotorulaspora mrakii and S. cerevisiae becoming the dominant species. These species though are reported as high risk spoilage microorganisms contrary to P. membranifaciens reported by [8].

Concluding this brief overview of yeast ecological studies in fermented table olives around the Mediterranean it becomes apparent that among Candida species, C. boidinii, C. diddensiae, C. famata (formerly D. hansenii), C. guilliermondii (formerly H. guilliermondii), and C. oleophila are the most prominent. Among Pichia species, P. membranifaciens, Pichia anomala (W. anomalus) and P. fermentans are the most prominent. From other species we tend to isolate more A. pullulans, Debaryomyces etchellsii, G. candidum, I. occidentalis, K. lactis, Rhodotorula spp., S. cerevisiae and Z. mrakii. These yeasts are plentiful for detailed studies on their role during table olive fermentations and could offer great opportunities for biotechnological tools as starter cultures.

4. BIOTECHNOLOGICAL APPLICATIONS OF TABLE OLIVES YEAST STRAINS

Yeast microbiota in olives is very heterogeneous and can be altered depending on the olive cultivar, region, type of fermentation process, salt concentration, pH, nutrients, oxygen and interactions with other microorganisms [5, 57, 64–69]. In table olive fermentations, yeasts are an important group of microorganisms that act as both desirable and spoilage microorganisms and it is important to evaluate their biodiversity in table olive fermentations [3, 4]. As a positive effect, some yeast species isolated from table olives, such as Debaryomyces, Pichia and Candida are known to include a considerable number of strains with a killer character and it has been found, that W. anomalus protects olives from unsaturated fatty acid oxidation and peroxide formation [3, 70, 71], relating also to the antioxidant activity of yeasts [7, 8]. Moreover, it has been reported that W. anomalus and S. cerevisiae isolated from diverse table olive fermentations, have phytase enzymes that are required for the degradation of phytate complexes [3, 41, 44]. On the other hand, W. anomalus is important yeast for olive fermentation, but it may also have a role in the deterioration of olives at the end of fermentation [72]. It has been reported that some olive-related yeast strains, such as W. anomalus, can produce enzymes that could cause softening of the olives as an unfavorable property [33].

According to Ref. [73], the combination of yeasts with LAB, has resulted to an improvement in growth of the LAB. As a result, the production of lactic acid was improved, thanks to the greater availability of the necessary nutrients by the yeasts activity and lysed cells. Indeed, recent studies have shown that the growth of LAB could be stimulated by the use of yeasts, as for example L. plantarum, improved its growth when D. hansenii was inoculated in olive brine [74]. Moreover, L. pentosus's performance was rapidly improved, with the use of S. cerevisiae as a starter in green table olive solutions [73]. In the same study, it has been proved that the production of lactic acid was increased, as well. Yeasts seem to be active microorganisms during the fermentation of table olives, synthesizing substances such as vitamins, amino acids and purines, or breakdown complex carbohydrates, which are essential for the growth of Lactobacillus species [75].

Therefore, selection of the most appropriate yeasts for starters should be firstly based on strains possessing the best enzymatic activities as mentioned earlier and secondly to their ability to predominate during fermentation. Moreover, it is needed to have a high resistance to salt and low pH values. Predictive microbiology is seemed to be a valuable tool for the discrimination and selection of the most promising strains, determining the influence of environmental variables on yeast growth [5]. A problem usually occurring in such applications is to find the appropriate methodology to manage such a large amount of data, which is necessary when researchers have to analyze several biochemical activities or growth data from a considerable number of strains.

Multivariate analysis techniques offer a viable approach in solving this setback. For instance, multivariate analysis approach to study growth and qualitative activity data of a number of yeasts isolated from Bella di Cerignola table olives [6] has been used. Principal Component Analysis (PCA) clearly differentiated and assisted the selection and discrimination of several W. anomalus and C. boidinii isolates with high global desirable activity levels [76].

However, in some cases, dominant yeasts could create products with milder taste and less self-life preservation [4, 15]. This problem was reported in [65] during fermentation of natural black olives at different NaCl levels and temperatures.

Moreover, an excessive growth of fermented yeast (7 log10 CFU mL-1) could produce a vigorous production of CO_2 resulting in penetrating olives and damaging the fruits [64]. The use of high levels of NaCl during fermentation (8% in the equilibrium) could privilege the growth of yeasts against LAB [4, 65]. In accordance with Ref. [33], it has been found that some strains of Rhodotorula minuta and D. hansenii in green table olive fermentations are having this ability. Finally, strains of R. glutinis, R. minuta and R. rubra could grow, form pellicles in olive brines and produce polygalacturonases causing a softening of olives kept in storage [77].

Moreover, yeasts present in packed olives can produce an excess gas (CO_2) leading to swollen containers, clouding of the brines, or produce off-flavors and off-odors [4]. Furthermore, it has been reported that yeasts identified currently as S. cerevisiae and P. anomala spoiled the olives through a combination of gas-pocket formation and softening [78]. Fortunately, yeasts from table olives are almost entirely nonpathogenic. The inhibitory effect of sorbic acid, benzoic acid and their salts on yeasts growth to stabilize table olive packing, were reported previously [5, 79, 80].

5. PROSPECTS FOR THE FUTURE

Currently, the scientific community has accumulated a substantial amount of knowledge regarding the ecology of fermented table olives, as well as the physiology, biochemistry, and genetics of yeasts isolated from the fermentation process. Furthermore, additional data will soon be obtained through the application of high throughput sequencing techniques. Although yeasts are commonly found in cover brines, current research has revealed that they are also present on the fruit epidermis. This discovery suggests that yeasts could be a valuable source of probiotic strains, adding to the numerous beneficial features they bring to the final product [43, 81]. The development of a novel and practical food product that incorporates the benefits of probiotic yeasts can enhance the overall value of the finished product, surpassing its existing merits. Based on the current research, it is evident that yeasts found in table olives possess a noteworthy technical and probiotic character. However, additional research is required regarding their physiology, ecology, and biochemistry. The study of yeasts at the molecular level is necessary during fermentation to determine their presence or absence in the final product, particularly when it involves probiotic bacteria. Without a doubt, the complete capabilities of yeasts associated with table olives have not yet been fully ascertained, and there are numerous obstacles that require further investigation, widespread distribution, and industrial utilization. The composition of the microorganisms present in olives differs to some extent depending on the geographical region, the specific variety of olive, and the method used for processing [5, 12, 33, 34, 76]. Due to the significance of yeasts in the end product and the need to explore and combine the variety of yeasts in table olives across locations as a geographical indication, it poses a difficulty. This may result in the emergence of novel table olives as Protected Designation of Origin (PDO) or Protected Geographical Indication (PGI) products, particularly for European Union (EU) nations. Starter cultures are crucial in olive fermentation as they effectively regulate the safety and quality of the end product. For comprehensive guidance on choosing yeasts as starting cultures for table olives, consult References [32, 82], which provide detailed reviews and precise instructions to follow. The recent use of Next Generation Sequencing (NGS) methods has contributed to the study of fungal populations in directly brined Alorena de Malaga green olives fermentations. This was achieved by the application of metabarcoding analysis, which has enhanced our understanding of the issue [83]. It is anticipated that there will be an increase in the number of publications in the following years. NGS technologies are generating vast quantities of data, which can often be challenging to analyze at a logical and practical level. Preliminary results have been announced for a significant endeavor including the organization, visualization, and investigation of bacterial communities in food using network analysis [84]. FoodMicrobionet is a platform that has been created by incorporating 17 bacterial studies on various food products such as dairy products, dairy starter cultures, raw and fermented meat, doughs, sourdoughs, and fermented vegetables. The main goal of this platform is to analyze nodes and network properties, and to develop an interactive web-based visualization. Through this method, the researcher can investigate the connections between Operational Taxonomic Units (OTUs) and samples to determine the fundamental and sample-specific bacterial populations. An analogous methodology will be extensively employed for the examination of yeast/fungal investigations across different matrices. If the aforementioned method is paired with a high throughput screening approach for fragrance production, similar to the one utilized in reference [85], it has the potential to yield fascinating outcomes with several novel connections. In a recent study, the bacterial community and its changes during the fermentation process of Sicilian table olives (cv. Nocellara Etnea) were examined, along with its impact on the synthesis of metabolites. In the coming years, we anticipate an increase in comprehensive research on the microbial ecology and biotechnology of table olives. This research will provide useful insights into the microbial processes involved in olive fermentations, with potential applications in other areas.

AT A GLANCE

The table olive is regarded as a customary fermented vegetable in Mediterranean countries, and its production and consumption are currently expanding globally. Yeasts play a crucial part in olive fermentation due to their dual function. Yeasts possess beneficial biochemical activities such as lipase, esterase, β-glucosidase, catalase, and synthesis of killer factors. These activities have important technological implications in the fermentation of this vegetable. However, spoiling activity may also be demonstrated. Recent research have indicated that yeasts derived from table olives could serve as a novel source of potential probiotics. Many yeast species commonly discovered in the preparation of table olives have been observed to have these qualities. Hence, the implementation of starting cultures technology will have a substantial impact, not only in regulating the safety and quality of the end product during olive fermentation, but also in promoting customer well-being.

REFERENCES

1. IOC (International Olive Oil Council), 2013. Statistical of Table Olive Production. Available from: http://www.internationaloliveoil.org/estaticos/view/132-world-table-olive-figures [Accessed: Jan 24, 2013].
2. Bonatsou S, Benítez A, Rodríguez-Gómez F, Panagou EZ, Arroyo-López FN. Selection of yeasts with multifunctional features for application as starters in natural black table olive processing. Food Microbiology. 2015;46:66-73
3. Arroyo-López FN, Romero-Gil V, Bautista-Gallego J, Rodríguez-Gómez F, JiménezDíaz R, García-García P, Querol A, Garrido-Fernández A. Yeasts in table olive processing: Desirable or spoilage microorganisms? International Journal of Food Microbiology. 2012;160:42-49
4. Garrido-Fernández A, Fernandez Diez MJ, Adams RM. Table Olives: Production and Processing. London: Chapman and Hall; 1997
5. Arroyo-López FN, Querol A, Bautista-Gallego J, Garrido-Fernández A. Role of yeasts in table olive production. International Journal of Food Microbiology. 2008;128:189-196
6. Bevilacqua A, Perricone M, Cannarsi M, et al. Technological and spoiling characteristics of the yeast microflora isolated from Bella di Cerignola table olives. International Journal of Food Science & Technology. 2009;44:2198-2207
7. Bautista Gallego J, Rodriguez-Gomez F, Barrio E, Querol A, Garrido-Fernandez A, Arroyo-Lopez FN. Exploring the yeast biodiversity of green table olive industrial fermentations for technological applications. International Journal of Food Microbiology. 2011;147:89-96
8. Silva T, Reto M, Sol M, et al. Characterization of yeasts from Portuguese brined olives, with a focus on their potentially probiotic behaviour. LWT – Food Science and Technology. 2011;44:1349-1354
9. Rodriguez-Gomez F, Bautista-Gallego J, Romero-Gil V, et al. Influence of yeasts on the oil quality indexes of table olives. Journal of Food Science. 2013;78:1208-1217
10. Kurtzman CP, Fell JW, Boekhout T. The Yeasts, A Taxonomic Study. 5th ed. Amsterdam: Elsevier; 2011
11. Kurtzman CP, Fell JW. The Yeast, A Taxonomic Study. 4th ed. Amsterdam: Elsevier; 1998
12. Arroyo-López FN, Durán-Quintana MC, Ruiz-Barba JL, Querol A, Garrido-Fernández A. Use of molecular methods for the identification of yeast associated with table olives. Food Microbiology. 2006;23:791-796
13. Flores CL, Rodriguez C, Petit T, Gancedo C. Carbohydrate and energy-yielding metabolism in non-conventional yeasts. FEMS Microbiology Reviews. 2000;24:507-529
14. Sanchez I, Palop L, Ballesteros C. Biochemical characterization of lactic acid bacteria isolated from spontaneous fermentation of 'Alamgro' eggplants. International Journal of Food Microbiology. 2000;59:9-17

15. Panagou EZ, Tassou CC. Changes in volatile compounds and related biochemical profile during controlled fermentation of cv. Conservolea green olives. Food Microbiology. 2006;23:738-746

16. Krajewska-Kulak E. Hydrolytic activity of Candida albicans and their susceptibility to antimycotics. Medical Science Monitor. 1998;4:1643-3750

17. Krajewska-Kulak E. Enzymatic biotypes of the yeast-like fungi strains and their susceptibilities to antimycotics isolated from ontocenosis of the urogenital system. Mikologia/Medical Mycology. 2002;9:2083-5744

18. Brasch J. Enzyme patterns of dermatophytes. Mycoses. 1994;37:1439-0507

19. Plomer-Niezgoda E, Baran E. Evaluation of extracellular activity of hydrolytic enzymes of selected mold fungi. Mikologia Lekarska. 1997;4:141-145

20. Bull AT, Goodfellow M, Slater JH. Biodiversity as a source of innovation in biotechnology. Annual Review of Microbiology. 1992;46:219-252

21. Abranches J, Morrais PB, Rosa CA, Mendonca-Hagler LC, Hagler HN. The incidence of killer activity and extracecullar proteases in tropical yeast communities. Canadian Journal of Microbiology. 1997;43:328-336

22. Bossi A, Bonizzato L, Zapparoli G. Acidic extracellular proteases from microrganisms of fairly acidic niche. Protein and Peptide Letters. 2006;13:737-741

23. Burden DW, Eveleigh DE. Yeasts-diverse substrates and production. In:Yeast Technology, Spencer JF, Spencer DM, editors. Berlin: Springer-Verlag; 1990. p. 199-227

24. Bordes F, Barbe S, Escalier P, Mourey L, André I, Marty A, Tranier S. Exploring the conformational states and rearrangements of Yarrowia lipolytica lipase. Biophysical Journal. 2010;99(7):2225-2234

25. Basaran P, Hang YD. Purification and characterization of acetyl esterase from Candida guilliermondii. Letters Applied Microbiology. 2000;30:167-171

26. Athenstaedt K, Daum G. Tgl4p and Tgl5p, two triacylglycerol lipases of the yeast Saccharomyces cerevisiae are localized to lipid particles. J BiolChem. 2005;280(45):37301-37309

27. Sakai N, Mano S, Nozaki K, Takaya H. Journal of the American Chemical Society. 1993;115:7033

28. Charoenchai C, Fleet GH, Henschke PA, Todd BEN. Screen of non-saccharomyces wine yeasts for the presence of extracellular hydrolytic enzymes. Australian journal of grape wine Research. 1997;3:2-8

29. Strauss RS, Pollack HA. Epidemic increase in childhood overweight 1986-1998. Journal of the American Medical Association. 2001;286:2845-2848

30. Kiehn E, Wong B, Edwards F, Armstrong D. Comparative recovery of bacteria and yeasts from Lysis centrifugation and a conventional blood culture system. Journal of Clinical Microbiology. 1983;18:300-304

31. Hurtado A, Reguant C, Bordons A, Rozes N. Lactic acid bacteria from fermented table olives. Food Microbiology. 2012;31:1-8

32. Bevilacqua A, Corbo MR, Sinigaglia M. Selection of yeasts as starter cultures for table olives: A step-by-step procedure. Frontiers in Microbiology. 2012;3:194

33. Hernández A, Martín A, Aranda E, Pérez-Nevado F, Córdoba MG. Identification and characterization of yeast isolated from the elaboration of seasoned green table olives. Food Microbiology. 2007;24:346-351

34. Rodriguez-Gomez F, Arroyo-Lopez FN, Lopez-Lopez A, Bautista-Gallego J, Garrido Fernandez A. Lipolytic activity of the yeast species associated with the fermentation/storage phase of ripe olive processing. Food Microbiology. 2010;27:604-612

35. Psani M, Kotzekidou P. Technological characteristics of yeast strains and their potential as starter adjuncts in Greek-style black olive fermentation. Springer Science+Business Media B.V. 2006;22:1329-1336

36. Caruso M, Fiore C, Contursi M, Salzano G, Paparella A, Romano P. Formation of biogenic amines as criteria for the selection of wine yeasts. World Journal of Microbiology and Biotechnology. 2002;18:159-163

37. Landete JM, Ferrer S, Pardo I. Biogenic amine production by lactic acid bacteria, acetic bacteria and yeast isolated from wine. Food Control. 2007;18:1569-1574

38. FAO/WHO. Health and Nutritional Properties of Probiotics in Food Including Powder Milk with Live Lactic Acid Bacteria. Córdoba, Argentina: Joint Food and Agricultural Organization of the United Nations and World Health Organization Expert Consultation Report; 2001

39. Sazawal S, Hiremath G, Dhingra U, Malik P, Deb S, Black RE. Efficacy of probiotics in prevention of acute diarrhoea: A meta-analysis of masked, randomised, placebo-controlled trials. The Lancet Infectious Diseases. 2006;6:374-382

40. Kourelis A, Kotzamanidis C, Litopoulou-Tzanetaki E, Scouras ZG, Tzanetakis N, Yiangou M. Preliminary probiotic selection of dairy and human yeast strains. Journal of Biological Research of Thessaloniki. 2010;13:93-104

41. Moslehi-Jenabian S, Lindegaard Pedersen L, Jespersen L. Beneficial effects of probiotic and food borne yeasts on humanhealth. Nutrients. 2010;2:449-473

42. Pennacchia C, Blaiotta G, Pepe O, Villani F. Isolation of Saccharomyces cerevisiae strains from different food matrices and their preliminary selection for a potential use as probiotics. Journal of Applied Microbiology. 2008;105:1919-1928

43. Etienne-Mesmin L, Livrelli V, Privat M, Denis S, Cardot JM, Alric M, BlanquetDiot S. Effect of a new probiotic Saccharomyces cerevisiae strain on survival of Escherichia coli O157:H7 in a dynamic gastrointestinal model. Applied and Environmental Microbiology. 2011;77:1127-1131

44. Olstorpe M, Schnüren J, Passoth V. Screening of yeast strains for phytase activity. FEMS Yeast Research. 2009;9:478-488

45. Ruiz Barba JL, Jiménez Diaz R. Availability of essential B-group vitamins to Lactobacillus plantarum in green olive fermentation brines. Applied and Environmental Microbiology. 1995;61:1294-1297

46. Gazi MR, Hoshikuma A, Kanda K, Murata A, Kato F. Detection of free radical scavenging activity in yeast culture. Bulletin of the Faculty of Agriculture of Saga University. 2001;86:67-74

47. Marquina D, Peres C, Caldas FV, Marques JF, Peinado JM, Spencer Martin J. Characterization of the yeast population in olives brines. Letters in Applied Microbiology. 1992;14:279-283

48. Marquina D, Toufani S, Llorente P, Santos A, Peinado JM. Killer activity in yeast isolated from olive brines. Advances in. Food Science. 1997;19:41-46

49. Fernández González MJ, Gárcia P, Garrido-Fernández A, Durán Quintana MC. Microflora of the aerobic preservation of directly brined green olives from Hojiblanca cultivar. Journal of Applied Bacteriology. 1993;75:226-233

50. Kotzekidou P. Identification of yeast from black olives in rapid system microtitre plates. Food Microbiology. 1997;14:609-616

51. Coton E, Coton M, Levert D, Casaregola S, Sohier D. Yeast ecology in French cider and black olive natural fermentations. International Journal of Food Microbiology. 2006;108:130-135

52. Hurtado A, Reguant C, Esteve-Zarzoso B, Bordons A, Rozès N. Microbial population dynamics during the processing of Aberquina table olives. Food Research. 2008;41:738-744

53. Hurtado A, Reguant C, Bordons A, Rozès N. Influence of fruit ripeness and salt concentration on the microbial processing of Arbequina table olives. Food Microbiology. 2009;26:827-833

54. Romo-Sánchez S, Alves-Baffi M, Arévalo-Villena M, Úbeda-Iranzo J, Briones-Pérez A. Yeast biodiversity from oleic ecosystems: Study of their biotechnological properties. Food Microbiology. 2010;27:487-492

55. Nisiotou AA, Chorianopoulos N, Nychas GJE, Panagou EZ. Yeast heterogeneity during spontaneous fermentation of black Conservolea olives in different brine solutions. Journal of Applied Microbiology. 2010;108:396-405

56. Aponte M, Ventorino V, Blaiotta G, Volpe G, Farina V, Avellone G, Lanza CM, Moschetti G. Study of green Sicilian table olive fermentations through microbiological, chemical and sensory analyses. Food Microbiology. 2010;27:162-170

57. Abriouel H, Benomar N, Lucas R, Gálvez A. Culture-independent study of the diversity of microbial populations in brines during fermentation of naturally fermented Alorena green table olives. International Journal of Food Microbiology. 2011;44:487-496

58. Muccilli S, Caggia C, Randazzo CL, Restuccia C. Yeast dy-namics during the fermentation of brined green olives treated inthe field with kaolin and Bordeaux mixture to control the olive fruit fly. International Journal of Food Microbiology. 2011;148:15-22

59. Alves M, Gonçalves T, Quintas C. Microbial quality and yeast population dynamics in cracked green table olives' fermentations. Food Control. 2012;23:363-368

60. Doulgeraki A, Hondrodimou O, Iliopoulus U, Panagou EZ. Lactic acid bacteria and yeast heterogeneity during aerobic and modified atmosphere packaging storage of natural black Conservolea olives in polyethylene pouches. Food Control. 2012;26:49-57

61. Barnett JA, Payne RW, Yarrow D. Yeasts: Characteristics and Identification. 2nd ed. Cambridge: University Press; 1990

62. Esteve-Zarzoso B, Belloch C, Uruburu F, Querol A. Identification of yeasts by RFLP analysis of the 5.8 rRNA gene and the two ribosomal internal transcribed spacers. International Journal of Systematic Bacteriology. 1999;49:329-337

63. Kurtzman CP, Robnett C. Identification and phylogeny of ascomycetous yeast from analysis of nuclear large subunit (26S) ribosomial DNA partial sequences. Antonie Van Leeuwenhoek. 1998;73:331-371

64. Duran-Quintana MC, Garcia-Garcia P, Garrido-Fernández A. Establishment of conditions for green table olive fermentation at low temperature. International Journal of Food Microbiology. 1999;51:133-143

65. Tassou CC, Panagou EZ, Katsaboxakis KZ. Microbiological and physicochemical changes of naturally black olives fermented at different temperatures and NaCl levels in the brines. Food Microbiology. 2002;19:605-615

66. Alvarez DME, Sánchez A, Lamarque AL. Naturally black olives: Comparison of three processes for fermenting cv. 'Farga' olives. Olivae. 2003;97:47-51

67. Chorianopoulos NG, Boziaris IS, Stamatiou A, Nychas GJE. Microbial association and acidity development of unheated and pasteurised green table olives fermented using glucose or sucrose supplements at various levels. Food Microbiology. 2005;22:117-124

68. Corsetti A, Perpetuini G, Schirone M, Tofalo R, Suzzi G. Application of starter cultures to table olive fermentation: An overview on the experimental studies. Frontiers in Microbiology. 2012;3:1-6

69. Tofalo R, Schirone M, Perpetuini G, Angelozzi G, Suzzi G, Corsetti A. Microbiological and chemical profiles of naturally fermented table olives and brines from different Italian cultivars. Antonie Van Leeuwenhoek. 2012;102:121-131

70. Llorente P, Marquina D, Santos A, Peinado JM, Spencer-Martins I. Effect of salt on the killer phenotype of yeasts from olive brines. Applied and Environmental Microbiology. 1997;63:1165-1167

71. Hernández A, Martín A, Córdoba MG, Benito MJ, Aranda E, Pérez-Nevado F. Determination of killer activity in yeasts isolated from the elaboration of seasoned green table olives. International Journal of Food Microbiology. 2008;121:178-188

72. Pitt JI, Hocking AD. Yeast. Fungi and Food Spoilage. New York: Springer; 2009

73. Segovia Bravo K, Arroyo-López FN, Garcia Garcia P, Duran Quintana MC, GarridoFernández A. Treatment of green table olive solutions with ozone. Effect on the polyphenol content on Lactobacillus pentosus and Saccharomyces cerevisiae growth. International Journal of Food Microbiology. 2007;114:60-68

74. Tsapatsaris S, Kotzekidou P. Application of a central composite design and response surface methodology to the fermentation of olive juice by Lactobacillus plantarum and Debaryomyces hansenii. International Journal of Food Microbiology. 2004;95:157-168

75. Viljoen BC. Yeast ecological interactions. Yeast–yeast, yeast bacteria, yeast–fungi interactions and yeasts as biocontrol agents. In: Querol A, Fleet H, editors. Yeasts in Food and Beverages. Berlin: Springer–Verlag; 2006. p. 83-110

76. Rodríguez-Gómez F, Romero-Gil V, Bautista-Gallego J, Garrido-Fernández A, Arroyo-López FN. Multivariate analysis to discriminate yeast strains with technological applications in table olive processing. World Journal of Microbiology and Biotechnology. 2012;28:1761-1770

77. Vaughn RH, Jakubczyk T, MacMillan JD, Higgins TE, Dave BA, Crampton VM. Some pink yeast associated with softening of olives. Applied Microbiology. 1969;18:771-775

78. Vaughn RH, Stevenson KE, Dave BA, Park HC. Fermenting yeast associated with softening and gas-pocket formation in olives. Applied Microbiology. 1972;23:316-320

79. Rodriguez de la Borbolla Alcalá JM, FernándezDíez MJ, González Cancho F. El empleo de ácidosórbico y sus sales en las aceitunasaderezadas. Grasas y Aceites. 1961;12:10-15

80. Marsilio V, Cichelli A. Influencia del sorbatopotásico y delbenzoatosódicosobre la estabilidad de las aceitunas de mesa ensalmuera. Grasas y Aceites. 1992;43:66-74

81. Domínguez-Manzano J, Olmo-Ruiz C, Bautista-Gallego J, Arroyo-López FN, Garrido-Fernández A, Jiménez-Díaz R. Biofilm formation on abiotic and biotic surfaces during Spanish style green table olive fermentation. International Journal of Food Microbiology. 2012;157:230-238

82. Bonatsu S, Tassou CC, Panagou EZ, Nychas G-JE. Table olive fermentation using starter cultures with multifunctional potential. Microorganisms. 2017;5:30. DOI: 10.3390/microorganisms5020030

83. Arroyo-López FN, Medina E, Ruiz-Bellido MA, Romero-Gil V, Montes-Borrego M, Landa BB. Enhancement of the knowledge on fungal communities in directly brined Alorena de Malaga green olive fermentations by metabarcoding analysis. PloS One. 2016;11(9). DOI: 10.1371/journal.pone.0163135

84. Parente E, Cocolin L, De Filippis F, Zotta T, Ferrocino I, O'Sullivan O, Neviani E, De Angelis M, Cotter PD, Ercolini D. FoodMicrobionet: A database for the visualisation and exploration of food bacterial communities based on network analysis. International Journal of Food Microbiology. 2016;219:28-37

85. Gamero A, Quintilla R, Groenewald M, Alkema W, Boekhout T, Hazelwood L. High-throughput screening of a large collection of non-conventional yeasts reveals their potential for aroma formation in food fermentation. Food Microbiology. 2016;60:147-159

86. Randazzo CL, Todaro A, Pino A, Pitino I, Corona O, Caggia C. Microbiota and metabolome during controlled and spontaneous fermentation of Nocellara Etnea table olives. Food Microbiology. 2017;65:136-148

CHAPTER 7

APPLYING SYSTEM IDENTIFICATION AND ADVANCED PROCESS CONTROL TECHNIQUES TO ENHANCE THE FERMENTATION PROCESS OF BAKER'S YEAST

1. INTRODUCTION: AN OVERVIEW

The microorganism Saccharomyces cerevisiae, commonly referred to as baker's yeast, is utilized in several applications including the generation of ethanol, glycerol, β-glucan, invertase enzyme, and mostly yeast itself. By utilizing molasses or glucose as the carbon source and employing batch or fed-batch operation, it is feasible to manufacture ethanol and the yeast itself on a large scale in anaerobic or aerobic circumstances, respectively [1]. In order to achieve a high volumetric efficiency, it is necessary to obtain cells with a high development phase in microorganisms, taking into account the process economics [2]. Research on S. cerevisiae is ongoing in the disciplines of biotechnology and genetics, as well as in research and development in the food and pharmaceutical industries [3]. Several studies on β-glucan production highlight the significance of S. cerevisiae production for its application in the pharmaceutical industry [4]. An in-depth analysis of production data is necessary to have a comprehensive understanding of the process [3]. Biocatalysts, which are microorganisms, can be generated in large quantities with strong enzyme activity by optimizing the operating conditions of a bioreactor. These factors include pH, temperature, dissolved oxygen (DO) concentration, air flow rate, agitation speed, and substrate concentration [2].

When discussing bioprocessing, it is important to note that the operating parameters of the systems undergo non-linear changes over time. Furthermore, they do not possess the mathematical models that precisely describe the intricate processes occurring during cell growth and the production of products [5, 6]. In addition, there are only a few online sensors available that can accurately measure state variables such as cell, substrate, and product concentrations. Additionally, the biocatalyst's activity is hindered by the inhibitory effects of the substrate, oxygen, or the product that is generated. According to Alpbaz et al., S. cerevisiae is highly responsive to alterations in the growth environment [7]. To ensure economic benefit, high-quality output, and safe operation, it is crucial to identify the optimal operating conditions and effectively regulate the operating parameters at the found optimum values, due to the limitations in bioprocesses [8]. Although there are installation, operation, and modeling studies available in the literature for batch and fed-batch operations, controlling them is extremely challenging due to the inherent biological processes and the dynamic nature of the entire process. The user's text is "[9]." The performance of the control system can be influenced by the tuning parameters of the controller and the parameters of the process model. Additionally, the selection of the control algorithm and the structure of the process model can also have an impact. The parameters of both parametric and nonparametric models are determined through dynamic assessments of various disturbance effects and system identification procedures. Parametric models are typically built using a polynomial model with discrete time intervals [10]. Typically, various types of inputs such as step, square wave, pseudorandom binary sequence (PRBS), impulse, pulse, and random signals are used as variables to analyze and collect input-output data from a process. These inputs can include flow rates of acid/base for pH, cooling/heating fluid flow rate for temperature, air flow rate for dissolved oxygen (DO), and substrate feed for substrate concentration [11]. Several system identification methods, including Biermann, Levenberg–Marquardt, genetic, least squares (LS), and recursive least squares (RLS), have been researched for the computation of model parameters [10, 11]. The parameters of nonparametric models are determined by analyzing the appropriate curves, such as the response curve and Bode diagrams [12, 13]. The current component of the system can be represented using an approximation structure and estimated parameters [3]. The closed-loop performance of the system during control is mostly determined by the discrepancy between the real process and the model. Hence, it is imperative that a specific model framework encompasses all the available data on operational circumstances and accurately approximates the system to a predetermined extent. Additionally, it should possess adaptability and facilitate rapid parameter estimation techniques [14]. The ARMAX model, which incorporates an autoregressive moving average structure and exogenous input polynomials, has been extensively utilized in the literature to accurately characterize process dynamics [2, 3, 10–12, 15–18]. The necessity for self-regulating controllers arises from the aspiration to manage processes with uncertain or gradually varying characteristics over time [11]. The foundation of this approach was derived from the self-tuning regulator (STR) pioneered by Åström and Wittenmark [19]. The primary aim of this strategy is to minimize the variability of the output variable, which includes pH, temperature, dissolved oxygen (DO), and substrate concentration. The STR, or Systematic Trading Rule, forecasts the future output variance and subsequently applies a control measure to ensure that the estimated variance is reduced to zero [11]. Nevertheless, the implementation of the STR technique has encountered several challenges, including the absence of real-time adjustment parameters, limitations in controlling nonminimum phase systems, and inadequate control over systems with variable or unknown time delays [2]. Subsequently, Clarke and Gawthrop made alterations to the STR, resulting in the self-tuning generalized minimal variance (ST-GMV) method, which was designed to address the aforementioned challenges [20]. ST-GMV is an adaptive algorithm that utilizes the GMV cost function and a predictive form of the process model. This formulation results in a more straightforward adjustment [21]. The ST-GMV approach is currently gaining popularity and is extensively utilized in various industrial applications [22, 23]. This chapter focuses on the dynamic analysis and system identification of the key operating parameters, namely pH and dissolved oxygen (DO), in the baker's yeast production process. Dynamic analysis was performed at the previously identified ideal temperature of 32°C for baker's yeast production. The purpose of this study was to understand the behavior of the process and collect data for use in the system identification stage. This investigation must be conducted in a process-specific manner because the data mostly relies on the physical structure of the process. This chapter specifically emphasized the effectiveness of the system identification process in control applications. In order to assess the impact of the process model structure and parameters on the performance of the controller, two opposing models, referred to as appropriate and unsuitable models, were employed in the ST-GMV simulation studies. The performance of the controllers was assessed based on their ability to track a steady set point trajectory under different noise situations for both controlled variables. Like many research in the literature, the simulation study results were conducted in the MATLAB environment and analyzed based on integrated squared errors.

2. INFLUENCE OF OPERATIONAL CONDITIONS ON THE GENERATION OF BAKER'S YEAST

Saccharomyces cerevisiae Undoubtedly, yeast is a highly significant microorganism that has been safely ingested by humans throughout history. Yeast cells require both nutrition and energy to facilitate their growth and the creation of products. In order to obtain essential nutrients and energy, they utilize sugars like glucose, maltose, and sucrose, as well as vitamins such as biotin, pantothenate, and inositol, along with minerals like copper (Cu), zinc (Zn), iron (Fe), molybdenum (Mo), and manganese (Mn) [22]. Elevated substrate concentrations can lead to suppression of microbe production as the substrate binds to a secondary, nonactive site on a specific kind of enzyme [22]. Similarly, ethanol production is also a result of insufficient oxygen in the given scenario. The phenomenon referred to as the Crabtree effect is considered unfavorable due to its negative impact on the yield of fermentative cell growth [22]. The inhibition of yeast production is mostly determined by the specific characteristics of the cell and substrate. The development of microorganisms in yeast production is inhibited when the concentration of glucose exceeds 200 g/L [2]. Another crucial prerequisite is the presence of oxygen. Yeast cells utilize oxygen in conjunction with sugar to facilitate growth without the creation of ethanol. In the context of beer and wine production, yeast will continue to thrive and produce ethanol when there is insufficient oxygen in the atmosphere. To achieve optimal yeast production, it is crucial to provide efficient oxygen and sugar transport to the growth medium while minimizing ethanol generation. Oxygen is a constraining substrate because it has a reduced solubility in water. At a temperature of 30°C, the amount of oxygen that may dissolve in water is 8 milligrams per liter [24]. The Pasteur effect refers to the phenomenon where a decrease in dissolved oxygen (DO) concentration leads to a decrease in substrate consumption and the rate of carbon dioxide generation [25]. When operating under aerobic circumstances, simply supplying the system with an oxygen source is insufficient. Ensuring a uniform dispersion of oxygen throughout the liquid medium is also a crucial factor [24]. The oxygen transfer from the gaseous phase to the microorganism in the feed medium plays a crucial role in setting the design and operational parameters of the bioreactor. The reactions that occur during yeast production depend on whether the medium state is aerobic or anaerobic.

$$\mathrm{C_6H_{12}O_6} + 6\mathrm{O_2} \xrightarrow{\text{Aerobic conditions}} 6CO_2 + 6\mathrm{H_2O} + 688\mathrm{kcal} \tag{1}$$

$$\mathrm{C_6H_{12}O_6} \xrightarrow{\textit{Anaerobic conditions}^2} C_2H_5OH + 6CO_2 + 56kcal \tag{2}$$

Carbon dioxide and water are produced in the medium when yeast can thrive in a favorable environment under aerobic circumstances. In order to generate yeast, it is necessary to have aerobic conditions [12]. Alternatively, ethanol synthesis takes place in the absence of oxygen [12]. Yeast cells may utilize ethanol as a carbon source, particularly during the diauxic phase [25]. Yeast has a preference for sugars over ethanol as a growth medium due to the poisonous nature of ethanol. During the process of cellular development, the yeast cell generates carbon dioxide. An additional dependable parameter that can be employed to characterize the condition of the yeast, as it does not rely on oxygen during ethanol generation, is the ratio of the carbon dioxide production rate (CPR) to the oxygen uptake rate (OUR). The respiratory quotient (RQ) is a determined ratio [2]. The pH of the growth medium is a crucial operational parameter as it directly impacts the enzymatic activity of microorganisms. Yeast exhibits resistance to acidic conditions while displaying high sensitivity to alkaline conditions. The growing medium should ideally have a pH ranging from 3 to 6 [24]. When the pH is not suitable for the microbe, cations and anions face challenges in traversing the cell membrane. Cell permeability disruption impacts enzyme activity and halts protein synthesis. In addition, cells become increasingly vulnerable to harmful chemicals. The pH of the growing medium significantly affects the substrate consumption and production output of the yeast. Yeasts, like other microorganisms, have certain temperature ranges for growth. These include a minimum temperature range of 5-25°C, a maximum temperature range of 40-50°C, and an optimum temperature range of 30-40°C [26]. The temperature has a significant impact on the activity of enzymes that are involved in the structures of microorganisms. Typically, the temperature of the growing medium significantly affects the development of microorganisms, respiration, and the production of products. According to chemical kinetics, an increase in temperature enhances the reaction rate. However, enzymatic reactions follow this pattern only up to a specific temperature. The temperature rise resulting from the heat created during the aerobic synthesis of baker's yeast is unfavorable in terms of product efficiency.

3. SYSTEM IDENTIFICATION

Before working on a system, it is necessary to know the system's upper and lower limits in order to be aware of the possible situations that can be faced. To control the system, firstly the relations between the input variables and the output variables must be obtained and the model of the system must be developed. While modeling the systems, it is possible to use input–output data obtained by experimental studies or mass-energy balances. However, it is difficult to obtain a model by using mass-energy balances in complex systems, and in some cases these balances may be insufficient to accurately identify the system. In such cases, it is more useful to create models using system identification methods from the experimental input–output data [27]. System identification might be described as a method based on giving a disturbance to the input variable and consequently obtaining the output variable data of the system and determining the model. The unknown parameters in the parametric model are found by an appropriate method using input–output data. Then, the model is compared with the experimental data and the fitness is tested.

3.1. Signals used in system identification

The first step of system identification is the selection of input signals that will affect the system. Step, square wave, sinusoidal wave, PRBS, impulse, pulse, and random signals are generally applied as inputs [11]. Often, discrete time models are used to describe the system [8–11]. For this purpose, input and output variable signals are sampled and recorded at a suitable time interval. This chapter was focused on step effect as input signal.

3.2. System models

Linear model of an open-loop discrete time system can be written in terms of u(t) as input variable and x(t) as output variable as shown in Eq. (3).

$$x(t) + a_1 x(t-1) + \ldots + a_{n_a} x(t-n_a) = b_0 u(t-1) + \ldots + b_{n_b} u(t-n_b) \tag{3}$$

Backward shift operator is defined as Eq. (4).

$$z^{-i} x(t) = x(t-i) \tag{4}$$

Here, x(t) represents the x value at time t, x(t - 1) represents the x value at time (t - Δt) for Δt = 1, and x(t - i) represents the x value at time (t - iΔt). Eq. (5) is defined as a discrete time transfer function.

x(t)=BAu(t)xt=BAut

$$x(t) = \frac{B}{A} u(t) \tag{5}$$

Here, polynomials of A and B can be written as Eqs. (6) and (7).

$$A = 1 + a_1 z^{-1} + \ldots + a_{n_a} z^{-n_a} \tag{6}$$

$$B = b_0 + b_1 z^{-1} + \ldots + b_{n_b} z^{-n_b} \tag{7}$$

Roots of polynomials A and B are poles and zeros of the system, respectively. If one of the poles or zeros of the system is placed outside the unit circle of the z-plane, system is defined as unstable or nonminimum phase, respectively [28]. In a self-tuning system, disturbance effects can occur in a wide variety of forms. The disturbance signal s(t) can be a part of the control system and is often treated as an additional disturbance factor at the output of the controlled process. In this case, the self-tuning controller will attempt to eliminate this disturbance effect. Such signals can be summed up in two groups as defined signals and random signals. In general, model of a constant random signal source is shown as follows.

$$s(t) = \frac{C}{A} e(t) \tag{8}$$

Here, C polynomial is defined as Eq. (9).

$$C = 1 + c_1 z^{-1} + c_2 z^{-2} + \ldots + c_{n_c} z^{-n_c} \tag{9}$$

Eq. (8) is also described as auto regressive moving average (ARMA) model. Whole system output can be written in various forms as Eqs. (10–12).

$$y(t) = x(t) + s(t) \tag{10}$$

$$y(t) = \frac{B}{A}u(t-1) + \frac{C}{A}e(t) \tag{11}$$

$$\begin{aligned} y(t) &= a_1 y(t-1) + \ldots + a_{n_a} y(t-n_a) = b_0 u(t-1) + \ldots + b_{n_b} u(t-n_b) + e(t) \\ &+ c_1 e(t-1) + \ldots + c_{n_c} e(t-n_c) \end{aligned} \tag{12}$$

This model is obtained by adding a control input to the ARMA-type signal, and the system is called ARMAX model.

3.3. Estimation of model parameters

3.3.1. LS method

Eq. (11) is written in matrix form, and transpose of data and parameter vectors are defined as (ϕT) and (θT), respectively.

$$\varphi^T = [-y(t-1), \ldots, -y(t-n_a), u(t-1), \ldots, u(t-n_b), e(t-1), \ldots, e(t-n_c)] \tag{13}$$

$$\theta^T = [a_1, \ldots, a_{n_a}, b_0, \ldots, b_{n_b}), c_1, \ldots, c_{n_c}] \tag{14}$$

Using Eqs. (13) and (14), output variable can be written as follows.

$$y(t) = \varphi^T \theta + e(t) \tag{15}$$

Eq. (15) is known as linear in the parameter model. There are N measurement values, where N is the number of data samples. Eq. (15) can be redefined as follows.

$$\begin{aligned} y(1) &= \phi T(1)\theta + e(1) \\ y(2) &= \phi T(2)\theta + e(2) \\ y(N) &= \phi T(N)\theta + e(N) \end{aligned} \tag{16}$$

Eq. (16) can be written in vector form as follows.

$$Y = \begin{bmatrix} y(1) \\ y(2) \\ . \\ . \\ y(N) \end{bmatrix}, \phi = \begin{bmatrix} \varphi^T(1) \\ \varphi^T(2) \\ . \\ . \\ \varphi^T(N) \end{bmatrix}, E = \begin{bmatrix} e(1) \\ e(2) \\ . \\ . \\ e(N) \end{bmatrix} \rightarrow Y = \phi\theta + E \tag{17}$$

The estimated parameter matrix ($\hat{\theta}$) is written as given in Eq. (18), with the output variable ($\hat{\theta}$) obtained from the model.

$$\hat{y}(t) = \varphi^T(t)\hat{\theta} \tag{18}$$

The difference between the measured output variable and the output variable calculated from the model is defined as the estimation error (ε(t)) and given in Eq. (19).

$$\varepsilon(t) = y(t) - \hat{y}(t) = \varphi^T(t)\hat{\theta} \tag{19}$$

Eq. (19) is rearranged for N measurements, and Eq. (20) is obtained.

$$\begin{aligned} \varepsilon(1) &= y(1) - \hat{y}(1) = \varphi^T(1)\hat{\theta} \\ \varepsilon(2) &= y(2) - \hat{y}(2) = \varphi^T(2)\hat{\theta} \\ &. \\ &. \\ \varepsilon(N) &= y(N) - \hat{y}(N) = \varphi^T(N)\hat{\theta} \end{aligned} \tag{20}$$

Eq. (20) can be written in vector form as follows.

$$\varepsilon(t) = Y - \phi\hat{\theta} \tag{21}$$

In order to calculate ($\hat{\theta}$) using LS method, cost function (J) given below in Eq. (22) must be minimized. ε value of Eq. (21) is written in Eq. (22) and must be rearranged as follows.

$$J = \sum_{t=1}^{N} \varepsilon^2(t) = \varepsilon^T\varepsilon = \|\varepsilon\|^2 \tag{22}$$

$$J = \left(Y - \phi\hat{\theta}\right)^T \left(Y - \phi\hat{\theta}\right) = Y^TY - \phi^T\hat{\theta}^TY - \left(Y^T\phi\hat{\theta} + \phi^T\phi\hat{\theta}^T\hat{\theta}\right) \tag{23}$$

To find the value that makes the Eq. (23) minimum, the derivative is taken, then equalized to zero, and rearranged and the predicted parameter vector calculated by the LS method is found as follows.

$$\frac{\partial J}{\partial\hat{\theta}} = -2\phi^T\left(Y - \phi\hat{\theta}\right) = 0 \tag{24}$$

$$\hat{\theta} = \left(\phi^T\phi\right)^{-1}\phi^TY\hat{\theta} = \left[\sum_{t=1}^{N} \varphi(t)\varphi^T(t)\right]^{-1} \sum_{t=1}^{N} \varphi(t)y(t) \tag{25}$$

Although the LS is a widely used method, it is not suitable for self-tuning and predictive control methods because the parameter calculation is not made in real time. In such control methods, the data and parameters must be solved at every t instant and updated.

3.3.2. RLS method

Real-time parameter estimation is possible with RLS method and can be easily applied to self-tuning and predictive control algorithms for calculating time-varying model parameters. In the RLS method, the new value of the output variable is calculated by using the model parameters based on past data and the new input–output variable values. The actual value (y(t)) is compared with this estimated value, and the error (ε(t)) is found. The model parameters calculated in the previous step are updated with the newly calculated model parameters [27]. If Eq. (25) is written for any t instant, the parameter calculation equation will be as follows.

$$\hat{\theta}(t) = \left(\phi^T(t)\phi(t)\right)^{-1}\phi^T(t)Y(t) \tag{26}$$

Description of terms in Eq. (26) is in terms of vector forms as follows.

$$Y(t) = \begin{bmatrix} y(1) \\ y(2) \\ . \\ . \\ y(t) \end{bmatrix}, \phi(t) = \begin{bmatrix} \varphi^T(1) \\ \varphi^T(2) \\ . \\ . \\ \varphi^T(t) \end{bmatrix} \tag{27}$$

If Eq. (26) is written for the next sampling time (t + 1), parameter calculation equation will be as follows.

$$\hat{\theta}(t+1) = \left(\phi^T(t+1)\phi(t+1)\right)^{-1}\phi^T(t+1)Y(t+1) \tag{28}$$

Terms φ(t+1) and Y(t+1) of Eq. (28) are written in vector form as follows.

$$\hat{\theta}(t+1) = \begin{bmatrix} \phi(t) \\ \phi^T(t+1) \end{bmatrix}, Y(t+1) = \begin{bmatrix} Y(t) \\ y(t+1) \end{bmatrix} \tag{29}$$

Using these equations, the terms in Eq. (28) are updated.

$$\phi^T(t+1)\phi(t+1) = \left[\phi^T(t) \;\; \varphi(t+1)\right]\begin{bmatrix} \phi(t) \\ \varphi^T(t+1) \end{bmatrix} = \phi^T(t)\phi(t) + \varphi(t+1)\varphi^T(t+1) \tag{30}$$

$$\phi^T(t+1)Y(t+1) = \left[\phi^T(t) \;\; \varphi(t+1)\right]\begin{bmatrix} Y(t) \\ y(t+1) \end{bmatrix} = \phi^T(t)Y(t) + \varphi(t+1)y(t+1) \tag{31}$$

Covariance matrix (P(t)) is defined and written in Eq. (30) as follows.

$$P(t) = \left[\phi^T(t)\phi(t)\right]^{-1} \tag{32}$$

$$P(t+1)^{-1} = P(t)^{-1} + \varphi(t+1)\varphi^T(t+1) \tag{33}$$

Using covariance matrix definition, Eqs. (26) and (28) can be rewritten as follows.

$$\hat{\theta}(t) = P(t)\phi^T(t)Y(t) \tag{34}$$

$$\hat{\theta}(t+1) = P(t+1)\phi^T(t+1)Y(t+1) \tag{35}$$

The term of φT(t+1)Y(t+1) in Eq. (31) is written in Eq. (35) and rearranged as follows.

$$\hat{\theta}(t+1) = P(t+1)\left[\phi^T(t)Y(t)\varphi(t+1)y(t+1)\right] = P(t+1)\phi^T(t)Y(t) + P(t+1)\varphi(t+1)y(t+1) \tag{36}$$

Eq. (37) is obtained using Eq. (34) as follows.

$$\phi^T(t)Y(t) = P(t+1)\hat{\theta}(t) - \varphi(t+1)\varphi^T(t+1)\hat{\theta}(t) \tag{38}$$

Eq. (33) is combined with Eq. (37) and rearranged as follows.

$$\phi^T(t)Y(t) = P(t+1)\hat{\theta}(t) - \varphi(t+1)\varphi^T(t+1)\hat{\theta}(t) \tag{38}$$

Eq. (38) is written in place at φT(t)Y(t) term of Eq. (36).

$$\begin{aligned}\hat{\theta}(t+1) &= P(t+1)\left[P(t+1)^{-1}\hat{\theta}(t) - \varphi(t+1)\varphi^T(t+1)\hat{\theta}(t)\right] + P(t+1)\varphi(t+1)y(t+1) \\ &= \hat{\theta}(t) + P(t+1)\varphi(t+1)\left[y(t+1) - \varphi^T(t+1)\hat{\theta}(t)\right]\end{aligned} \tag{39}$$

Estimation error at time (t + 1) is defined as Eq. (40), and model parameter vector at time (t + 1) is found as Eq. (41).

$$\varepsilon(t+1) = y(t+1) - \varphi^T(t+1)\hat{\theta}(t) \tag{40}$$

$$\hat{\theta}(t+1) = \hat{\theta}(t) + P(t+1)\varphi(t+1)\varepsilon(t+1) \tag{41}$$

Matrix inversion is applied to Eq. (33), and future value of covariance matrix is obtained as Eq. (42).

$$P(t+1) = P(t) - \frac{P(t)\varphi(t+1)\varphi^T(t+1)P(t)}{1 + \varphi^T(t+1)P(t)\varphi(t+1)} \tag{42}$$

RLS method consists of Eqs. (40–42), and the algorithm used is given below.

At time t + 1,

- Form ϕ(t+1) vector using new data.
- Calculate ε(t+1) using Eq. (40).
- Calculate P(t+1) using Eq. (42).
- Update model parameter vector ($\hat{\theta}$) (t+1) using Eq. (41).
- Return to step 1.

4. ST-GMV CONTROL

Cost function of STR, defined as the difference between the set point and the measured value for an input–output model, is as follows.

$$J(u,t) = \Xi\left\{(y(t+k) - r(t+k))^2\right\} \tag{43}$$

where y is the output variable, r is the set point, u is the manipulated variable (input), Ξ is the expectation, and k is the default time delay [16]. It is possible to minimize this cost function by choosing u(t) which can be defined as an appropriate control output at time t. At the next sampling time step (t + Δt), a new situation occurs between y and r, and u will need to get a new value. If the default time delay is smaller than the time delay to be encountered in the real system, then the control output will try to remove the noise components before being transmitted to the system with the time delay in the real system. This would result in large feedback gains, resulting in an unrealizable controller that would make the system unstable. On the other hand, if the default time delay is greater than the time delay of the real system, then the lowest possible noise value will not be obtained since the highest rate for manipulation is not provided [16]. Clarke and Gawthrop set out the ST-GMV method using the control cost of the STR of Aström and Wittenmark to remove the difficulties in the STR altogether [29]. ST-GMV control is a one-step ahead optimal control strategy. The cost function of this technique is expressed by the following equation.

$$J(u,t) = \Xi\left\{(y(t+k) - r(t+k))^2 + \lambda(u(t))^2\right\} \tag{44}$$

This type of controller design can internally stabilize the system, and the stability depends on the selected λ values. ST-GMV algorithm has a good set point tracking characteristic and has the ability to control nonminimum phase systems. If the default time delay is implemented within the generalized system, then the control signal compensates the pseudo output φ(t) accordingly and directs the feed-forward path.

Using Eq. (44), ST-GMV method relies on maintaining closed-loop stability by taking λ as small as possible while maintaining a minimum output change to stay reasonably close to the expectation. Cost function can be generally expressed as follows.

$$J(u,t) = \Xi\left\{\phi^2(t+k)\right\} \tag{45}$$

ST-GMV method uses a system pseudo output φ(t + k) given by the following equation to minimize the cost function expressed in general by Eq. (45).

$$\phi(t+k) = Py(t+k) + Qu(t) - Rr(t) \tag{46}$$

Here, r(t) is the set point, P, Q, and R are the transfer functions with backward shift operator (z-k). Pseudo output of the system includes a feed-forward feed term (Q) and filters (P, R) of output and the set point. ST-GMV algorithm uses the feed-forward polynomial Q to prevent output noise removal problem before signal transmission. φ(t + k) term of Eq. (46) can be expressed using Eq. (11) with the implementation of default time delay as follows.

$$\phi(t+k) = \frac{PB+QA}{A}u(t) + \frac{PC}{A}e(t+k) - Rr(t) \tag{47}$$

According to this equation, cost function to be minimized given by Eq. (45) will be the pseudo output variation. ST-GMV control algorithm divides the system into parts. For this, firstly, the error term is fragmented to include past, current, and future data.

$$\frac{PC}{A}e(t+k) = Ee(t+k) + z^{-k}\frac{G}{A}e(t) \tag{48}$$

Both sides of Eq. (48) are multiplied by A and rearranged as follows.

$$PC = AE + z^{-k}G \tag{49}$$

Polynomials are written as follows.

$$A = 1 + a_1z^{-1} + \ldots + a_{n_a}z^{-n_a} \tag{50}$$

$$B = b_0 + b_1z^{-1} + \ldots + b_{n_b}z^{-n_b} \tag{51}$$

$$C = 1 + c_1z^{-1} + \ldots + c_{n_c}z^{-n_c} \tag{52}$$

$$E = 1 + e_1z^{-1} + \ldots + e_{k-1}z^{-(k-1)} \tag{53}$$

$$G = g_0 + g_1z^{-1} + \ldots + g_{n_g}z^{-n_g} \tag{54}$$

$$P = 1 + p_1z^{-1} + \ldots + p_{n_p}z^{-n_p} \tag{55}$$

AE term of Eq. (49) is expressed in terms of ARMAX model including offset as follows.

$$Ay(t+k) = Bu(t) + Ce(t+k) + d \tag{56}$$

If both sides of Eq. (54) is multiplied by E and written in Eq. (49), Eq. (55) is obtained.

$$PCy(t+k) = BEu(t) + CEe(t+k) + Ed + Gy(t) \tag{57}$$

Both sides of Eq. (55) are divided to C, and Eq. (56) is obtained.

$$Py(t+k)=\frac{BE}{C}u(t)+Ee(t+k)+\frac{Ed}{C}+\frac{G}{C}y(t) \tag{58}$$

Eq. (56) is combined with Eq. (46) and rearranged as follows.

$$\phi(t+k)=\frac{1}{C}[(BE+QC)u(t)+Gy(t)-CRr(t)+Ed]+Ee(t+k) \tag{59}$$

Eq. (57) is the sum of current and future terms. Current terms can be expressed as follows and represents the best φ(t + k) estimation made by using the data until time t.

$$\phi(t+k|_t)=\frac{1}{C}[(BE+QC)u(t)+Gy(t)-CRr(t)+Ed] \tag{60}$$

The second term is the estimation error caused by the noise source, e (t + 1), e (t + 2), ..., e (t + k). The second term cannot be removed using control signal u(t) as mentioned before.

$$Ee(t+k)=\phi(t+k)+\phi(t+k|_t) \tag{61}$$

So, J is minimized by equalizing Eq. (58) to zero.

$$\phi(t+k|_t)=0\rightarrow\frac{[(BE+QC)u(t)+Gy(t)-CRr(t)+Ed]}{C}=0 \tag{62}$$

Eq. (60) is rearranged using following definitions.

$$F=BE+QC \tag{63}$$

$$H=CR \tag{64}$$

ST-GMV control law can be expressed as follows.

$$Fu(t)+Gy(t)-Hr(t)+Ed=0 \tag{65}$$

Calculation of input variable using ST-GMV control law is made using following equation.

$$u(t)=\frac{Hr(t)-Gy(t)-Ed}{F} \tag{66}$$

Application of ST-GMV algorithm consists of following steps [16, 29]:

- Apply a PRBS to the system as a forcing function and obtain the plant output.
- Estimate F, G, H from Eq. (63), implementing the RLS algorithm.
- Employ Eq. (64) to evaluate the control signal.
- Apply the control signal.

- The system output is obtained.
- Return to step 1.

5. MATERIAL AND METHODS

5.1. Microorganism, inoculum preculture, and growth medium

S. cerevisiae NRRL Y-567 was obtained from NRRL-Agricultural Research Service Culture Collection. Preculture and growth media consist of 2% glucose, 0.6% yeast extract, 0.3% K_2HPO_4, 0.335% $(NH_4)_2SO_4$, 0.376% NaH_2PO_4, 0.052% $MgSO_4 \cdot 7H_2O$, and 0.0017% $CaCl_2 \cdot 4H_2O$ which were sterilized by autoclaving under 1.2 atm at 121°C for 20 min. Microorganisms were incubated for 8 hours at 32°C at 120 rpm, and inoculum ratio of 1:10 was used for scale enlargement.

5.2. Experimental system

In order to observe the change of DO and pH over time during baker's yeast production using a 2-L working volume of laboratory-scale bioreactor which was operated continuously, the input–output data were recorded and the ARMAX model parameters were determined by RLS method written in MATLAB. Experimental system is given in Figure 1. Most suitable parametric model was estimated using different values of α (covariance matrix), λ (forgetting factor), and order of parametric model. During experiments, DO and pH were measured with a WTW Oxi 340 with polarographic DO sensor and WTW pH340i pH meter, respectively. The DO and pH probes immersed in the bioreactor measure the online DO and pH values of the growth medium, and these values were converted to the electrical signal with DO, and pH meters reach the I/O card in the computer via the carrier interface modules. The signals arriving to card are interpreted by the algorithm written in Visual Basic in the ADVANTECH VISIDAQ package program and was sent to the system online. Operating conditions of the bioreactor was given in Table 1.

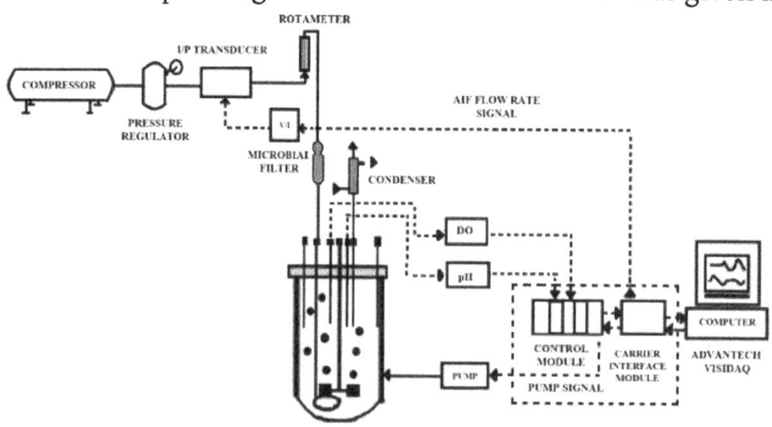

Figure 1. Experimental system.

Table 1. Operating conditions of the bioreactor.

Temperature (°C)	Air flow rate (vvm)	Cooling water flow rate (mL/min)	Cooling water temperature (°C)	Agitation speed (rpm/min)
32	1	55	21	600

6. RESULTS AND DISCUSSION

6.1. Dynamic analysis

6.1.1. Step input given to air flow rate

In baker's yeast production, the oxygen concentration must not fall below the critical value (0.7 mg/L); therefore, firstly manipulated variable must be selected to control the DO [12]. For that purpose, the air flow rate was chosen as the manipulated variable for the control of the DO. However, the effect of air flow rate on the DO in the bioreactor was investigated in order to observe effective control of this variable. For this purpose, while the system was in steady state at 1 mg/L DO for 0.5 L/min air flow rate, the positive step input was given to air flow rate as 3.4 L/min, and the change in DO over time was observed. In this case, DO was increased to 3 mg/L as can be seen from Figure 2.

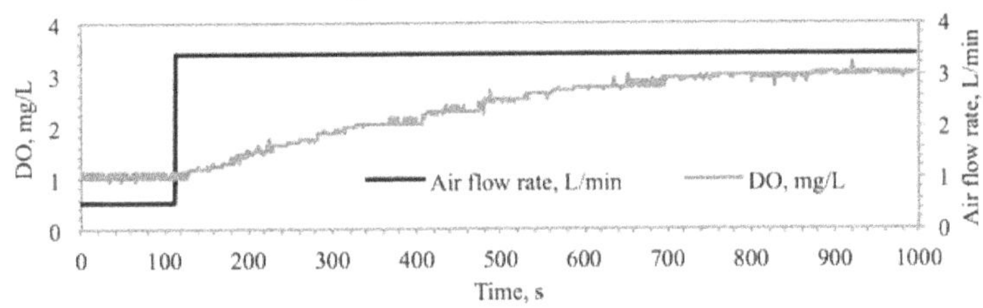

Figure 2. Positive step input given to air flow rate from 0.5 to 3.4 L/min.

6.1.2. Step input given to base flow rate

During the yeast growth, due to the degradation process of glucose in aerobic conditions to save the chemical energy in ATP molecules cause an increase in the concentration of H+ ions resulted with pH decrease. Decreasing the pH of the medium affects not only the cell division, but also the cleavage rate and the production of many products from yeast and the activity of enzymes. For this purpose, it is necessary to determine the manipulated variable in order to achieve pH control. Therefore, the base flow rate was selected as the manipulated variable for pH control. However, in order to observe effective control of this variable, the effect of the base flow rate on the pH in the bioreactor was investigated. The bioreactor was settled at pH 3.90 under the specified operating conditions, and then, microorganism was fed to the bioreactor. Positive step input was given to base flow rate from 0.26 to 1.41 mL/min with 0.05 M NaOH solution. The acid

(H2SO4) flow rate was kept constant at 0.22 mL/min. The change in pH value over time under such an effect is shown in Figure 3.

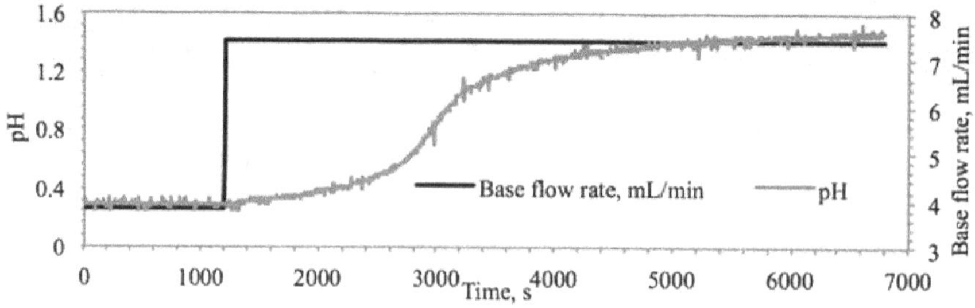

Figure 3. Positive step input given to base flow rate from 0.26 to 1.41 mL/min.

6.2. System identification results

6.2.1. Determination of model parameters for controlled variable of DO

In order to find the most appropriate ARMAX model using data of the manipulated variable air flow rate and the controlled variable DO obtained from dynamic analysis, the various forgetting factors (0.96–1), the initial value of the covariance matrix (1100,1000,10,000), and the order of the model (na = 2 nb = 1, na = 2 nb = 2, na = 3 nb = 1, na = 3 nb = 2) were run with the RLS algorithm, and integrated square error (ISE) was used for comparison. The models with the lowest and highest ISE values were used in the ST-GMV control algorithm to demonstrate the effect of the model structure on the control performance which will be explained in the next section. The compared values in terms of estimation performance are given in Table 2.

Table 2. Estimated performance criteria as a result of positive step from 0.5 to 3.4 L/min for air flow rate and ISE values.

Estimation performance criteria na = 2, nb = 1		na = 2, nb = 2	Order of the model na = 3, nb1	na = 3, nb = 2	
λ	P	ISE	ISE	ISE	ISE
0.96	1	1.8720	1.2037e + 36	1.5937	1.9718e + 61
	100	1.6297	3.8648e + 05	1.4280	2.2419e + 34
	1000	1.6110	9.9163e + 23	1.4124	8.5341e + 04
	10,000	1.6080	7.4019e + 25	1.4089	1.7541e + 25
0.97	1	2.0045	1.9599	1.7180	1.6972
	100	1.7544	1.7441	1.5485	1.5402
	1000	1.7337	1.7304	1.5306	1.5278
	10,000	1.7308	1.7295	1.5267	1.5254

Estimation performance criteria na = 2, nb = 1		Order of the model			
		na = 2, nb = 2	na = 3, nb1	na = 3, nb = 2	
λ	P	ISE	ISE	ISE	ISE
0.98	1	2.1602	2.1154	1.8570	1.8365
	100	1.9039	1.8920	1.6852	1.6752
	1000	1.8803	1.8772	1.6641	1.6610
	10,000	1.8776	1.8765	1.6597	1.6585
0.99	1	2.3344	2.2891	2.0055	1.9853
	100	2.0702	2.0559	1.8307	1.8186
	1000	2.0427	2.0395	1.8052	1.8018
	10,000	2.0404	2.0392	1.8002	1.7991
1	1	2.4483	2.4060	2.1171	2.0993
	100	2.1663	2.1490	1.9384	1.9243
	1000	2.1336	2.1304	1.9077	1.9040
	10,000	2.1317	2.1305	1.9020	1.9010

The estimation performance criterion λ value shows a decrease in ISE values between 0.96 and 0.97, but an increase in ISE values is observed after 0.97 of λ. The initial values of the covariance matrix between 1 and 10,000 resulted in a decrease in ISE values (Table 2).

Consequently, the most suitable ARMAX model was obtained with the order na = 3 nb = 2, the forgetting factor of 0.97, and the initial value of the covariance matrix of 1000. In the least successful ARMAX model case, the order was na = 2, nb = 1, the forgetting factor was 1, and the initial value of the covariance matrix was 1. At the end of this approach, it was decided that the type of ARMAX model to be developed for GMV control was given in Eq. (67).

$$y(t) - 0.1928y(t-1) + 0.0338y(t-2) - 0.0168y(t-3) = 0.1853u(t-1) + 0.5794u(t-2) + e(t-1) \quad (67)$$

As a conclusion, the most suitable ARMAX model was obtained with the model order as na = 3, nb = 2, forgetting factor of 0.97, and initial value of the covariance matrix as 1000, and the RLS estimation of DO in the growth media by alterations with the air flow rate is given in Figure 4.

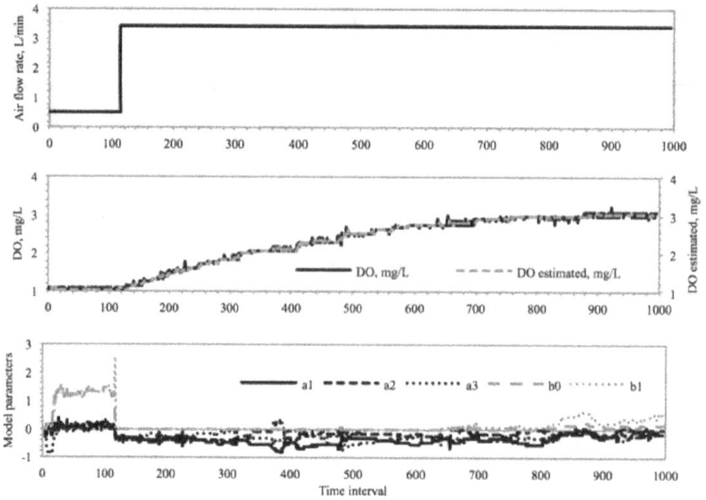

Figure 4. Experimental system identification estimation of DO (na = 3 nb = 2, λ = 0.97, P = 1000).

6.2.2. Determination of model parameters for controlled variable of pH

In order to find the most appropriate ARMAX model using data of the manipulated variable base flow rate and the controlled variable pH obtained from dynamic analysis, the various forgetting factors (0.96–1), the initial value of the covariance matrix (1, 100, 1000, 10,000), and the order of the model (na = 2, nb = 1; na = 2, nb = 2; na = 3, nb = 1, na = 3, nb = 2) were run with the RLS algorithm and ISE values of the prediction were used for comparison. The models with the lowest and highest ISE values were used in the ST-GMV control algorithm, which will be explained in the next section. The compared values in terms of estimation performance are given in Table 3.

Table 3. Estimated performance criteria as a result of positive step from 0.26 to 1.412 mL/min for base flow rate and ISE values.

Estimation performance criteria na = 2, nb = 1		Order of the model			
		na = 2, nb = 2	na = 3, nb = 1	na = 3, nb = 2	
λ	P	ISE	ISE	ISE	ISE
0.96	1	1.2954	1.2948	1.1136	1.2699
	100	1.2394	1.2278	1.0715	1.0658
	1000	1.2096	1.2029	1.0521	1.0479
	10,000	1.1952	1.1943	1.0410	1.0383

Estimation performance criteria		Order of the model			
na = 2, nb = 1		na = 2, nb = 2	na = 3, nb = 1	na = 3, nb = 2	
0.97	1	1.3644	1.3631	1.1858	1.1851
	100	1.3160	1.3062	1.1474	1.1423
	1000	1.2815	1.2731	1.1260	1.1204
	10,000	1.2620	1.2595	1.1117	1.1091
0.98	1	1.4439	1.4428	1.2706	1.2699
	100	1.4020	1.3940	1.2351	1.2312
	1000	1.3670	1.3558	1.2149	1.2077
	10,000	1.3405	1.3373	1.1961	1.1929
0.99	1	1.5302	1.5292	1.3676	1.3669
	100	1.4910	1.4852	1.3326	1.3298
	1000	1.4602	1.4468	1.3159	1.3079
	10,000	1.4257	1.4213	1.2930	1.2887
1	1	1.6207	1.6202	1.4828	1.4820
	100	1.5827	1.5785	1.4484	1.4460
	1000	1.5582	1.5441	1.4355	1.4276
	10,000	1.5171	1.5108	1.4104	1.4044

ISE values were raised with the increase of the estimation performance criterion λ. For the same forgetting factor values, the ISE values decrease with the increase of covariance matrix initial value. The lowest ISE was obtained when the initial value of the covariance matrix was 10,000. As the order of polynomial A increases, the ISE values decrease, and as the order of polynomial B increases, a significant change in ISE values cannot be observed.

$$y(t) - 0.2235y(t-1) + 0.1077y(t-2) - 0.0737y(t-3) = 1.2723u(t-1) + 1.921u(t-2) + e(t-1) \quad (68)$$

As a result, the most suitable ARMAX model was obtained with the order na = 3 nb = 2, the forgetting factor of 0.96, and the initial value of the covariance matrix of 10,000. In the least successful ARMAX model case, the order was na = 2, nb = 1, the forgetting factor was 1, and the initial value of covariance matrix was 1. At the end of this screening, it was decided that the suitable ARMAX model structure in order to develop the GMV control algorithm for pH control with by manipulating the base flow rate is given in Eq. (68).

As a conclusion, the most suitable ARMAX model was obtained with the model order of na = 3, nb = 2, forgetting factor as 0.96, and initial value of the covariance matrix of 10,000, and the RLS estimation is given in Figure 5.

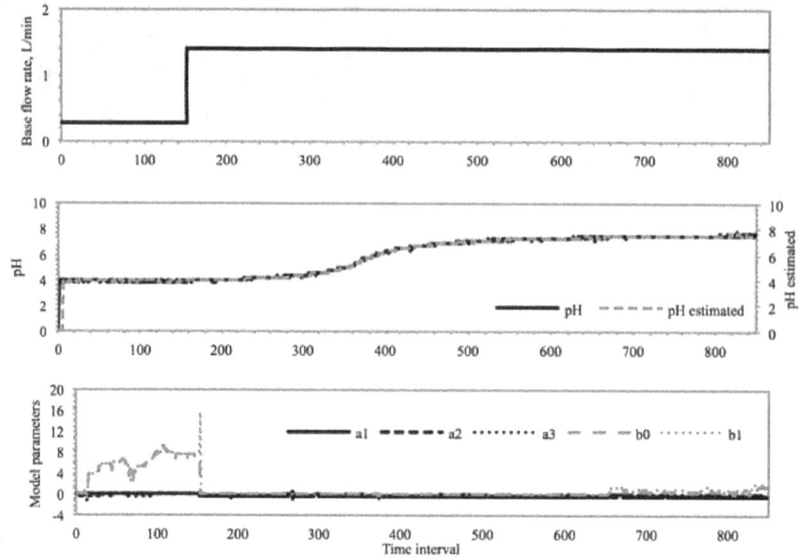

Figure 5. Experimental system identification estimation of pH (na = 3 nb = 2, λ = 0.96, P = 10,000).

6.3. ST-GMV control applications of baker's yeast production

The suitable and unsuitable ARMAX models of the yeast production process expressed the relationship between the controlled variables of DO and pH, with the manipulated variables of air flow rate and base flow rate in system identification results section. After this step, ST-GMV control performances were evaluated with the suitable and unsuitable ARMAX models determined for each controlled variable in the case of constant set point trajectory for various noise levels. The control performance criterion was selected as ISE and values were evaluated for the ST-GMV control simulations of both DO and pH control cases. By this way, how much the system identification step, including the determination of model structure and model parameter settings, has affected the success of process control is demonstrated by using control simulations.

6.3.1. DO control

In the baker's yeast production process, in which DO was controlled variable and the air flow rate was selected as the manipulated variable, the most suitable and unsuitable ARMAX models obtained from system identification have been used in ST-GMV control algorithm. In the case of positive step input from 0.5 to 3.4 L/min for the air flow rate, the order of the most suitable model was na = 3 nb = 2, λ = 0.97, and P = 1000. By the same way, the model that does not identify the system (largest ISE value) was found as na = 2 nb = 1, λ = 1, and P = 1. When the both suitable and unsuitable obtained models were used in the ST-GMV control algorithm with the controller parameters of P = 1, Q = 0.9975, R = 2.0885 in the presence of two different noises. It was observed that the suitable model with calculated ISE values of 50.14 and 52.37 was definitely able to identify the system as expected and provided a good control (Figure 6) in contrast to unsuitable model with ISE values of 502 and 609.56, respectively (Figure 7).

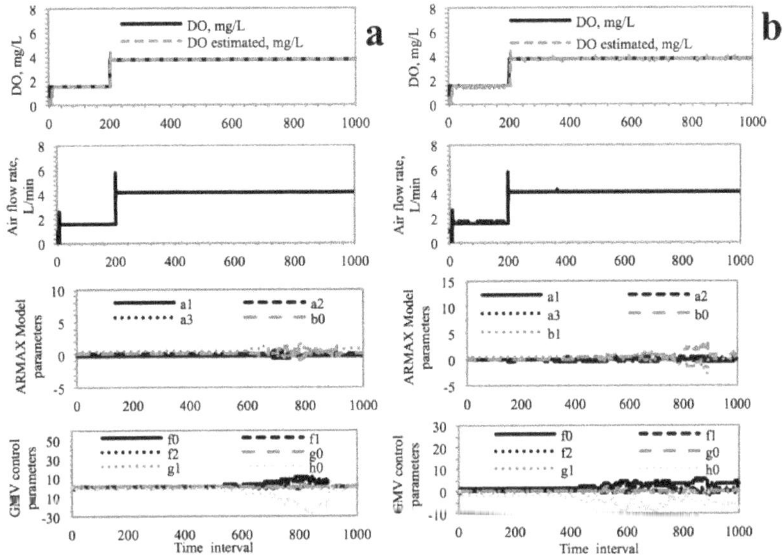

Figure 6. ST-GMV control simulation results with the most appropriate ARMAX model at the desired constant DO set point (a) having 0.005 noises, (b) having 0.05 noises.

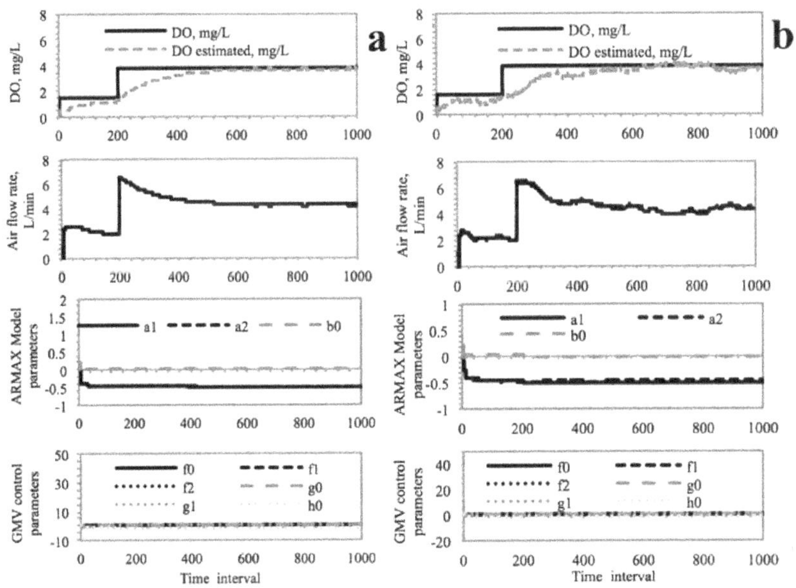

Figure 7. ST-GMV control simulation results with unsuitable ARMAX model at the desired DO set point (a) having 0.005 noises, (b) having 0.05 noises.

6.3.2. pH Control

In the baker's yeast production process, in which the pH value was controlled variable and the base flow rate was selected as the manipulated variable, the most suitable and unsuitable ARMAX models obtained from system identification have been used in the ST-GMV control algorithm. In the case of positive step input from 0.26 to 1.41 mL/min for the base flow rate, the order of the most suitable model was na = 3, nb = 2, λ = 0.96, and P = 10,000. By the same way, the model that does not identify the system (largest ISE value) was found as na = 2 nb = 1, λ = 1, and P = 1. When the both suitable and unsuitable obtained models were used in the ST-GMV control algorithm with the controller parameters of P = 1, Q = 0.9375, R = 1.1885 in the presence of two different noises. It was observed that the suitable model with calculated ISE values of 110.2 and 113.51 was definitely able to identify the system as expected and provided a good control (Figure 8) in contrast to unsuitable model with ISE values of 245.69 and 213.69, respectively (Figure 9). ST-GMV control simulation results are summarized in Table 4.

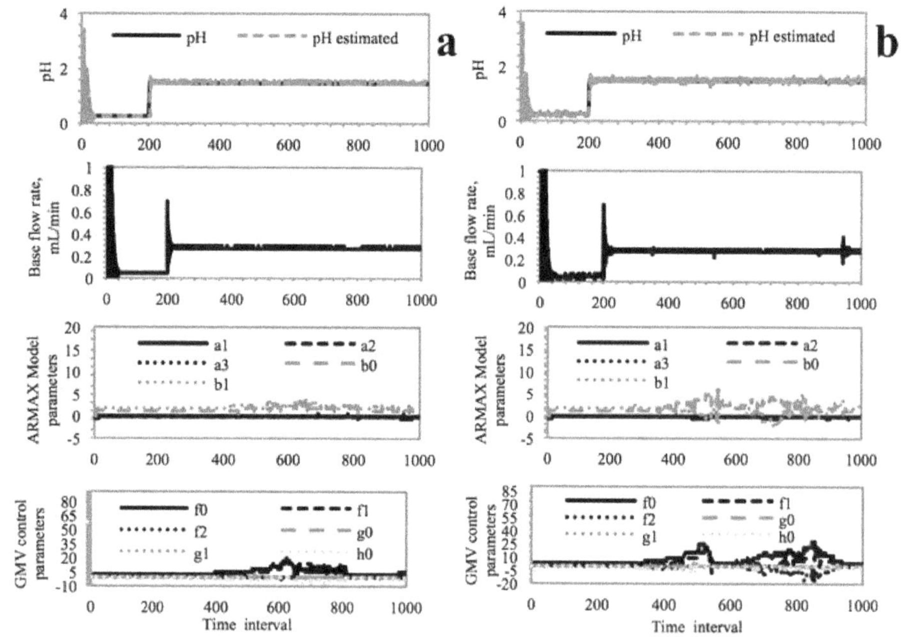

Figure 8. ST-GMV control simulation results with the most appropriate ARMAX model at the desired constant pH set point (a) having 0.005 noises, (b) having 0.05 noises.

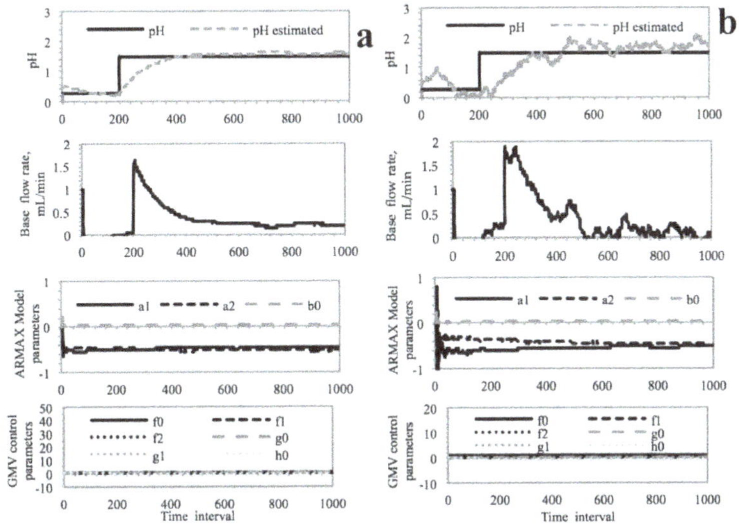

Figure 9. ST-GMV control simulation results with unsuitable ARMAX model at the desired pH set point (a) having 0.005 noises, (b) having 0.05 noises.

Table 4. Theoretical ST-GMV control performance summary with suitable and unsuitable ARMAX models.

ontrolled variable	Noises	Control with suitable model ISE	Control with unsuitable model ISE
DO	e = 0.005	50.1361	502.0056
	e = 0.05	52.3745	609.5629
pH	e = 0.005	110.2026	245.6958
	e = 0.05	113.5143	213.6950

7. CONCISE SUMMARY

Comprehending the dynamic behavior of biotechnological processes, where living cells are employed as biocatalysts, is now one of the most difficult problems. This is because numerous biochemical events occur simultaneously. The operation of the process in batch mode will be challenging due to the presence of time-varying parameters. In this scenario, the estimating approach is becoming increasingly significant in accurately representing the actual behavior of the process through mathematical models. There are multiple ways and approaches available for this aim. The subsequent crucial stage is choosing the most suitable approach for system identification. It also impacts the accuracy of estimating the dynamic behavior of the operation. The RLS algorithm was used to easily carry out system identification studies in the production process of baker's yeast. These studies successfully identified variations in dissolved oxygen (DO) and pH, which were manipulated by adjusting the air flow rate and the base flow rate, respectively. The production process was conducted in a batch operational mode with aerobic conditions, using the S. cerevisiae microorganism. The prediction error, as measured by the Integrated Squared Error (ISE), indicates that the estimation performance was satisfactory. Choosing the appropriate model structure is essential for accurately representing the behavior of a process. The study determined that the order of the model for both polynomial-type ARMAX model structures is na = 3 and nb = 2, based on an analysis of various model orders. As the degree of the polynomial A increases, the discrepancy between the observed value and the estimated value diminishes, which is advantageous. Nevertheless, the rise in B polynomial does not exhibit any noteworthy alteration. The forgetting factor was determined to be 0.96 and 0.97. The initial value of the covariance matrix was not as influential as the forgetting factor. A value of 1000 was deemed suitable for all experiments. Theoretical ST-GMV control of dissolved oxygen (DO) and pH was effectively executed using the most appropriate ARMAX models derived from system identification. By increasing the noise level in the theoretical ST-GMV control, it becomes possible to accomplish successful control under the constant set point condition using the given models. Furthermore, the effectiveness of a controller that utilizes inappropriate models diminishes as the levels of noise increase. In conclusion, effective control can only be achieved through a robust system identification process.

AT A GLANCE

Several researchers have studied the fermentation process of Saccharomyces cerevisiae in order to improve product quality and production while reducing costs. To obtain desirable products, it is necessary to adjust operating parameters such as pH, dissolved oxygen (DO) concentration, temperature, substrate type and concentration, agitation speed, and air flow rate. At this juncture, system identification and advanced control approaches arise to offer solutions. The pH and dissolved oxygen (DO) levels of the growing medium were analyzed dynamically under aerobic conditions in a batch bioreactor. This was done by applying step inputs to the base flow rate and square wave inputs to the air flow rate. The link between the controllable and manipulated variables in baker's yeast production was determined using system identification, based on the input-output data of the process and a linear Auto Regressive Moving Average with eXogenous (ARMAX)-type model. The model parameters were estimated utilizing the recursive least squares (RLS) technique. Thc optimal parametric model was identified by estimations using various beginning values of the covariance matrix, forgetting factor, and order of the ARMAX model. The ARMAX model was used to implement self-tuning generalized minimum variance (ST-GMV) control for pH and DO regulation. The performance requirements for modeling and control studies were determined based on the values of integrated square error (ISE).

REFERENCES

1. Akay B, Ertunç S, Kahvecioğlu A, Hapoğlu H, Alpbaz M. Adaptive control of S. cerevisiae production. Food and Bioproducts Processing. 2002;80(1):28-38. DOI: 10.1205/096030802753479089
2. Hitit ZY, Boyacioglu H, Ozyurt B, Ertunc S, Hapoglu H, Akay B. Self-tuning GMV control of glucose concentration in fed-batch baker's yeast production. Applied Biochemistry and Biotechnology. 2014;172(8):3761-3775. DOI: 10.1007/s12010-014-0794-5
3. Boyacıoğlu H, Ertunç S, Hapoğlu H. Modelling of baker's yeast production. International Journal of Secondary Metabolite (IJSM). 2017;4(1):10-17. DOI: 10.21448/ijsm.252053
4. Huang B, Guo J, Yi B, Yu X, Sun L, Chen W. Heterologous production of secondary metabolites as pharmaceuticals in Saccharomyces cerevisiae. Biotechnology Letters. 2008;30(7):1121-1137. DOI: 10.1007/s10529-008-9663-z
5. Cinar A, Parulekarb SJ, Undey C, Birol G. Batch Fermentation: Modelling: Monitoring, and Control. New York: CRC press; 2003. DOI: 10.1201/9780203911358.ch7
6. Bailey JE, Ollis DF. Biochem. Eng. Fund: New York: Mc Graw Hill; 1986
7. Alpbaz M, Bursali N, Ertunç S, Akay B. Application of a statistical technique to the production of Saccharomyces cerevisiae (baker's yeast). Biotechnology and Applied Biochemistry. 1997;26(2):91-96. DOI: 10.1111/j.1470-8744.1997.tb00452.x
8. Bursali N, Akay B, Ertunc S, Hapoglu H, Alpbaz M. New tuning method for generalized predictive control of the production of S. cerevisiae. Food and Bioproducts Processing. 2001;79(1):27-34. DOI: 10.1205/09603080151123335
9. Ertunc S, Akay B, Hapoglu H, Alpbaz M. A new computer algorithm approach to identification of continuous-time batch bioreactor model parameters. Computational Science–ICCS 2005. 2005;154-147. DOI: 10.1007/11428831_19
10. Bursali N, Ertunc S, Akay B, Pamuk V, Hapoğlu H, Alpbaz M. The effect of the tuning parameters on the performance of the parametric and nonparametric model based control methods for growth medium temperature of baker's yeast production. Food and Bioproducts Processing. 2001;79(4):242-249. DOI: 10.1205/096030801753252315
11. Akay B, Ertunc S, Boyacioglu H, Hapoglu H, Alpbaz M. Discrete system identification and self-tuning control of dissolved oxygen concentration in a stirred reactor. Korean Journal of Chemical Engineering. 2011;28(3):837-847. DOI: 10.1007/s11814-010-0450-x
12. Ertunc S, Akay B, Boyacioglu H, Hapoglu H. Self-tuning control of dissolved oxygen concentration in a batch bioreactor. Food and Bioproducts Processing. 2009;87(1):46-55. DOI: 10.1016/j.fbp.2008.04.003
13. Dorf RC, Bishop RH. Modern Control Systems. New Jersey: Pearson; 2011
14. Warwick K, Rees D. Industrial digital control systems (no. 37). In: IET; 1988
15. Soderstrom T, Stoica P. System Identification. Hertfordshire: Prentice hall international; 1989
16. Ertunc S, Akay B, Bursali N, Hapoğlu H, Alpbaz M. Generalized minimum variance control of growth medium temperature of baker's yeast production. Food and Bioproducts Processing. 2003;81(4):327-335. DOI: 10.1205/096030803322756411
17. Camcıoğlu Ş, Özyurt B, Zeybek Z, Hapoğlu H. Experimental application of one step ahead advanced pH control to water-based paint wastewater treatment. Journal of the Faculty of Engineering and Architecture of Gazi University. 2016;31(3):655-664. DOI: 10.17341/gummfd.52934

18. Camcıoğlu Ş, Ozyurt B, Hapoglu H. pH control of paper wastewater treated with electrocoagulation method. Anadolu University Journal of Science and Technology–A Applied Sciences and Engineering. 2015;16(2):107-115

19. Åström KJ, Wittenmark B. Adaptive Control:Courier Corporation; 2013.

20. Clarke DW, Gawtthrop PJ. Self-tuning control. IEEE Proc. 1979;126(6):633-640

21. Olmstead DR. Computer Control of Fermentation Processes. Florida: CRC Press; 1989

22. Shuler ML, Kargi F. Bioprocess Engineering: Basic Concepts. New Jersey: Prentice Hall; 1992

23. Sales KR, Billings SA. Self-tuning control of non-linear ARMAX models. International Journal of Control. 1990;51(4):753-769. DOI: 10.1080/00207179008934096

24. Perry J-G. Introduction to Food Biotechnology. Florida: CRC Press; 2002

25. Furukawa K, Heinzle E, Dunn IJ. Influence of oxygen on the growth of Saccharomyces cerevisiae in continuous culture. Biotechnology and Bioengineering. 1983;25(10):2293-2317. DOI: 10.1002/bit.260251003

26. Walker G. Yeast Physiology and Biotechnology. Chichester: Wiley; 1998

27. Wellstead PEE, Zarrop MB. Self-Tuning Systems: Control and Signal Processing. New York: John Wiley & Sons Inc; 1991

28. Hapoğlu H, Balas CE, Wardle AP. Self-tuning and conventional control of an industrial scale packed distillation column. Turkish Journal of Engineering and Environmental Sciences. 1998;22(1):1-8

29. Karagöz AR, Hapoğlu H, Alpbaz M. Generalized minimum variance control of optimal temperature profiles in a polystyrene polymerization reactor. Chemical Engineering and Processing: Process Intensification. 2000;39(3):253-262. DOI: 10.1016/S0255-2701(99)00104-X

CHAPTER-8

THE MODE OF ACTION OF PROBIOTIC YEAST AND ITS IMPACT ON RUMINANT NUTRITION

1. INTRODUCTION: AN OVERVIEW

Ruminants possess the ability to consume various kinds of feed, which is broken down by microbial biomass. This process leads to enhanced metabolism and, thus, has a direct effect on the productivity of dairy animals. The microbiota in the gastrointestinal tract (GIT) significantly influences the productivity, health, and overall welfare of dairy animals [1–3]. The diversity and activity of ruminal gastrointestinal tract (GIT) microorganisms play a crucial role in the process of meal digestion. The manner in which nutrients are digested in the gastrointestinal tract (GIT) of ruminants has a significant influence on their growth, health, and production [4]. The gastrointestinal tract (GIT) contains a diverse range of microorganisms that play a crucial role in maintaining optimal health, physiology, and productivity of animals [1]. The current gut microbiota controls food safety by eliminating infections, engaging with organisms, and competing for resources in the gastrointestinal tract [5]. The microbiota in the gastrointestinal tract helps to stimulate the immune system, which serves as a protective barrier against pathogenic diseases. Furthermore, it inhibits the growth of harmful and disease-causing bacteria in the gut [6]. Various treatments have been employed to improve the microbiota of the gastrointestinal tract, leading to subsequent impacts on the production capacity and growth efficiency of dairy animals. Currently, the utilization of probiotics to promote the microbiota of the gastrointestinal tract has emerged as a beneficial and cost-effective approach to improve the health and productivity of animals. A probiotic is a living microbe that positively affects the host by enhancing the microbial flora in its intestinal [7]. Several microorganisms have been approved as probiotics for use in the diet of ruminants to enhance nutrient utilization and animal performance [8]. Young calves, hens, and pigs show improved outcomes when given bacterial probiotics, but adult ruminants benefit from yeast/fungal probiotics [9]. In recent decades, Saccharomyces cerevisiae has been employed as a preventative supplement for livestock to combat diarrhea and other gastrointestinal issues [10]. In addition, they provide advantages in productivity, alleviate digestive issues, and promote improved animal health in a cost-efficient manner [11]. Supplementing the diet with yeast culture has a beneficial impact on the amount of food consumed by ruminants, leading to improved growth [12] and increased efficiency in production [13]. Additionally, it has a beneficial impact on the concentration of fat in milk [13–15] and the level of milk urea nitrogen [16]. Furthermore, it reduces the generation of lactate [17], enhances the growth of beneficial bacteria [13], minimizes rumen acidosis [18], improves the breakdown of hemicelluloses, and increases the digestibility of essential nutrients such as NDF and ADF [19–21]. One additional benefit of using S. cerevisiae is that the cost-benefit ratio of supplementing dairy cattle with S. cerevisiae is 4:1 [22]. This chapter elucidates the impact of yeast on ruminant animals and presents recommendations for the creation of a probiotic yeast tailored to certain animal breeds.

2. PROBIOTIC YEAST

Yeasts are eukaryotic microorganisms that differ from bacteria in terms of their structure and activity [23]. Yeasts are organisms that can survive with or without oxygen and vary in terms of their habitat, morphology, reproductive behavior, nutrients they consume, and their strong resistance to many antibiotics, including sulfamides and other antibacterial compounds [24]. The yeast cells possess an inherent and genetically determined capacity to resist external factors. This resistance is immutable and cannot be transferred to other microbial species. The yeast cell is larger than bacteria, with dimensions of 5 × 10 μm compared to 0.5 × 5 μm. The investigation of antagonistic yeast to hinder bacterial pathogenicity during early development primarily involves the following steps: (1) competing for nutrients, (2) inducing pH changes in the medium, (3) generating high concentrations of ethanol, (4) secreting antibacterial compounds and releasing antimicrobial compounds (such as toxins or "mycocins"). Nevertheless, the efficacy of probiotic organisms is considered to be peculiar to certain populations due to differences in the microbial ecology of the gastrointestinal tract, feeding patterns, and the intricate interplay between the host animal and the bacteria. Since the majority of probiotic yeast strains available in the market are of Western or European origin, it is crucial to investigate new probiotic strains that are local. Yeast cells synthesize numerous crucial fermentation products, as well as various essential minerals and enzymes, which render them valuable and extremely nutritious as a feed supplement for ruminants [25–27]. Additionally, it offers enhanced productivity, decreased digestive issues, and increased health in a cost-efficient manner.

3. INVESTIGATING THE RUMINAL GUT MICROBIOLOGY TO IDENTIFY A NEW PROBIOTIC STRAIN WITH SPECIFIC TARGET PROPERTIES

Molecular approaches have revolutionized the investigation of the rumen ecology [10]. Gaining a more comprehensive knowledge of rumen microbiology is a crucial prerequisite for the identification and cultivation of a novel yeast strain that can influence certain functional microorganisms. The stomach of ruminants is composed of four compartments: reticulum, rumen, omasum, and abomasum [28]. The rumen is a chamber devoid of oxygen that contains a wide range of microorganisms, including as bacteria, archaea, fungus, and single-celled ciliated protozoa (Figure 1) [29].

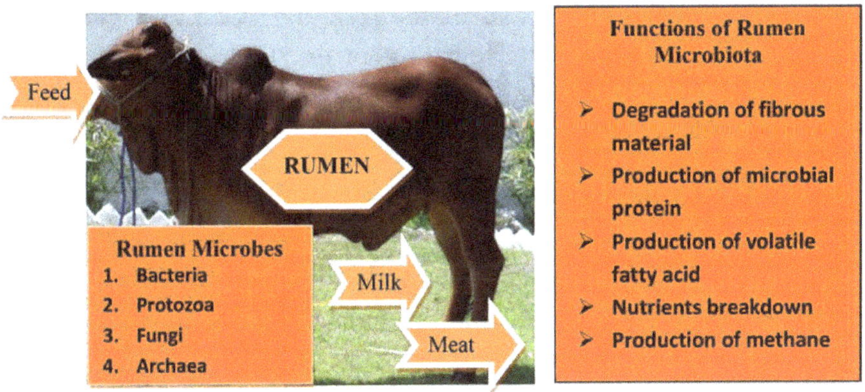

Figure 1. Estimated rumen microbial ecosystems.

This microbial ecosystem has been utilized to enhance the process of feed digestion. Bacteria are abundant microorganisms found in the rumen [30]. The majority of bacteria are linked to feed, whereas some are independent, adhering to mucous membranes and forming associations with fungus and protozoa. The predominant morphologies of rumen bacteria include cocci, rods, spirochetes, budding forms, and filamentous structures. The size of rumen bacteria ranges from 1 to 2 μm. Most of the bacterial species found in the rumen are Gram-negative. The composition of this microbial community is shaped by various factors, including as the host species, age, seasons [31], feed type, geographical location, and if the animal has undergone any therapy [32]. The equilibrium in rumen microbial flora is essential for efficient feed utilization and has the potential to enhance productivity [33]. The composition of the rumen microbial community is directly influenced by the specific type of feed being consumed [29]. Ruminants have the ability to consume various types of feed supplies that are broken down by microbial biomass, resulting in improved metabolism [34]. This ultimately affects the efficiency of dairy animals. Beneficial microbial supplements can be used to manage the feed microbial flora. Several studies on ruminant species have focused on the management and modulation of ruminal fermentation to enhance animal performance. Through this line of research, we will employ a technique to modify the rumen fermentation in order to enhance the utilization of nutrients and increase the productivity of animals. The prohibition of antibiotics as animal growth promoters in the European Union in 2006 has led to a higher need for alternative feed additives that can be utilized to control ruminal fermentation and enhance animal production [31, 35, 36]. The manipulation of the rumen population to enhance nutrition use. Metabolism can be manipulated by controlling the nutrition, administering antibiotics, and introducing certain microbial inoculants. The diet-shift effect, such as a high-forage diet, raises the rumen pH, leading to enhanced stability and survival of cellulolytic and hemicellulolytic bacteria and protozoa. Conversely, a high-concentrate diet that reduces the pH of the rumen leads to a drop in cellulolytic and hemicellulolytic bacteria, an increase in amylolytic bacteria, and a decrease in rumen protozoa. Microbial inoculants have the potential to improve the stability and vitality of the microbial community in the rumen and hindgut.

4. ROLE OF PROBIOTIC YEAST IN RUMINANT NUTRITION

The balance in rumen microbial flora plays a crucial role in feed utilization and could result in better animal productivity [37]. Several hypotheses concerning the mode of action of probiotic yeast in animal nutrition have been proposed; however, a majority of them emphasize positive effects by modifying rumen microbial population. The first and most widely supported mode of action is that the yeast stimulates the growth of bacteria (cellulolytic, amylolytic, and proteolytic) and protozoa [38, 39]. The rumen dissolved oxygen (O2) can be measured in situ [40]. Loesche [41] found that a majority of rumen microbial flora are highly sensitive to O2. Probiotic yeasts remove oxygen from rumen and provide a better anaerobic environment for bacterial growth [42]. Sixteen liters of oxygen can enter inside rumen daily, by the mean of feeding, rumination, and salivation [43]. Inside rumen, yeast cells use oxygen for their metabolic process. Freshly ingested feed particles have sugars and small oligosaccharides. Probiotic yeast metabolizes these small particles and produces peptides, polypeptides, and amino acids. This respiratory activity of probiotic yeast lowers the oxidation-reduction potential inside rumen [44]. A negative change in the redox potential (-20 mV) has been observed in rumen with probiotic yeast addition [36]. This change gives a better anaerobic condition inside rumen [33]. Aforementioned environment helps in the protection of rumen bacteria from damage by oxygen and stimulation of growth of cellulose degrading bacteria [45, 46]. These conditions will also be helpful in the cellulose degrading process (cellulose digestion). Respiratory-deficient mutants of probiotic yeast cannot stimulate bacterial growth. As we mention earlier that O2 scavenging property of yeasts is very important for growth of rumen microbial biomass, hence, this O2 scavenging property should be considered when probiotic yeast is selected for ruminants (Figure 2).

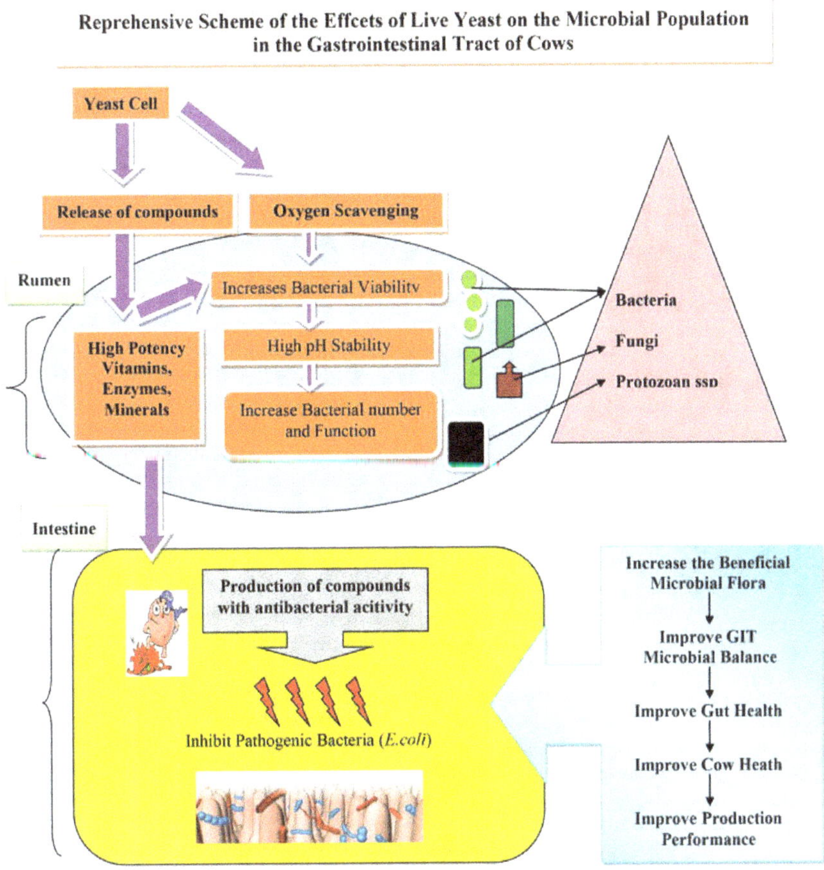

Figure 2. Representative scheme of effect of live yeast on the microbial flora of the gastrointestinal tract in ruminants: the probiotics yeast can improve the composition of the intestinal microbiota through the production of antimicrobial substances which inhibit the pathogenic bacteria ultimately improving the gut health. Although there is a difference between the probiotic colonization microbiota and the target rumen microbiota, many researchers suggested that there is a relationship between the GIT microbiota and other tissues of the host body.

Probiotic yeasts have beneficial effects on the lactate-metabolizing bacterial species. S. cerevisiae provides different growth factors essential for the growth of lactic acid fermenting bacterial species, such as Megasphaera elsdenii or Selenomonas ruminantium. In dairy animals, a reduction of lactic acid concentration was seen inside the rumen with live yeast addition (Figure 3).

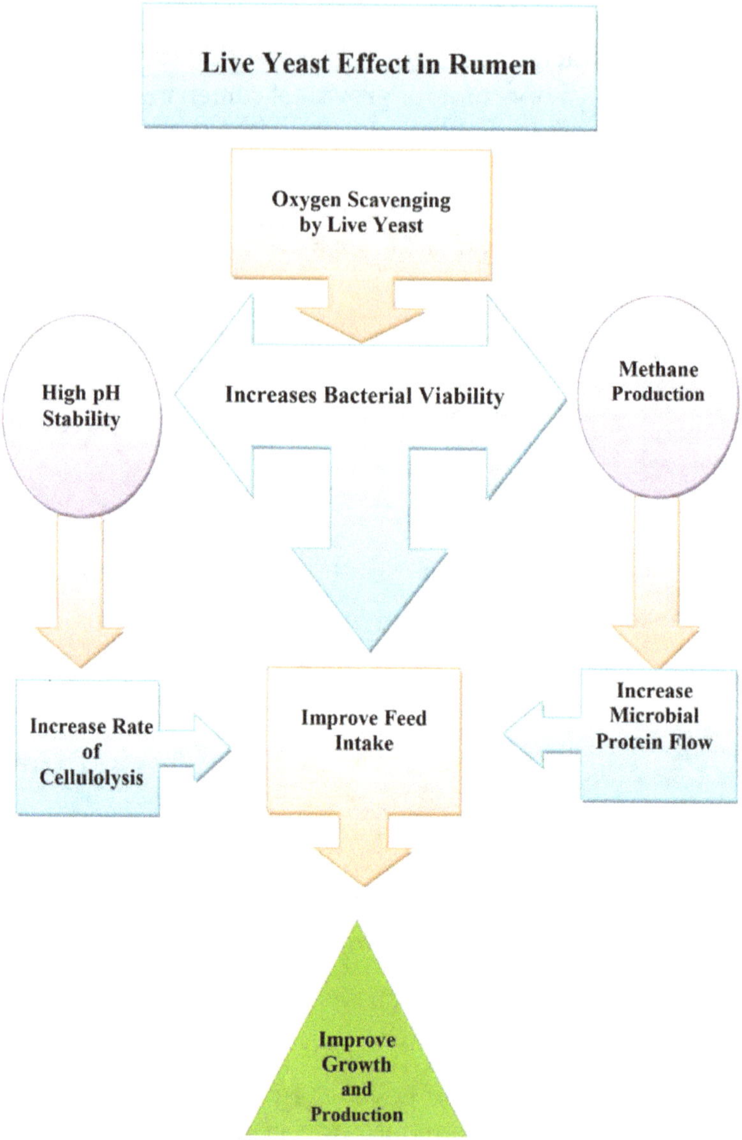

Figure 3. A scheme describing the mode of action of yeast culture. Adopted from Ref. [21].

4.1. Role of probiotics in the establishment of rumen and hindgut micro-flora establishment

The key of the rumen development is to provide supporting conditions to the microbiota to ensure its optimal establishment. It has been well studied that live yeast can help in establishing different types of micro-flora in neonate by positively modulating rumen colonization, by important functional microbial population. A newborn ruminant digestive system is sterile but with passage of time when they contact with their mother and other animals, they get microbes from their saliva and feces [10]. In contrary, the mother and her young connection are more common in small-scale farming systems. On the other hand, in intensive dairy farming systems, the neonate is alienated from the mother and is fed on solid feeding that provides a negative situation in the development of rumen microflora [47]. The early maternal separation has a negative effect on the rumen colonization by important microbial species. This negative situation leads to poor rumen microbial development making the neonate to suffer from different digestive diseases like diarrhea. Different diseases of digestive system are most important factors of low income heifers rearing. It has been well studied that live yeast culture can help the establishment of key microbial communities (Bacteroides-Prevotella and Clostridium coccoides-Eubacterium rectale group) in neonate by removing the oxygen from rumen. The rate of cellulose degrading microflora population was greater in lambs fed on S. cerevisiae addition as compared to the control [10]. Live yeast could be used as a nutrition tool for maturation of the rumen microbial ecosystem, which can result in a positive effect on animal performance, and health both before and after weaning, with an increase in grain intake and reduced frequency of diarrhea. S. cerevisiae had the ability to provide different types of organic acids or vitamins, those stimulating ruminal populations of cellulolytic bacteria and lactic acid utilizing bacteria (LUB) [48]. The cellulose degrading microbial population was also much stable in the animals fed on yeast addition because protozoa comes in rumen only once the bacterial species are present inside the rumen. It has been also noted that protozoa appeared earlier in those animals who fed on S. cerevisiae addition [49]. Amylolytic bacterial population is also affected by yeast in the rumen [38]. It is because the protozoan concentrations are proliferated and are able to store starch and postpone bacterial fermentation [50]. Proteolytic bacterial activity was highest in the yeast supplemented animals. Proteins in the feed are quickly broken down into peptides, amino acids and ammonia (NH3) by different protozoa and fungi inside rumen [51]. Some NH3 is converted into microbial protein (MP) and some ammonia is used by the animal in the form of urea. An important portion of rumen ammonia is excreted and represents an indication of nitrogen (N) loss of the dietary nitrogen intake (20–25%) [30]. Amino acids and peptides issued from dietary proteins cannot be directly slipped in the animal intestine, if the diet has highly nutritious value. The same effect on ammonia concentration was observed with daily yeast culture supplementation in adult ruminants [52]. In vitro findings tell that probiotic yeast could alter the growth and activities of protein-degrading bacteria, which ultimately enhanced the protein digestion inside rumen [53]. The mode of action of yeast can be explained by a fight between live S. cerevisiae cells and different bacterial species for energy utilization [54]. A study on 14 dairy cows field trials addition of yeast strain in the diet revealed that the soluble nitrogen of the diet was a key factor to drive the production parameters to the probiotics yeast

[55]. However, with other yeast strains, no significant effect was observed on the concentration and fraction of microbial nitrogen in dairy cattle [56]. Further study is needed to investigate the effect of probiotic yeast on the nitrogen microbial metabolism [10]. Many studies have shown that increased feed intakes are driven by increased flow of absorption nitrogen [43, 57]. This step stems simultaneously from the proliferation and stimulation of viable cell counts of anaerobic bacteria population. A higher ammonia nitrogen concentration measured for vessel in which live yeast was added compared to autoclaved yeast suggest that, the live yeast stimulated the proteolytic activity of the rumen bacterial species that ultimately influenced rumen fermentation [58]. It was noted that digestibility of crude protein was significantly higher in animals fed on the mixed fungal (yeast and Aspergillus) supplementation and it is suggesting that fungal supplementation might promote proteolytic activities by supplying some types of stimulatory factors [59]. Many studies have shown that animals fed on the yeast supplementation have been associated with higher concentration of ammonia nitrogen, which might suggest that proteolytic bacterial activity has been stimulated by yeast culture [60, 61]. The second proposed mechanism is that yeast cell provides the soluble growth factors such as, organic acids, branched-chain volatile fatty acids, vitamins, and amino acids, which have a positive effect in stimulating cellulolytic, proteolytic and lactic acid utilizing bacteria [59, 62].

4.2. Effect of yeast and yeast cultures on rumen fiber digestion

Fiber is non-digestible polysaccharides (a complex form of carbohydrate) [63]. In nutrition, the term fiber defines as a component of plant that is not digestible by mammalian enzyme [64]. Cellulose, hemicellulose, and lignin are the primary components of fiber. Cellulose and hemicelluloses constitute 15–70% of most ruminant diet [65]. Cellulose is the most abundant carbohydrate in plant cell wall. Chemically, cellulose is made up of linear chains of sugar molecules. In cellulose, glucose molecules are linked together in a β-1,4 links, and this linkage can only be digested by microbial cellulolytic enzymes (Table 1).

Table 1. The effects of various probiotic yeast strains on ruminant performance.

Strain	Diet type and dose	Animal type	Effect	References
Saccharomyces cerevisiae, QAU03, (locally isolated yeast from *Sahiwal* cow dung sample)	3 kg concentrate feed, 8 kg silage and 20 kg fodder 3.13×10^{07} (CFU/g) yeast	Lactating cows	Increase fiber digestibility and improve milk and its fat contents 1. Improve gastrointestinal tract microbial balance 2.	[21]

Strain	Diet type and dose	Animal type	Effect	References
Saccharomyces cerevisiae Yea-sacc (1026) (Alltechinc., Nicholasville, KY)	1 g/kg as fed High concentration or low concentration diets 2.5 × 10^9 (CFU/g)	Dairy Holstein heifers	Improved feed efficiency of HC-fed heifers. 1. Yeast culture increased dry matter digestibility in HC- and LC-fed heifers 2.	[21]
Saccharomyces cerevisiae CNCM l-1077, Pasteur institute), (Levucell Sc),	1 × 10^{10} CFU/head/day Total mixed ration (TMR)	Non-lactating dairy Holstein cows	Enhanced rumen fermentation1. Lower CH 2.$_4$ emissions	[48]
Saccharomyces cerevisiae, Yea-sacc (1026) (Alltechinc., Nicholasville, KY)	4.5 × 10^9 CFU High starch low starch diet	Holstein heifers	Positive effect on DM, NDF, ADF, and hemicellulose digestibility 1.	[21]
Dry yeast (CNCM-1077, Levucell Sc 20. Sc, Lallemand, animal nutrition)	0.5 g/hd/day basal diet consisting	Holstein dairy cows in late lactation	Decreased time spent in subacute rumen acidosis 1.	[49]
Levucell SC 10 ME	Maize silage, 1 × 10^{10} CFU/g yeast	Holstein dairy cows in Non-lactation	Lower the risk of rumen acidosis 1. Increased fiber degradation of low quality maize silages 2.	[18]
Saccharomyces cerevisiae, Yea-sacc (1026) (Alltechinc., Nicholasville, KY)	Balanced TMR or pasture 2.5 × 10^9 CFU/g yeast	Multiparous dairy cows	Improve the metabolic status 1.	[38]

Cellulose makes up about 40% of plant cell walls. Hemicellulose also can only be digested by microbial enzymes because it also has β-1,4 linkages. Hemicellulose has a strong negative effect on fiber degradation because of close association with lignin [66]. The rumen is an important part of the ruminant's stomach because cellulose is broken down into simple sugar that can be used by the animal body inside rumen. The rumen represents a mobile, self-sustaining fermentation system for plant material [67, 68]. It is a complex microbial ecosystem that contain many types of microorganisms such as, bacteria (1010–1011 cells per ml), protozoa (104–106 per ml) and fungi species (103–105 zoospores per ml) [69, 70].

4.2.1. Rumen fibrolytic bacteria

Rumen bacteria (1011 viable cells per ml) dominate the fermentation, both in terms of numbers and metabolic processes. The rumen bacteria are 99.5% obligatory anaerobic. In rumen, 200 species with many subspecies of bacteria are present. There are different kinds of bacteria in the rumen, which aid in fermentation process [71]. Fibrobacter and Ruminococcus are the main rumen fibers degrading bacteria in cattle [72, 73]. Fibrobacter succinogenes is a Gram-negative and rod-shaped anaerobe first isolated from the cattle [74]. Despite their important role, cellulose degrading bacteria are argued to only comprise of 0.3% of the total bacteria population inside rumen [75]. Rumen bacteria are classified into fibrolytic, amylolytic, pectinolytic, proteolytic, lipolytica, lactate using bacteria and hydrogen-using bacteria. Amylolytic bacteria ferment starch while fibrolytic bacteria involve in the fermentation of fiber. Different bacterial populations dominate the rumen fermentation depending on the type of feed. Cattle that are fed on high-fiber diet will have a ruminal bacterial population that is high in fibrolytic bacteria especially Ruminococcus ssp. Rumen bacteria are mainly involved in the fermentation of fiber, starch, and sugar in the feed.

4.2.2. Rumen fibrolytic fungi

Ruminal anaerobic fungi, an emerging group of animal probiotics, account for approximately 8% of the total rumen microbial biomass in ruminants [76]. Rumen fungi have a crucial role in the degradation of fiber material [77–80]. The fungi have an important role in fiber digestion because of the vegetative thallus rhizoids. The rhizoids have a more penetrating capability to plant cell wall as compared to the bacteria and protozoa. Degradation of lignin of the plant cell wall is a main characteristic of rumen fungi [81, 82]. Fungi degraded 37–50% of barley straw. The fungi fibrolytic activity enhanced by hydrogen-utilizing methanogens decreases the cruel effect of hydrogen [76, 83]. Fungi play an active and significant role in fiber digestion of low quality roughages by breaking the beta-1,4 linkages between lignin and hemicelluloses inside the plant cell [84]. Fungi have a positive role in fiber degradation as evidenced by producing a wide array of potential hydrolytic enzymes [79, 85, 86].

4.2.3. Rumen protozoa

In vitro studies have suggested that 19–28% of the total cellulosic activity in fiber digestion can be attributed to protozoa [87]. However, digestion seems to be limited to very susceptible tissue, for instance, mesophyll cells [88]. Studies have demonstrated that defaunation (removal of protozoa) reduces the rate of fiber/cell wall degradation digestion [89, 90]. However, in the absence of protozoa, there is an increased requirement for non-protein nitrogen (NPN) because of an increased bacterial population. A reduction of N may therefore result in the reduction in fiber digestion [91].

4.3. Mode of action of probiotic yeast in the post-ruminal GIT

The GIT inhabits multifarious microbial diversity that helps in generating impactive response regarding nutritious, health, physiology, and productivity of animals [1]. The existing gut microbiota regulates food safety through shedding of pathogens, interaction with organisms, and resource competition in the GIT [5]. The physiological, anatomical, and immunological status of the host is strongly dependent upon microbiota of GIT which facilitates essential functions to host. GIT microflora aids in the stimulation of immune system that acts as a barrier against infectious pathogens. It also restrains the injurious and pathogenic bacteria gut colonization [6]. The microflora that resides in GIT mostly belongs to Bacteroides, Bifidobacterium, Clostridium, Eubacterium, Fusobacterium, and Lactobacillus families. Among all of the intestinal microbiota, Enterococcus and Escherichia coli represent the least contribution (upto 1%) whereas, anaerobes show dominancy over microaerophiles and facultative anaerobes by 1000:1 [92]. Lactobacillus and Bifidobacteria are marked as predominant flora which counts for 90% of the total population in GIT. The fluctuating flora represents their existence in trace, i.e., less than (0.01%) that is usually considered as more diversified and pathogenic ones [93]. The GIT microbiota protects the host from pathogen, which produces digestive diseases like diarrhea. The performance of the calf is directly related to the efficient growth along with the improved health status [94]. Gut microbial flora play an important role in the growth and health of the animal. Probiotics put beneficial effects on the health of gut by improving its microbial balance. They have antidiarrheal capability and enhance the growth performance of animals [94, 95]. The intestinal microbiota of cattle performs its vital role in the fermentation process. They help in methane emission by means of fermentation both from rumen and large intestine [96]. The microbial diversity in the GIT of the dairy cattle has lot of impact on the productivity and well-being of the cattle [1–3, 97]. There is no direct evidence that yeast or fungal extracts affect digestion or metabolism in the lower gut. However, the potential for such effects should not be ignored [98]. This improvement can be due to either the effect of mannan-oligosaccharides (MOS, a component of yeast cell wall) on the immune modulation or direct effect of yeast on the reduction of pathogenic bacteria and toxic metabolites. According to the findings of Heinrichs et al. [99], MOS has an ability to bind selected pathogen by blocking the microbial lecithin and preventing the pathogens from colonization in host GIT. As noted, previous inquiries regarding feeding direct fed microbial products (DFM) to ruminant animals focused on its potential beneficial effects on the post-ruminal GIT (Figure 4).

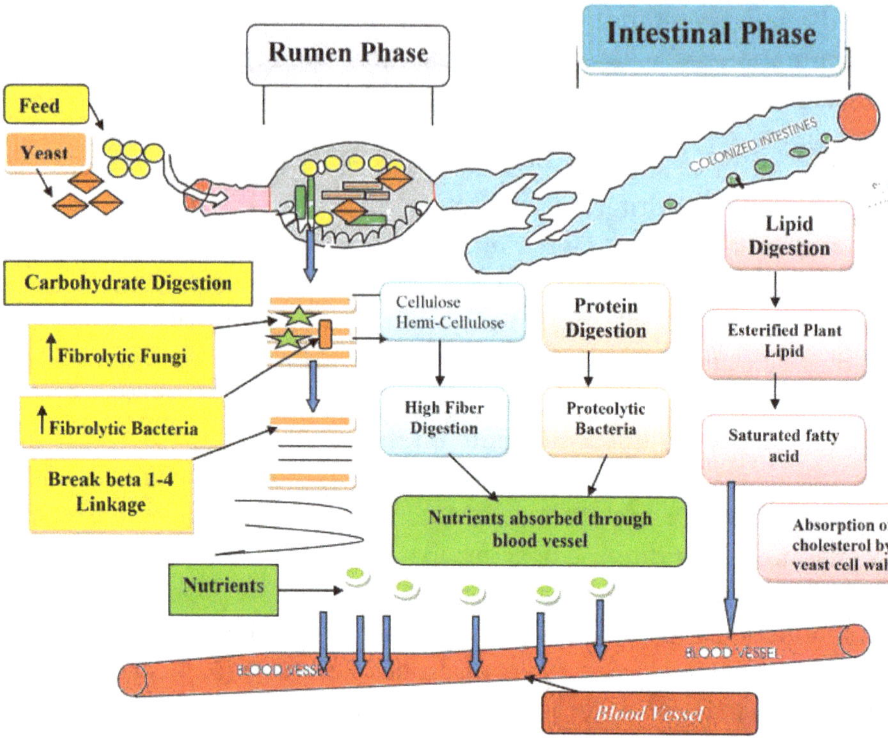

Figure 4. Simple scheme proposed to explain mode of action of probiotic yeast in rumen and post-ruminal GIT: live yeast improves carbohydrates, protein, and lipid digestion rate by improving the production of cellulolytic, hemicellulolytic and proteolytic, and lipolytic bacteria and fungi.

5. EXPERIMENTAL PROOFS

5.1. Experiment 1: effect of probiotic yeast on the growth performance and fecal biomarkers of dairy heifers

Poor growth performance in growing animals is associated with imbalanced nutrition. The use of probiotic yeast would minimize the expenditure of replacement heifers with optimum growth rate. In our experiment, young animals fed on the diet supplemented with yeast culture gain more weight than non-supplemented animals. In this experimental study, eight dairy heifers (87 ± 5 kg and 6–7 months) were divided into two equal groups of four animals each (control and probiotic) following completely randomized design [100]. During the trial, heifers in control group were offered control diet (NRC recommended diet), while in the probiotic group fed with control diet plus commercial available probiotic yeast (Yea-Sac1026; 5 g/animal) for a period of 120 days. Results reveal that dairy heifers fed on the probiotic feed gained significantly ($P < 0.05$) higher average weights than dairy heifers fed on control feed (Figures 5 and 6) [34].

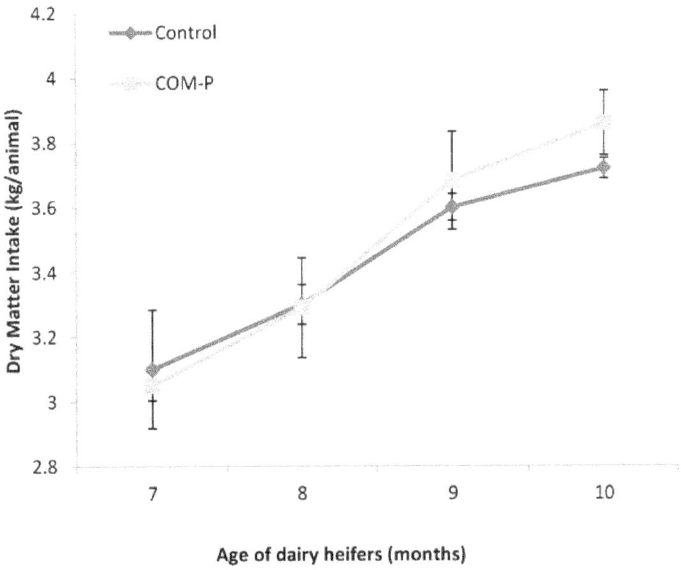

Figure 5. Average monthly dry matter intake pattern (kg) of dairy heifers fed on control feed (control, ♦; no yeast) or commercial probiotic feed (COM-P, ■; control feed plus commercial yeast).

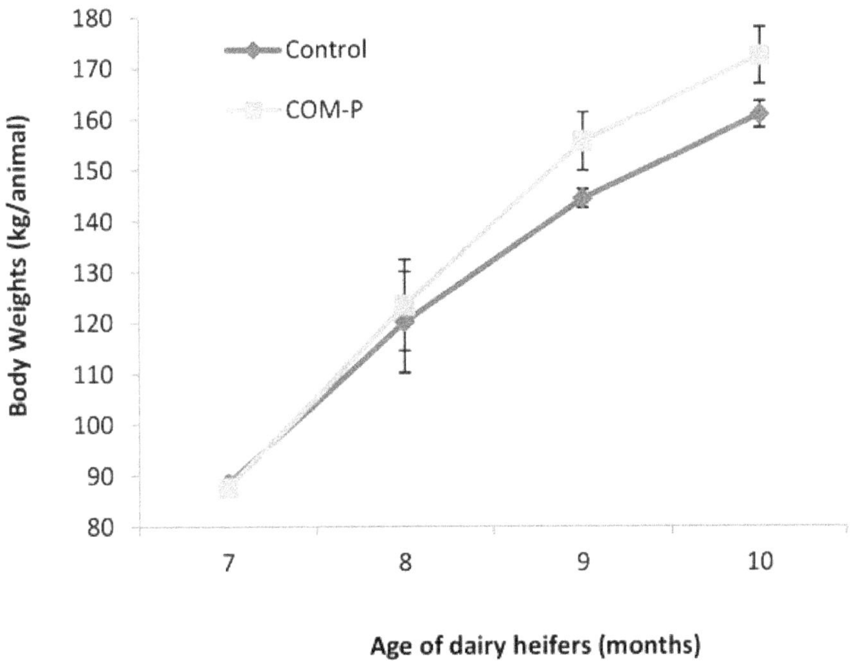

Figure 6. Average monthly growth pattern (kg) of dairy heifers fed on control feed (control, ♦; no yeast) or commercial probiotic feed (COM-P, ■; control feed plus commercial yeast).

Probiotic yeast decreases the pathogenic bacteria and increase the beneficial bacteria in present study (Figure 7) [34].

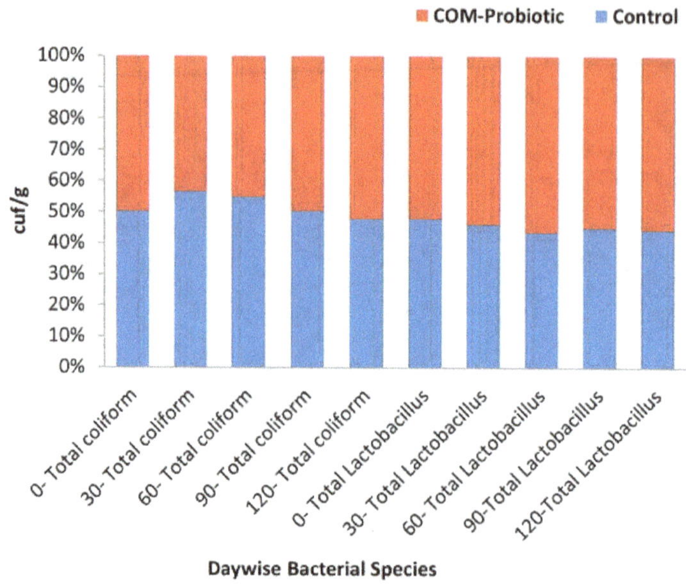

Figure 7. Total E. coli and Lactobacillus count (CFU/g) in the ruminal gut of dairy heifers fed on control feed (control, ♦; no yeast) or commercial probiotic feed (COM-P, ■; control feed plus commercial yeast).

5.2. Experiment 2: development of indigenous probiotic yeast for local breed

From the aforementioned discussion, we found that an important step to establish a breed probiotic strain is that the origin of the isolated strain should be animal based for their better adhesion and colonization in the animal GIT. We hypothesize that, breed-specific probiotic yeast gives better results in terms of milk production. From this line of research, we conduct an experimental study to develop the indigenous probiotic yeast for local breed and to evaluate its effect on the lactating animals. A S. cerevisiae strain was isolated from dung samples of the dairy animals. After careful assessment of its probiotic test, that yeast strain (animal probiotic) was further used in same dairy cattle feed (Figure 8) [100].

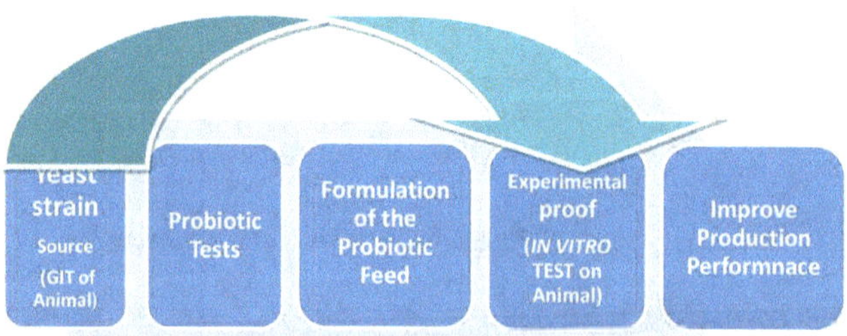

Figure 8. Flow sheet of development of indigenous probiotic yeast for local breed.

5.3. Experiment 3: impact of indigenously isolated Saccharomyces cerevisiae probiotics on milk production and gut microbial species of lactating cows

Nine lactating dairy cattle of mix breed (Sahiwal and Sahiwal × Jersey) at their first and second lactation (producing 4–5 l/day) were used for the experiment. Cows were fed a concentrate feed, maize silage, and oat fodder. The neutral detergent fiber digestibility was improved in the presence of 3.13 × 1007 CFU/g of our laboratory produced live yeast and milk production was improved by 0.7 kg/d in the laboratory produced probiotic fed dairy cows (Figure 9).

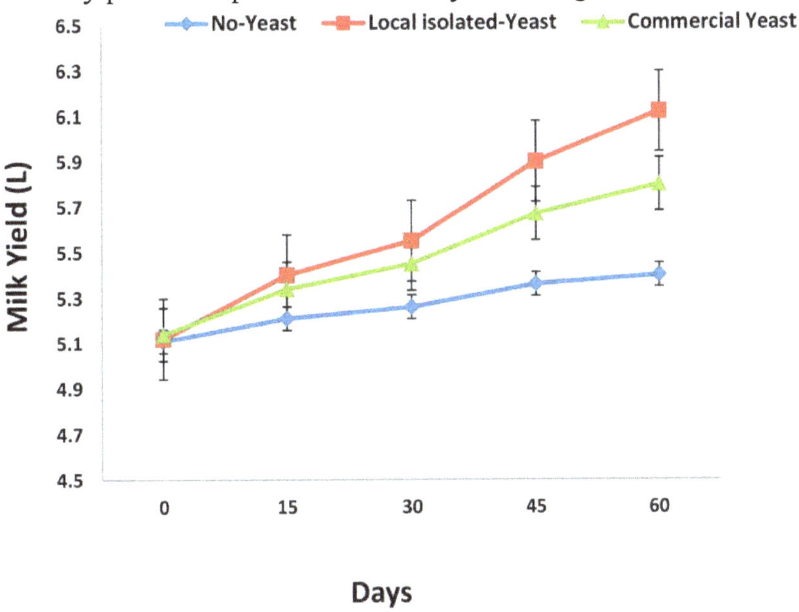

Figure 9. Effect of yeast on milk yield (Means ± SEM) in lactating dairy cattle fed on the diet supplemented with no yeast (control, ♦), laboratory yeast (LAB-Y, ■) or commercial yeast (COM-Y,).

We assumed that improved performance is might be due to cellulolytic activity of the S. cerevisiae which was isolated from dairy animal dung sample [101]. This activity enhanced the cellulose digestion rate and helped in the milk synthesis [14]. Yeast culture significantly ($P < 0.05$) increased the fiber digestibility, resulting in an increased supply of absorbed nutrients for milk synthesis in our experiment [101]. Results of the ruminal gut microflora showed that the average beneficial Lactococcus species (CFU/g) counts were increased while pathogenic Enterococcus species (CFU/g) counts were lower in laboratory produced yeast (LAB-Y) fed lactating cows than other groups which lead to improve GIT microbial balance. The economic efficiency of LAB-Y fed group that was 4.7% better than the control group, fed no yeast culture. It can be concluded that indigenous isolated probiotic yeast strain improves the production performance, gut health, and well-being of lactating dairy cattle in cost-effective manner. Locally isolated yeast strain may be adopted well in the cattle gut than exotic probiotics [100].

6. DIFFICULTIES IN THE PROCESS OF DEVELOPING APPROPRIATE PROBIOTIC YEAST

Traditionally, as ruminates live in different parts of the world, hence, different yeast strains may exhibit different effects upon the ruminal fermentation depending on their location. Therefore, we should identify new yeast strains for getting best results for the rumen fermentation in their own living condition. For getting positive results in the ruminants, probiotic strains should be breed specific. Latest knowledge related to modes of action of probiotic yeast and its beneficial effects on rumen fermentation, may aid in selection of new breed-specific strains which act on specific key target microorganisms (Cellulolytic, hemicellulolytic bacteria and fungi) and areas of rumen fermentation. Inside the rumen fluid, certain probiotic yeast candidate cannot remain active for longer periods of time. On the other hand, probiotic strains viability and stability are the more advanced technological challenges faced by the livestock industry holders. There is a strong interaction between the host animal and microbial population. To overcome these challenges, further empirical studies are needed on the study of probiotic candidates as well as the ruminal gut microbiota activities to enhance the information of the host-specific interactions. Then the goal is to apply the knowledge of ruminal gut health normal microbial species composition in comparison with microbiota present during disease to select the right breed-specific probiotic strain (Figure 10).

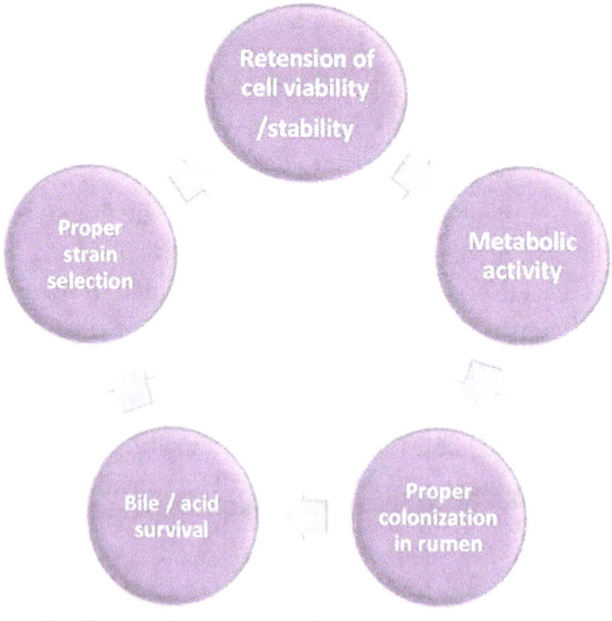

Figure 10. Challenges in preparation of suitable probiotics yeast.

7. CONCISE SUMMARY

The addition of probiotics to animal feed has potential consequences for the future of the cattle business. Nevertheless, the primary obstacle encountered by rumen microbiologists is the development of an economically efficient bioactive diet for dairy animals. Searching for new probiotic strains will be the primary focus of research and development in the future livestock markets worldwide. The focused uses of particular strains may hold significant potential for treating numerous chronic diseases in animals. This would result in the promotion of both economic and ecological farming practices. Probiotic strains from the same ecological origin may be more compatible with the rumen microbiome, resulting in higher productivity. Hence, the forthcoming feed supplements may need to be tailored to various breeds. In recent times, there has been a global surge in customer demand for food products that are safe and promote good health. Therefore, the use of probiotics not only improves productive efficiency but also helps decrease the likelihood of ruminant gastrointestinal tract (GIT) carrying human pathogens and reduces the excretion of harmful substances like nitrogen-based chemicals and methane. The S. cerevisiae has been granted the Generally Recognized as Safe (GRAS) designation by the Food and Drug Administration (FDA), making it suitable for inclusion in animal diets. When deciding whether to employ a feed additive for an experiment, it is important to evaluate elements such as predicted response, net profit, ongoing research, and field responses. Fermented yeast culture is a cost-effective product with numerous advantages for ruminants. An important advantage of probiotic yeast is its lack of antibiotic resistance genes. Furthermore, it possesses the capacity to colonize in the gastrointestinal tract (GIT), counteract enterotoxin, and withstand bile salt and gastric acid, hence enhancing the health and productivity of dairy animals. Typically, the cost of animal feed is expensive. However, using probiotic yeast as a nutritional strategy can improve the digestibility of the diet and enhance the performance of dairy animals in a cost-effective way. Further empirical research is required to examine the influence of yeast cells in the gastrointestinal tract of dairy animals. Future studies should also investigate the behavior of yeast cells in the digestive environment. We anticipate the advancement of novel probiotic strains tailored to specific breeds. This development is expected to shift the role of rumen microbiologists from passively observing the nutritional aspects to actively driving progress in ruminant nutrition in the upcoming year.

AT A GLANCE

The primary objective of yeast supplementation is to address rumen microbial dysbiosis, hence improving nutrition use and ultimately promoting animal growth and productivity. Yeast enhances the rumen ecology through two mechanisms: firstly, by directly producing digesting enzymes and acting as a growth stimulant, and secondly, by increasing the growth and functionality of beneficial bacteria.

Yeasts possess the capacity to generate metabolites that promote development, such as rumen acetogens, as well as antimicrobial chemicals that hinder the growth of potential pathogens. The effect of yeast probiotics on animals is influenced by various interacting elements, including as animal breed, dosage, type, diet, strain, physiological stage, and feeding system. During times of elevated global feed costs, probiotic yeast offers a valuable nutritional approach that can improve the digestibility of diets and enhance the performance of ruminants in a cost-effective manner. Numerous yeast culture-based products are readily accessible on a global scale, however, their efficacy as a probiotic dietary supplement in a specific breed is largely uncertain. Hence, the investigation of the novel indigenous probiotic strain is quite intriguing in this particular situation. Probiotic strains that come from the same ecological origin may be more compatible with the rumen microbiome, resulting in optimal outcomes. Furthermore, the use of a probiotic yeast tailored to a certain breed is a cost-effective and practical solution for farmers to mitigate the consequences of malnutrition.

REFERENCES

1. Guarner F, Malagelada JR. Gut flora in health and disease. The Lancet. 2003;361:512-519
2. Dowd SE, Callaway TR, Wolcott RD, Sun Y, McKeehan T, Hagevoort RG, Edrington TS. Evaluation of the bacterial diversity in the feces of cattle using 16S rDNA bacterial tag-encoded FLX amplicon pyrosequencing (bTEFAP). BMC Microbiology. 2008;8:125
3. Eckburg PB, Bik EM, Bernstein CN, Purdom E, Dethlefsen L, Sargent M, Gill SR, Nelson KE, Relman DA. Diversity of the human intestinal microbial flora. Science. 2005;308:1635-1638
4. Fuller R. Probiotics in man and animals. The Journal of Applied Bacteriology. 1989;66:365-378
5. Nurmi E, Rantala M. New aspects of Salmonella infection in broiler production. Nature. 1973;241:210-211
6. Cebra JJ. Influences of microbiota on intestinal immune system development. The American journal of clinical nutrition. 1999;69(5):1046-1051
7. Chaucheyras-Durand F, Chevaux E, Martin C, Forano E. Use of yeast probiotics in ruminants: Effects and mechanisms of action on rumen pH, fibre degradation, and microbiota according to the diet. In: Probiotic in Animals. InTech Open Book Publisher; 2012
8. Grochowska S, Nowak W, Mikula R, Kasprowicz-Potocka M. The effect of Saccharomyces cerevisiae on ruminal fermentation in sheep fed high-or low-NDF rations. Journal of Animal and Feed Sciences. 2012;21:276-284
9. Musa H, Wu S, Zhu C, Seri H, Zhu G. The potential benefits of probiotics in animal production and health. Journal of Animal and Veterinary Advances. 2009;8:313-321
10. Chaucheyras-Durand F, Walker ND, Bach A. Effects of active dry yeasts on the rumen microbial ecosystem: Past, present and future. Animal Feed Science and Technology. 2008;145(1):5-26
11. Huber JT. Probiotics in Cattle. London: Chapman and Hall; 1997. p. 162-186
12. Robinson PH, Erasmus LJ. Effects of analyzable diet components on responses of lactating dairy cows to Saccharomyces cerevisiae based yeast products: A systematic review of the literature. Animal Feed Science and Technology. 2009;149:185-198

13. Poppy GD, Rabiee AR, Lean IJ, Sanchez WK, Dorton KL. A meta-analysis of the effects of feeding yeast culture produced by anaerobic fermentation of Saccharomyces cerevisiae on milk production of lactating dairy cows. Journal of Dairy Science. 2012;95:6027-6041

14. Dolezal P, Dolezal J, Szwedziak K, Dvoracek J, Zeman L, Tukiendorf M, Havlicek Z. Use of yeast culture in the TMR of dairy Holstein cows. Iranian Journal of Applied Animal Science. 2012;2:51-56

15. Ferraretto LF, Shaver RD, Bertics SJ. Effect of dietary supplementation with live-cell yeast at two dosages on lactation performance, ruminal fermentation, and total-tract nutrient digestibility in dairy cows. Journal of Dairy Science. 2012;95:4017-4028

16. Sabbia JA, Kalscheur KF, Garcia AD, Gehman AM, Tricarico JM. Soybean meal substitution with a yeast-derived microbial protein source in dairy cow diets. Journal of Dairy Science. 2012;95:5888-5900

17. Hristov AN, Varga G, Cassidy T, Long M, Heyler K, Karnati SK, Corl B, Hovde CJ, Yoon I. Effect of Saccharomyces cerevisiae fermentation product on ruminal fermentation and nutrient utilization in dairy cows. Journal of Dairy Science. 2010;93:682-692

18. Guedes CM, Gonçalves D, Rodrigues MAM, Dias-da-Silva A. Effects of a Saccharomyces cerevisiae yeast on ruminal fermentation and fibre degradation of maize silages in cows. Animal Feed Science and Technology. 2008;145:27-40

19. Lesmeister KE, Heinrichs AJ, Gabler MT. Effects of supplemental yeast (Saccharomyces cerevisiae) culture on rumen development, growth characteristics, and blood parameters in neonatal dairy calves. Journal of Dairy Science. 2004;87:1832-1839

20. Lehloenya KV, Krehbiel CR, Mertz KJ, Rehberger TG, Spicer LJ. Effects of propionibacteria and yeast culture fed to steers on nutrient intake and site and extent of digestion. Journal of Dairy Science. 2008;91:653-662

21. Lascano GJ, Heinrichs AJ, Tricarico JM. Substitution of starch by soluble fiber and Saccharomyces cerevisiae dose response on nutrient digestion and blood metabolites for precision-fed dairy heifers. Journal of Dairy Science. 2012;95:3298-3309

22. Hutjens MF. Feed additives. Veterinary Clinics of North America: Food Animal Practice. 1991;7:525

23. Faria-Oliveira F, Ferreira C, Puga S. Yeast: World's Finest Chef. Intech Open Access Publisher; 2013

24. Stone CW. Yeast Products in the Feed Industry: A Practical Guide for Feed Professionals. Cedar Rapids, IA, USA: Diamond V Mills Inc; 1998

25. Yalcin S, Yalcin S, Can P, Gurdal AO, Bagci C, Eltan O. The nutritive value of live yeast culture (Saccharomyces cerevisiae) and its effect on milk yield, milk composition and some blood parameters of dairy cows. Asian Australas Journal of Animal Sciences. 2011;24:1377-1385

26. Sontakke U. Benefits of Saccharomyces cerevisiae as a Feed Additive in Ruminants. Vol. 1714. Karnal: National Dairy Research Institute; 2012

27. Szucs JP, Suli A, Halasz T, Arany A, Bodor Z. Effect of live yeast culture Saccharomyces cerevisiae on milk production and some blood parameters. Scientific Papers Animal Science and Biotechnologies. 2013;46:40-44

28. Jouany J-P, Morgavi D. Use of 'natural'products as alternatives to antibiotic feed additives in ruminant production. Animal. 2007;1(10):1443-1466

29. Carberry CA, Kenny DA, Han S, McCabe MS, Waters SM. Effect of phenotypic residual feed intake and dietary forage content on the rumen microbial community of beef cattle. Applied and Environmental Microbiology. 2012;78:4949-4958

30. Fonty G, Chaucheyras-Durand F. Effects and modes of action of live yeasts in the rumen. Biologia. 2006;61:741-750

31. Weinberg ZG. Effect of lactic acid bacteria on animal performance. Indian Journal of Biotechnology. 2003;2:378-381

32. Wcimer P, Waghorn GC. Effect of her on populations of three species of ruminal cellulolytic bacteria in lactating cows. Journal of Daily Science. 1999;82:122-134

33. Dehority BA, Orpin CG. Development of, and natural fluctuations in, rumen microbial populations. In: Hobson PN, Stewart CS, editors. The Rumen Microbial Ecosystem. London, United Kingdom: Blackie Academic and Professional; 1997. p. 196-245

34. Shakira G, Atiya A, Ahmed I. Effects of dietary supplementation of yeast (Saccharomyces cerevisiae) culture on growth performance, blood parameters, nutrient digestibility and fecal flora of dairy heifers. The Journal of Animal and Plant Science. 2015;25:53-59

35. Dehority B, Orpin C. Development of, and natural fluctuations in, rumen microbial populations. In: The Rumen Microbial Ecosystem. Springer; 1997. p. 196-245

36. Weimer P, Waghorn G, Odt C, Mertens D. Effect of diet on populations of three species of ruminal cellulolytic bacteria in lactating dairy cows. Journal of Dairy Science. 1999;82:122-134

37. Santra A, Karim S. Rumen manipulation to improve animal productivity. Asian Australasian Journal of Animal Sciences. 2003;16:748-763

38. Arakaki L, Stahringer R, Garrett J, Dehority B. The effects of feeding monensin and yeast culture, alone or in combination, on the concentration and generic composition of rumen protozoa in steers fed on low-quality pasture supplemented with increasing levels of concentrate. Animal Feed Science and Technology. 2000;84:121-127

39. El-Ghani AA. Influence of diet supplementation with yeast culture (Saccharomyces cerevisiae) on performance of Zaraibi goats. Small Ruminant Research. 2004;52:223-229

40. Hillman K, Lloyd D, Williams A. Use of a portable quadrupole mass spectrometer for the measurement of dissolved gas concentrations in ovine rumen liquor in situ. Current Microbiology. 1985;12:335-339

41. Loesche WJ. Oxygen sensitivity of various anaerobic bacteria. Applied Microbiology. 1969;18:723-727

42. Rose AH. Live Yeast, A Microorganism for all Species: A Theoretical Look at its Mode of Action. Nicholasville, Kentucky: Alltech Technical Publications; 1987. p. 113-118

43. Newbold C, Wallace R, Chen X, McIntosh F. Different strains of Saccharomyces cerevisiae differ in their effects on ruminal bacterial numbers in vitro and in sheep. Journal of Animal Science. 1995;73:1811-1818

44. Wallace R. Ruminal microbiology, biotechnology, and ruminant nutrition: Progress and problems. Journal of Animal Science. 1994;72:2992-3003

45. Dawson K, Newman K, Boling J. Effects of microbial supplements containing yeast and lactobacilli on roughage-fed ruminal microbial activities. Journal of Animal Science. 1990;68:3392-3398

46. Jouany JP, Mathieu F, Sénaud J, Bohatier J, Bertin G. Influence of protozoa and fungal additives on ruminal pH and redox potential. South African Journal of Animal Science. 1999;29:65-66

47. Fonty G, Gouet P, Jouany J-P, Senaud J. Establishment of the microflora and anaerobic fungi in the rumen of lambs. Journal of General Microbiology. 1987;133:1835-1843

48. Chaucheyras F, Fonty G, Bertin G, Gouet P. Effects of live Saccharomyces cerevisiae cells on zoospore germination, growth, and cellulolytic activity of the rumen anaerobic fungus, Neocallimastix frontalis MCH3. Current Microbiology. 1995;31:201-205

49. Chaucheyras-Durand F, Fonty G. Yeasts in ruminant nutrition. Experiences with a live yeast product. Kraftfutter. 2002;85:146-150

50. Enjalbert F, Garrett J, Moncoulon R, Bayourthe C, Chicoteau P. Effects of yeast culture (Saccharomyces cerevisiae) on ruminal digestion in non-lactating dairy cows. Animal Feed Science and Technology. 1999;76:195-206

51. Wallace RJ, Onodera R, Cotta MA, editors. The Rumen Microbial Ecosystem. London: Chapman and Hall; 1997. p. 283-328

52. Kumar U, Sareen VK, Singh S. Effect of Saccharomyces cerevisiae yeast culture supplement on ruminal metabolism in buffalo calves given a high concentrate diet. Animal Science. 1994;59:209-215

53. Beev G, Todorova P, Tchobanova S. Yeast cultures in ruminant nutrition. Bulgarian Journal of Agricultural Science. 2007;13:357-374

54. Chaucheyras-Durand F, Masseglia S, Fonty G. Effect of the microbial feed additive Saccharomyces cerevisiae CNCM I-1077 on protein and peptide degrading activities of rumen bacteria grown in vitro. Current Microbiology. 2005;50:96-101

55. Sniffen C, Chaucheyras-Durand F, De Ondarza M, Donaldson G. Predicting the impact of a live yeast strain on rumen kinetics and ration formulation. 2004:53-59.

56. Putnam D, Schwab C, Socha M, Whitehouse N, Kierstead N. Effect of yeast culture in the diets of early lactation dairy cows on Ruminal fermentation and passage of nitrogen fractions and amino acids to the small intestine 1, 2. Journal of Dairy Science. 1997;80:374-384

57. Kamel H, Sekine J, El-Waziry A, Yacout M. Effect of Saccharomyces cerevisiae on the synchronization of organic matter and nitrogen degradation kinetics and microbial nitrogen synthesis in sheep fed Berseem hay (Trifolium alexandrinum). Small Ruminant Research. 2004;52:211-216

58. Oeztuerk H. Effects of live and autoclaved yeast cultures on ruminal fermentation in vitro. Journal of Animal and Feed Sciences. 2009;18:142-150

59. Wiedmeier R, Arambel M, Walters J. Effect of yeast culture and Aspergillus oryzae fermentation extract on ruminal characteristics and nutrient digestibility. Journal of Dairy Science. 1987;70:2063-2068

60. Kung L, Kreck E, Tung R, Hession A, Sheperd A. Effects of a live yeast culture and enzymes on in vitro ruminal fermentation and milk production of dairy cows. Journal of Dairy Science. 1997;80:2045-2051

61. Moallem U, Lehrer H, Livshitz L, Zachut M, Yakoby S. The effects of live yeast supplementation to dairy cows during the hot season on production, feed efficiency, and digestibility. Journal of Dairy Science. 2009;92:343-351

62. Newbold C, Wallace R, McIntosh F. Mode of action of the yeast Saccharomyces cerevisiae as a feed additive for ruminants. British Journal of Nutrition. 1996;76:249-261

63. Callaway E, Martin S. Effects of a Saccharomyces cerevisiae culture on ruminal bacteria that utilize lactate and digest cellulose. Journal of Dairy Science. 1997;80:2035-2044

64. Tungland B, Meyer D. Nondigestible oligo and polysaccharides (dietary fiber): Their physiology and role in human health and food. Comprehensive Reviews in Food Science and Food Safety. 2002;1:90-109

65. Moore KJ, Hatfield RD. Carbohydrates and Forage Quality. Madison: ASA-CSSA-SSSA; 1999. p. 229-280

66. Hobson PN, Stewart CS. The Rumen Microbial Ecosystem. Springer Science and Business Media; 1997

67. Fernando SC, Purvis HT, Najar FZ, Sukharnikov LO, Krehbiel CR, Nagaraja TG, Roe BA, DeSilva U. Rumen microbial population dynamics during adaptation to a high-grain diet. Applied Environmental Microbiology. 2010;76:7482-7490

68. Flint HJ. The rumen microbial ecosystem-some recent developments. Trends in Microbiology. 1997;5:483-488

69. Shakira G, Mirza I, Latif A. Scope of common DNA based methods for the study of rumen bacterial population. Bangladesh Journal of Animal Science. 2013;41:141-146

70. Kamra D. Rumen microbial ecosystem. Current Science. 2005;89:124-135

71. Cho SJ, Cho KM, Shin EC, Lim WJ, Hong SY, Byoung RC, Jung MK, Sun ML, Yong HK, Kim H, Yun HD. 16S rDNA analysis of bacterial diversity in three fractions of cow rumen. Journal of Microbiology and Biotechnology. 2006;16:92-101

72. Kobayashi Y, Shinkai T, Koike S. Ecological and physiological characterization shows that Fibrobacter succinogenes is important in rumen fiber digestion – review. Folia Microbiologia (Praha). 2008;53:195-200

73. Kim M, Morrison M, Yu Z. Evaluation of different partial 16S rRNA gene sequence regions for phylogenetic analysis of microbiomes. Journal of Microbiological Methods. 2011;84:81-87

74. Hungate RE. The anaerobic mesophilic cellulolytic bacteria. Bacteriological Reviews. 1950;14:1-49

75. Brulc JM, Yeoman CJ, Wilson MK, Miller ME, Jeraldo P, Jindou S, Goldenfeld N, Flint HJ, Lamed R, Borovok I, Vodovnik M. Cellulosomics, a gene-centric approach to investigating the intraspecific diversity and adaptation of Ruminococcus flavefaciens within the rumen. PLoS One. 2011;6(10):25329

76. Orpin CG, Joblin KN. The rumen anaerobic fungi. In: The Rumen Microbial Ecosystem. London, UK: Elsevier Applied Science Publishers; 1988. p. 129-150

77. Theodorou MK, Longland AC, Dhanoa MS, Lowe SE, Trinci AP. Growth of Neocallimastix sp. strain R1 on Italian ryegrass hay: Removal of neutral sugars from plant cell walls. Applied and Environmental Microbiology. 1989;55:1363-1367

78. Paul SS, Kamra DN, Sastry VRB, Sahu NP, Agarwal N. Effect of administration of an anaerobic gut fungus isolated from wild blue bull (Boselaphus tragocamelus) to buffaloes (Bubalus bubalis) on in vivo ruminal fermentation and digestion of nutrients. Animal Feed Science and Technology. 2004;115:143-157

79. Lee SS, Choi CK, Ahn BH, Moon YH, Kim CH, Ha JK. In vitro stimulation of rumen microbial fermentation by a rumen anaerobic fungal culture. Animal Feed Science and Technology. 2004;115:215-226

80. Thareja A, Puniya AK, Goel G, Nagpal R, Sehgal JP, Singh PK, Singh K. In vitro degradation of wheat straw by anaerobic fungi from small ruminants. Archiver Animal Nutrition. 2006:412-417

81. Mountfort DO, Asher RA, Bauchop T. Fermentation of cellulose to methane and carbon dioxide by a rumen anaerobic fungus in a triculture with Methanobrevibacter sp. strain RA1 and Methanosarcina barkeri. Applied and Environmental Microbiology. 1982;44:128-134

82. Akin DE, Benner R. Degradation of polysaccharides and lignin by ruminal bacteria and fungi. Applied and Environmental Microbiology. 1988;54:1117-1125

83. Joblin K. Physical disruption of plant fibre by rumen fungi of the Sphaeromonas group. In: The Roles of Protozoa and Fungi in Ruminant Digestion. Armidale, NSW: Penambul; 1989. p. 259-260

84. Tripathi VK, Sehgal JP, Puniya AK, Singh K. Hydrolytic activities of anaerobic fungi from wild blue bull (Boselaphus tragocamelus). Anaerobe. 2007;13:36-39

85. Williams AG, Orpin CG. Polysaccharide-degrading enzymes formed by three species of anaerobic rumen fungi grown on a range of carbohydrate substrates. Canadian Journal of Microbiology. 1987;33:418-426

86. Paul SS, Kamra DN, Sastry VR, Sahu NP, Kumar A. Effect of phenolic monomers on biomass and hydrolytic enzyme activities of an anaerobic fungus isolated from wild nil gai (Baselophus tragocamelus). Letters in Applied Microbiology. 2003;36:377-381

87. Gijzen HJ, Lubberding HJ, Gerhardus MJT, Vogels GD. Contribution of rumen protozoa to fibre degradation and cellulase activity in vitro. FEMS Microbiology Letters. 1988;53:35-43

88. Akin DE. Histological and physical factors affecting digestibility of forages. Agronomy Journal. 1989;81:17-25

89. Bonhomme A. Rumen ciliates: Their metabolism and relationships with bacteria and their hosts. Animal Feed Science and Technology. 1990;30:203-266

90. Yang CM, Varga GA. The effects of continuous ruminal dosing with dioctyl sodium sulphosuccinate on ruminal and metabolic characteristics of lactating Holstein cows. British Journal of Nutrition. 1993;69:397-408

91. Ushida K, Jouany JP. Effect of defaunation on fibre digestion in sheep given two isonitrogenous diets. Animal Feed Science and Technology. 1990;29:153-158

92. Mestecky J, Russell M. Passive and active protection against disorders of the gut. Veterinary Quarterly. 1998;20:83-87

93. Tournut J. The digestive flora of the pig and its variations. Recueil De Medecine Veterinaire. 1993;169:645-652

94. Soberon F, Raffrenato E, Everett R, Van Amburgh M. Preweaning milk replacer intake and effects on long-term productivity of dairy calves. Journal of Dairy Science. 2012;95:783-793

95. Donovan D, Franklin S, Chase C, Hippen A. Growth and health of holstein calves fed milk replacers supplemented with antibiotics or enteroguard 1, 2. Journal of Dairy Science. 2002;85:947-950

96. Johnson KA, Johnson DE. Methane emissions from cattle. Journal of Animal Science. 1995;73:2483-2492

97. Engelbrektson A, Kunin V, Wrighton KC, Zvenigorodsky N, Chen F, Ochman H, Hugenholtz P. Experimental factors affecting PCR-based estimates of microbial species richness and evenness. The ISME Journal. 2010;4:642-647

98. Elghandour MM, Salem AZ, Castañeda JS, Camacho LM, Kholif AE, Chagoyán JC. Direct-fed microbes: A tool for improving the utilization of low quality roughages in ruminants. Journal of Integrative Agriculture. 2015;14:526-533

99. Heinrichs A, Jones C, Heinrichs B. Effects of mannan oligosaccharide or antibiotics in neonatal diets on health and growth of dairy calves. Journal of Dairy Science. 2003;86:4064-4069

100. Ghazanfar S. Study on the effects of dietry supplmentation of Saccharomyvces cerevisiae on performance of dairy cattle and heifers. [PhD thesis]. Islamabad. Pakistan: Quaid-e-Azam University; 2016

101. Ghazanfar S, Qubtia M, Hassan F, Afzal M, Ahmed I, et al. Effect of indigenously isolated Saccharomyces cerevisiae probiotics on milk production, nutrient digestibility, blood chemistry and fecal microbiota in lactating dairy cows. The Journal of Animal and Plant Sciences. 2017 (Accepted)

SECTION 3

ENERGY PRODUCTION

CHAPTER-9

PHENOTYPIC ANALYSIS OF YEASTS FOR BIOETHANOL PRODUCTION

1. INTRODUCTION: AN OVERVIEW

In recent decades, recurrent crises in world oil have resulted in serious economic crises, leading to the search for alternative fuels [1]. In 1930, Brazil presented the first National Congress on Industrial Applications of Alcohol that was aimed at establishing the infrastructure for the production and use of bioethanol. This led Brazil to start production in the early twentieth century, while other countries started the production of fuel from grains, using its potential for bioenergy production. The remainder of residues with high protein content is a source of nutrition in agriculture, as well as being a rich source of sustainability [2].

The three major world powers producing bioethanol are Brazil (sugarcane), the USA (corn), and China (wheat and corn), where Brazil is the largest producer through a direct source of sugar, as production by grains requires an additional step with the liquefaction and hydrolysis of the starch. Estimates indicate that around 85% of all bioethanol worldwide is the responsibility of Brazilian and North American production, as well as inferior productions in all parts of the continents, as shown in Figure 1 [3].

In Brazil, the consumption of bioethanol has been relatively low in comparison to the usage of conventional fuels, which has resulted in the country's continued dependence on imported oil [4]. In response to this, the national government initiated the National Alcohol Program (ProÀlcool) in the 1970s. This program aimed to promote better air quality and advancements in alternative energy sources [5].

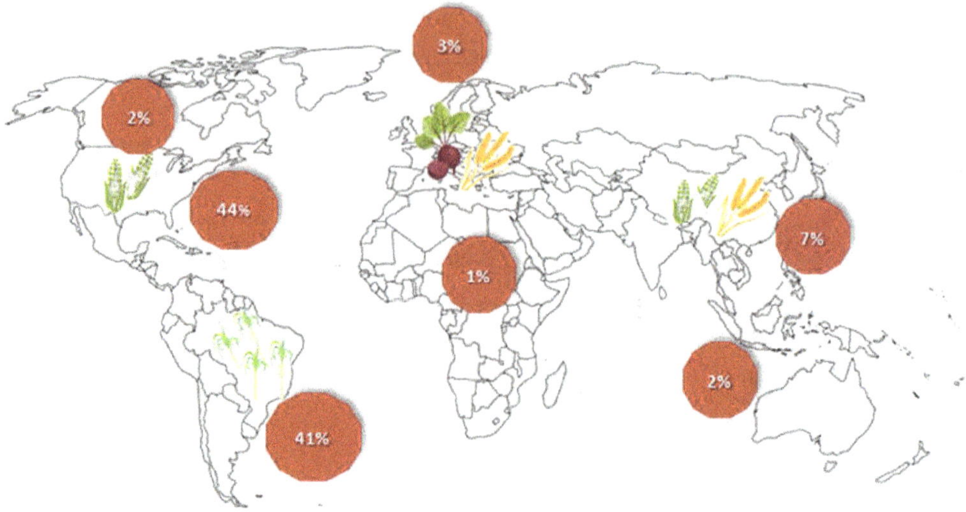

Figure 1. Global distribution of production and percentage of production per continent: Americas: South America—Brazil 43% (sugarcane); North America: USA 44% (corn) and Canada 2%; Europe 3% (vegetables and wheat); Africa 1% and Oceania 1%.

The ProÀlcool initiative witnessed a substantial surge in bioethanol production, soaring from 500 million liters at the program's inception to around 13 billion liters annually [6]. Subsequently, Brazil has been identified as a prospective bioethanol producer, with a robust strategy for domestic use [6, 7]. Brazil's bioethanol production for the 2016-2017 harvests is anticipated to be 33.2 billion liters [8]. This production level has remained consistent since 1986. The whole production is derived from sugarcane, which is cultivated using large-scale technology that involves the use of novel cane types, good climate conditions, fertile soil, and innovative agricultural techniques [4, 6, 9]. Brazilian bioethanol production is based on first-generation methods, in which sucrose, a natural sugar extracted from sugarcane, is fermented by yeast to produce ethanol [10]. The sugarcane plant utilized for Brazilian bioethanol production is obtained from the hybridization of 637 species from the Saccharum genus, which belongs to the Poaceae family and the Andropogoneae tribe. This plant is native to hot temperate climates and is defined by its stem and straw form [11, 12]. The stem serves as the source of sugarcane juice, which is then utilized in the manufacturing of sugar and bioethanol. Bagasse consists of post-grind materials and waste, namely the dried, green leaves of plants. These leaves are used in second-generation processes to produce bioethanol by fermentation [13]. Fermentation, a biochemical and biological process, has been recognized since ancient times. Its purpose is to convert sugar into ethanol (both anhydrous and hydrated), carbon dioxide, succinic acid, as well as volatile acids and esters [14]. The Brazilian fermentation method stands out for its distinctiveness and uniqueness, primarily because it is predominantly fed-batch in most states. This entails brief fermentation cycles and cell treatments with sulfuric acid [10]. This procedure utilizes cane juice as the primary substance, resulting in a finished product with a concentration of 9-12% (v/v) and an efficiency rate of 90-92% [15]. The bioethanol yield is directly influenced by the sugar content in the must, which is a combination of molasses, water, and sugarcane juice. The ratio of bioethanol generated to the amount of raw material utilized varies accordingly. The process begins with the action of the invertase exoenzyme, which breaks down the sugar sucrose (a disaccharide) into glucose and fructose (structural monosaccharides). These monosaccharides are then absorbed by facultative aerobic microorganisms. Under anaerobic conditions, these microorganisms undergo the pyruvic acid cycle, where the enzymes pyruvate, decarboxylase, and alcohol dehydrogenase work together to produce bioethanol and its byproducts at the end of fermentation [9]. During the national fermentation process, an important step is the centrifugation and sulfuric acid wash of the yeasts at the end of each fermentation cycle. This is done to reduce the likelihood of contamination [10]. After completing the therapy, the cells are reintroduced into the fermenters as a fresh inoculum for the next cycle. This process is done twice every day for a period of 6-9 months, as seen in Figure 2 [16].

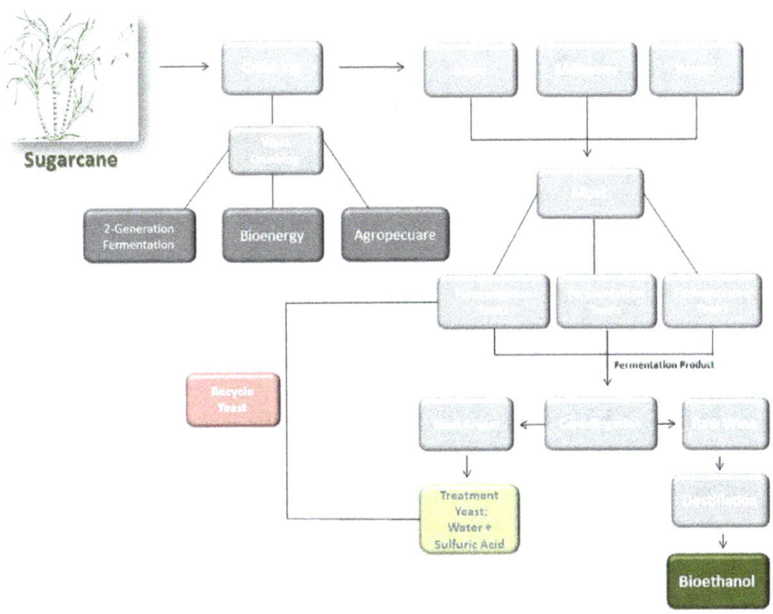

Figure 2. Simplified scheme of a fed-batch fermentation process with the recycling of yeast cells in Brazilian distillers by fermenting of sugarcane.

The fermentation process involving grains, particularly in the USA and China, is characterized by a high carbohydrate content. This necessitates the liquefaction and hydrolysis stage of the raw material, during which starch molecules are broken down into fermentable sugars. This breakdown enables fermentation to take place, as depicted in Figure 3 [17]. One notable feature of grain fermentations is the absence of recycled yeast cells, as seen in Brazilian mills. This is because the concentration of residues and fermentable products is preserved for distillation, resulting in a shorter fermentation process compared to the cell recycle method [18].

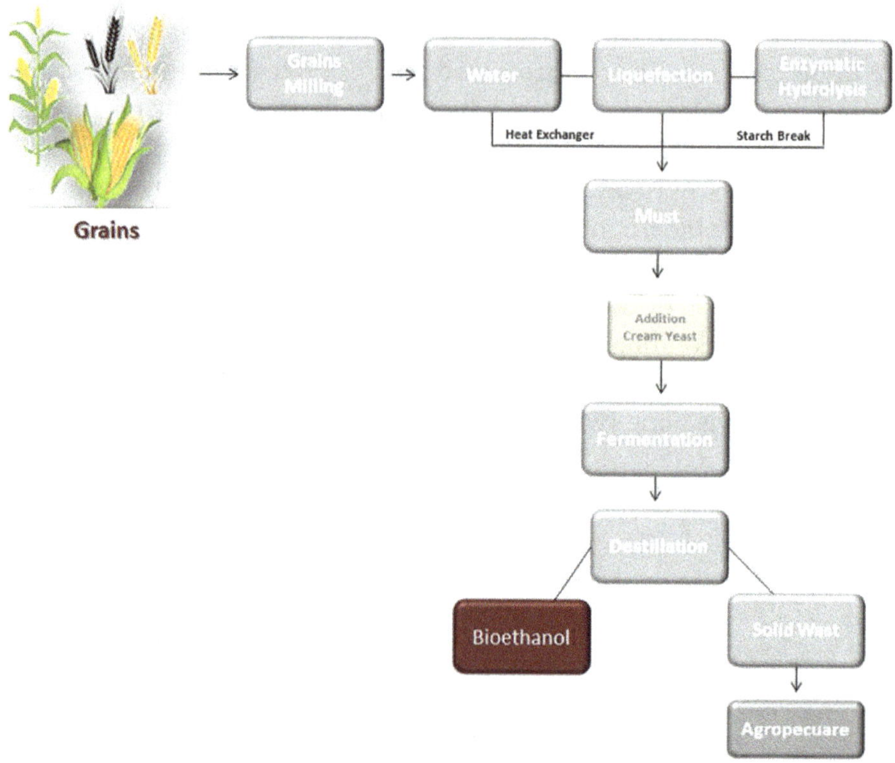

Figure 3. Simplified scheme of grains fermentation process with the liquefaction and hydrolysis in North American's and Chinese's distillers by fermenting of corns and wheat.

Significant changes are also observed when comparing the Brazilian and North American fermentation processes. In the fermentation of sugarcane, we have a lower concentration of solid residues, a concentration of larger yeast cells, and a much shorter time for bioethanol production, 6–12 h. The US process, that is derived for approximately 45–60 h, presents the advantages of a final concentration of bioethanol of 12–18%, against the 7–12% of the Brazilian process, and the raw material, that comes from corn plantations, lasts approximately one year, as opposed to the sugarcane harvest and its losses with rains that last around 200–240 day per year [19].

The main microorganisms used for the fermentation process are yeasts, such as Saccharomyces sp., Schizosaccharomyces sp., Kluyveromyces sp., among others

[13]. Currently, the most used yeast in the sugar and alcohol sector, for fermentation processes in the production of bioethanol is the specie Saccharomyces cerevisiae [20]. The methodologies used for the identification of yeasts based on morphology, biochemical characteristics, and sexual reproduction require the evaluation of 70–90 tests to obtain the identification of species. Macroscopic and microscopy features may be the first method of identification of S. cerevisiae yeasts, as presented in Figure 4 [21, 22].

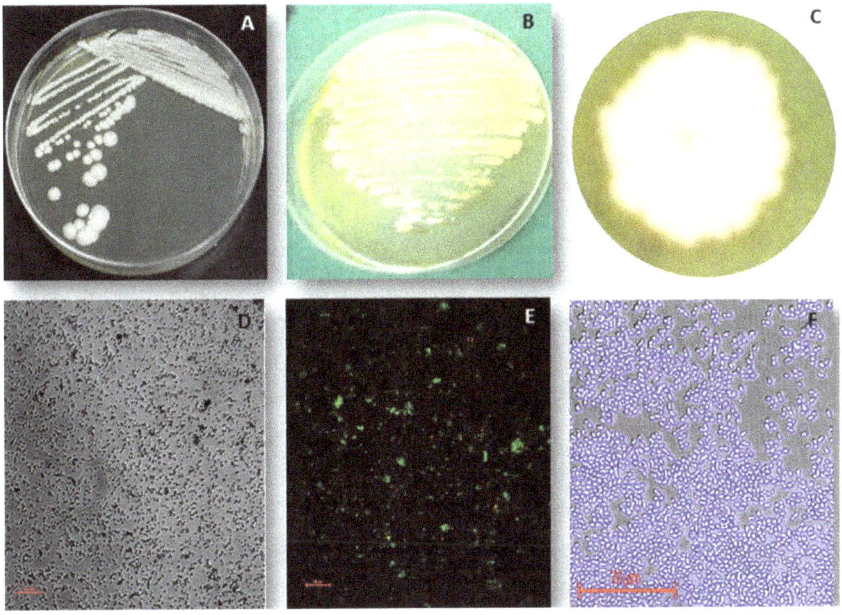

Figure 4. Macroscopic and microscopy of yeasts Saccharomyces cerevisiae. Macroscopies (A), (B), and (C) of PE-2 isolates, grown in YEPD (yeast extract-peptone-dextrose growth medium) solid with creamy and yellowish-white culture characteristics. Microscopies: (D) clear field microscopy of PE-2 isolate; (E) PE-2 isolate with FITC (fluorescein isothiocyanate fluoroforo) and Propidium Iodide cell tags; (F) PE-2 isolate with Calcofluor cellular target; Microscopy presence of oval yeasts with budding presence, with size of approximately 4–8 µm. Microscopies were performed in IN Cell Analyzer, objective of 20× and diameter 70 µm.

Molecular approaches have been created as substitutes for traditional procedures in order to identify and characterize yeasts. These techniques have the benefit of creating a separate representation of the genes, which enables rapid and precise identification of yeast species [23]. Novel molecular approaches are being developed to characterize and analyze polymorphisms in wild yeasts, owing to their high mutation potential [24]. Various genetic techniques, including as DNA analysis, electrophoretic karyotyping, rRNA sequencing, rDNA restriction analysis, and polymerase chain reaction (PCR) and restriction fragment length polymorphism (RFLP), have been employed to differentiate Saccharomyces sp.

strains within the strict sense group [25]. Melo Pereira's studies [26] created two novel pairs of specific primers for the HO gene of Saccharomyces bayanus, S. cerevisiae, and Saccharomyces pastorianus. These primers enable rapid PCR amplification, allowing for the accurate identification of these species in under 3 hours. Guillamón [27] and Oliveira [28] used ribosomal DNA RFLP of ITS1, ITS2, and 5.8S to identify various yeast species that were isolated during wine fermentation. They were also able to examine the range of yeast species present during spontaneous fermentation. S. cerevisiae is a yeast that can grow in media containing simple sugars and disaccharides. It has a high capacity for genetic alterations and exhibits strong resilience to adverse conditions in the growth medium. These characteristics make it crucial for fermentation operations [29]. Certain strains of S. cerevisiae possess the ability to be extremely productive, exerting control over the entire fermentation process during the harvest period. This enables efficient and consistent fermentations, leading to reduced costs and improved fermentation performance (with a high ethanol production capacity), as well as maintaining high viability throughout the process [30]. Research suggests that S. cerevisiae has the ability to thrive in various conditions, making it a valuable resource for studying the phenotypic characteristics of the Saccharomyces sp. species evolution [31, 32]. A recent study has demonstrated that the interplay between ambient factors and organisms can impact the recognition of distinct traits in S. cerevisiae [33, 34]. Saccharomyces cerevisiae is extensively utilized and grown in industrial fermentation processes because of its remarkable ability to adapt to various environmental conditions, including varying sugar and ethanol concentrations, pH levels, oxygen levels, resistance to contaminants, salt stress, protein stress, temperature fluctuations, and osmotic pressure [35]. Gao et al. [36] suggest that utilizing thermophilic bacteria is advantageous in processes involving simultaneous saccharification and fermentation (SSF). This is because SSF may operate at temperatures ranging from 45 to 50°C, leading to increased bioethanol production. This phenomenon arises due to the yeast experiencing reduced harm as the temperature rises, resulting in a decreased likelihood of microbial contaminations. Therefore, it is preferable for the thermostable yeast fermentation to take place at the enzyme's optimal temperature in order to maximize the process of ethanol synthesis. Consequently, enhancing thermotolerance in yeast leads to reduced production expenses, higher ethanol yield in the simultaneous saccharification and fermentation (SSF) system, and less risk of contamination. Understanding the fermentative yeasts is crucial for managing and overseeing alcoholic fermentation, particularly when seeking certain traits of dominance and resistance in bioethanol yeast production. The purpose of this chapter is to separate the primary phenotypic traits of S. cerevisiae yeasts during alcoholic fermentation, in order to potentially select novel strains with distinct phenotypic characteristics that are resistant and well-suited for bioethanol production.

2. PHENOTYPIC PARAMETERS OF S. CEREVISIAE IN ALCOHOLIC FERMENTATION

2.1. Flocculation test

Cell flocculation of yeast strains such as S. cerevisiae is called cell aggregation and sedimentation in liquid media [37]. Cells have the characteristics of agglomeration at the end of each fermentation process, which makes it an interesting and divergent phenomenon in the industry, as shown in Figure 5 [38]. Studies point out some divergences in flocculation, which can be a phenomenon of cooperative protection mechanism found in cells during adverse factors in the fermentation cycle [39]. It also facilitates the separation of yeasts at the end of the fermentation by sedimentation, thus helping the collection, centrifugation, and cellular treatment, contributing to the new inoculum stage, for a next cycle in the fermenter [40].

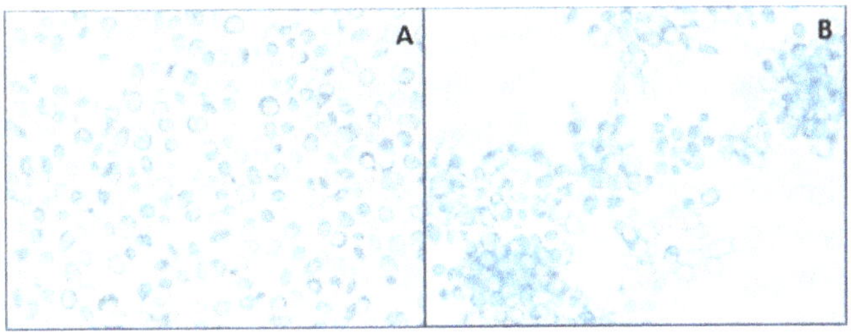

Figure 5. Characterization of aggregative power of Saccharomyces cerevisiae yeast cells, grown in YEPD liquid medium and stained with lactophenol. (A) Cells in normal condition without aggregative power. (B) Cells in condition of cellular aggregation.

The principle of cell adhesion is initiated through the recognition of mannose chains, located on cell surfaces, by lectin-like proteins, and agitation is necessary for the beginning of flocculation [41]. Two hypotheses for flocculation are well established: (I) sensitively to proteinases; (II) inhibited by saccharides, suggesting the existence of a protein that recognizes these sugars [42].

These two hypotheses classify two flocculation groups in yeasts that are distinguished by the inhibition of sugar: the first group, called New Flo phenotype, characterized by the inhibition of mannose, glucose, maltose, and sucrose with the exception of galactose; and the second group, the Flo1, is inhibited by mannose, but not by glucose, maltose, sucrose, and galactose, and its action is normally bound to a gene [43].

It is believed that these two distinct phenotypes are caused by two different proteins of the lectin type. Furthermore, the physical-chemical interaction in the cells surface may be involved in the aggregation process, where there is a correlation between flocculation and electrophoretic mobility of yeast cells on certain stress conditions. Other studies reported the correlation of hydrophobicity in the process of flocculation [44].

Another hypothesis for flocculation is the action of a dominant gene family (FLO1, FLO5, FLO8, FLO9, FLO10, and FLO11), where they encode a yeast cell wall protein that acts directly on the cell aggregation [45]. The proteins encoded by these FLO genes share a cellular/modular organization in three domains: an amino-terminal responsible for carbohydrate binding, a central domain, and a carboxyl-terminal domain containing a glycosylphosphatidylinositol anchor sequence [46]. However, the central domain contains tandem repeat regions of DNA sequence that can drive recombination reactions within and between FLO genes, resulting in new generations of FLO alleles, thus conferring yeast cells a wide diversity in the flocculation phenomenon [47].

2.2. Sensitivity test temperature and ethanol

The environmental adversities occurring in a fermentation cycle, such as the decrease of nutrients by sugar consumption, temperature changes, pH changes, risk of contamination, phenolic compounds, and the concentration of ethanol by its own production occurs in different forms and some of them were completely studied [48]. An understanding of the cellular mechanisms of protection to the multiphysical and chemical stresses that the yeast undergoes during fermentation cycle is fundamental for the selection of ideal yeast [49].

Temperature elevations result in reduced fermentation efficiency in S. cerevisiae, due to the high fluidity in the membranes, caused by the altered composition of fatty acids in the adverse response [50]. As one of the stress factors known in the fermentative cycle in yeast, temperature change restricts ethanol production and induces the accumulation of proteins bound to tolerance stress [51].

In the first-generation fermentation cycle, yeasts require a temperature of 30°C, whereas, in the production of second-generation bioethanol, where cellulose enzymes start the process by saccharification, yeasts require a higher temperature of 45–50°C [52]. The efficacy of the fermentation is decreased at high temperatures, because it causes damage to the yeast cell, such as the rupture of the protein structure or the loss of function, thus preventing cell proliferation, decreasing viability during the process, and leading to cell death [53]. This temperature control in the fermentative cycles is a problem for the plants in tropical countries, where the ambient temperature is already naturally high and cooling systems are necessary for the total control of this temperature [54].

In the bioethanol production, the process temperature must be stabilized at around 30°C (the cell growth temperature), which is reaching 40°C [55, 56].

Thus, thermotolerant yeast strains may be a promising approach to a profitable fermentation process, as is the case of simultaneous saccharification and fermentation that requires high temperatures to increase ethanol yield [36].

Osmotolerance can be an important factor in the production of ethanol for its adaptation strategy employed in all cell types by accumulating compatible solutes (sulfite), resulting in a decrease in the potential of intracellular water [57]. As sulphite and sulfite-generating compounds have long been used as antimicrobial agents in alcoholic fermentation, tolerance to sulfite in yeast is another desired characteristic for the production of bioethanol from sugarcane juice [58].

The high levels of ethanol in the fermentation medium are considered as negative parameters in the process conditions, because at the same time that the production is essential, the accumulation of ethanol by this production generates an acidification of the medium, leading to irreversible damages in the yeast membrane, thereby decreasing cell viability [59].

The true physiological and ecological relevance of ethanol tolerance in S. cerevisiae is its ability to generate mechanisms that protect the cell from chemical and physical damage at high levels of ethanol [60]; this is usually observed in a typical fermentation environment, where there is a large amount of sugars, leading later to ethanol production [30]. This stage generally occurs by stationary phase cells and its tolerance to the ethanol produced is only controlled by the integrity of the yeast membrane in contact with the ethanol accumulation, which is composed of chitin, glucans, glycoproteins, fatty acids, and ergosterol [61].

However, S. cerevisiae is resistant to ethanolic stress for its capacity of modifying the conformation of its membrane in the increase of fatty acids and ergosterol when coming in contact with the adverse environment, thus neutralizing the damages caused, mainly in relation to its viability [62, 63]. The accumulation of ethanol can also affect the structural compliance of the cellular proteins causing the inefficiency of its actions, such as the decrease of the activity of glycolytic enzymes: pyruvate kinase and hexokinase, besides altering the absorption of glucose, maltose, and amino acid. In some cases, there may occur cellular extravasation of essential cellular components [64].

In industrial fermentations, a high capacity of production is observed by the accumulation of ethanol in the medium, indicating a positive assimilation of residual sugar, which is measured by the visualization of cellular proliferation in the presence of the gradual levels of ethanol produced during the fermentative process [65]. Tolerance and ethanol characteristics of the main industrial strains of S. cerevisiae studied are described in Table 1 [30, 66–69].

Table 1. Characteristics of temperature tolerance and ethanol for major industrial strains worldwide used for the production of bioethanol.

Strains	Group	Origin	Feedstocks	Temperature tolerance (°C)	Ethanol tolerance (%)
ZTW1	Industrial (fuel ethanol)	China	Grains	55	18
YJ5329	Industrial (fuel ethanol)	China	Grains	55	18
PE-2	Industrial (fuel ethanol)	Brazil	Sugarcane	40	15
CAT-1	Industrial (fuel ethanol)	Brazil	Sugarcane	40	15
AT-3	Industrial (fuel ethanol)	USA	Grains	40	14
E r O h red	Industrial (fuel ethanol)	USA	Grains	40	15

Although this assay is routinely used in industries as large-scale screenings, its actual importance in ethanol resistance in yeast is not elucidated, due to divergent of actions that this process can cause, for example, the negative side acting in the decrease of the cell viability, and positive the increase in resistance to contaminating microorganisms in the fermentation process [70].

The metabolic pathways correlated to the expression of genes responsive to high levels of heat stress and ethanol stress include heat shock proteins (HSPs) and also metabolic enzymes such as trehalose, which is directly involved in tolerance in S. cerevisiae [71]. HSPs play a role in folding and refolding, transport, and degradation of intracellular proteins, triggered by stress in fermentation process and located in the cytoplasm, nucleus, and mitochondria, acting immediately in response to an accumulation of denatured proteins, activating the transcription factors of thermal shock (HSF), and leading to a positive regulation of thermotolerance gene expression [72].

The interactions of multiple genes at loci for cellular functions under heat and ethanol stresses are essential [73]. HSPs are known as chaperones ensuring the functional and structural conformation of the yeast, on the action of genes such as SSA1, SSA2, SSA3, and SSA4 which are expressed together with the HSP genes HSP12, HSP26, HSP30, HSP31, and HSP150 which were also found active at high stress levels [74] and interactions between chaperones of different types are widely encountered [75].

However, the inference of several chaperones shows an effective activity in neutralizing the stress, with the activation of the functional chaperones specific to more complex structures in the yeast cell walls [76], which have as a main function to repair of these denatured proteins to maintain cell viability [77].

In addition to serving as chaperones, HSPs have numerous other functions, for example, Hsp30p is characterized as a hydrophobic plasma membrane protein that acts on the regulation of H+-ATPase, Hsp31p, and Hsp32p functions as hydrolases and peptidase, and Hsp150 is characterized as a protein in supporting the cell wall stability and remodeling [78]. HSPs and chaperone-mediated genomic regulation are also linked to glucose metabolism, which are indispensable tools for stress tolerance in yeast metabolism, especially with storage of carbohydrates, such as trehalose [79].

Trehalose is a compound that acts to prevent the influx of excess salts resulting in irreversible dehydration of cells; therefore, yeasts are capable of accumulating trehalose up to 15% in a stress environment [80]. The trehalose acts by reducing the permeability of the membrane thereby rendering it hydrophobic, due to some regulatory genes such as TPS1, TPS2 and, TSL1, as well as acting in the remodeling of proteins under stress conditions [81]. Cells incapable of accumulating trehalose presented depreciated growth, leading to a significant decrease in cell viability during fermentation stresses [82].

2.3. Assimilation of sugars

Sugarcane juice is one of the main means used in the production of bioethanol, which is derived from the break of fermentable sugars such as sucrose, glucose, and fructose in contact with fermenting microorganisms such as yeast S. cerevisiae [83]. Yeast consumes the sugars in the medium in a complex and highly regulated manner, the principle of fermentation, where the sucrose is consumed first, followed by glucose and fructose, and finally maltose, this assimilation of sugars can occur simultaneously between the breaks of sugars, which is the standard process for sequential uptake of the glucose repression pathways or the catabolite repression pathway [84].

Glucose and sucrose may trigger beneficial effects on cells, including stimulation of cell proliferation, mobilization of storage compounds such as glycogen and trehalose, as well as decreased resistance to cell stress [85]. In contrast, negative impacts due to lack of glucose in the process can lead to several problems such as decreased or blocked fermentations, instability of cellular viability and low ethanol production [86], where the break of sugars, sucrose into simple sugars (glucose) occurs by an intracellular enzyme known as invertase, located in wall the yeast industries [87].

Microorganisms that possess the ability to assimilate the highest amount of sugars are indicated for the production of bioethanol, examples are shown in Table 2 characterizing the main strains of S. cerevisiae worldwide used in industries for the production of bioethanol [30, 66–69].

Table 2. Characteristics of the assimilation of residual sugars and ethanol production for large industrial strains of Saccharomyces cerevisiae yeasts used in the production of bioethanol.

Strains	Group	Origin	Feedstocks	Assimilation sugar (%)	Production bioethanol (%)
ZTW1	Industrial (fuel ethanol)	China	Grains	65	28
YJ5329	Industrial (fuel ethanol)	China	Grains	60	33
PE-2	Industrial (fuel ethanol)	Brazil	Sugarcane	51	22
CAT-1	Industrial (fuel ethanol)	Brazil	Sugarcane	52	26
AT-3	Industrial (fuel ethanol)	USA	Grains	42	18
E r O h red	Industrial (fuel ethanol)	USA	Grains	75	30

The assimilation of sugars in the fermentation process is not exclusively the fermentation of sugarcane. Currently, new technologies are available to produce ethanol from vegetables such as potatoes, cassava, beets, cereals such as corn, and there are also studies showing the production of green bioethanol in algae fermentations [88].

This type of fermentation is due to the breakdown of starch, carried out by the action of the enzyme glucoamylase, acting directly on the conversion of starch to glucose, by breaking the successive bonds of the nonreducing end of the glucose finally producing straight chains [89]. The process of producing ethanol from starch involves two main steps: enzymatic hydrolysis as the main step and habitual fermentation as the second step [90].

2.4. Second-generation bioethanol

All adverse parameters studied for the first-generation fermentation process have been highly researched to reach an ideal model of production of second-generation bioethanol, which is characterized by being profitable and environmentally sustainable [91].

Second-generation bioethanol production starts from the lysis of the raw material (sugarcane bagasse, vinasse, and residues from the milling of grains). The main step is characterized by a pretreatment where the breakdown of the cellulose-hemicellulose-lignin complex allows the production of fermentable sugar levels for a subsequent fermentation, demonstrated in Figure 6 [92].

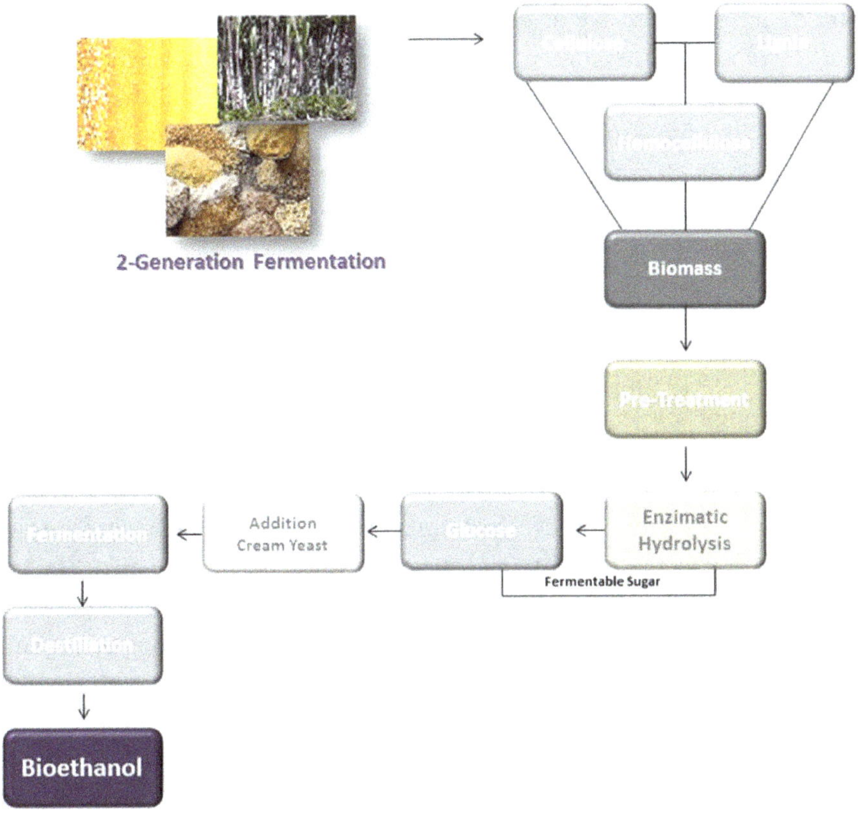

Figure 6. Simplified scheme of second-generation fermentation process with steps of biomass and hydrolysis for bioethanol production worldwide.

The hydrothermal and lime pretreatments are the most used, known for making the method more effective in preparing the biomass bioconversion step [93], a strong advantage for the sugarcane bagasse. They can be carried out under conditions of low temperature and pressure, resulting in lower sugar degradation, whereas in the saccharification the pretreatment is observed with high temperatures and difficult breaks of carbohydrate chains, resulting in a lower amount of sugars [94].

The fermentation of lignocellulose hydrolysates for bioethanol production presents two main problems: first, the fermentation of xylose that requires a low and controlled oxygenation; second, the removal of microbial inhibitors, which can contaminate the process [95]. Furthermore, these yeasts present a certain tolerance limited to ethanol [96].

Genetic manipulation in the metabolism of xylose in yeast fermentation has advanced and pioneering studies on glucose transporters that mediate xylose uptake, allyl-xylitol-reductase genes, xylitol dehydrogenase and xylulokinase have been expressed, which allows a better assimilation and fermentation of xylose [97].

The main concern in this step is that the balanced supply of NADP (enzyme nicotinamide adenine dinucleotide phosphate) and NADPH (enzyme nicotinamide adenine dinucleotide phosphate oxidase) has to be constant to avoid the production of xylitol. The path is the reduction of NADPH production by blocking the oxidative pentose phosphate cycle in xylose assimilation [98]. Cellulose hydrolysates present different inhibitors from lignin derivatives and sugar degradation, resulting in high amounts of acetic acid, intrinsically necessary for the deconstruction of biomass [99].

All adversities of typical fermentation first generation associated with pH, temperature, elevation of ethanol concentrations, and temperature among other stress factors present in large-scale fermentations are seen together in the adversity challenges of second-generation fermentation [100].

2.5. Progress and outlook

Historically, first-generation fermentation has mostly been employed in the context of large-scale industrial applications. While the process is widely recognized, it is not fully understood. Each new process launched reveals observable changes in the fundamental characteristics and behavior of the yeasts employed. Saccharomyces cerevisiae possesses the quality of adaptability to many environments, resulting in a multitude of behavioral reactions during fermentation. With each new cycle, changes are unavoidable, making it crucial to thoroughly monitor all fermentation parameters to ensure the success of the process. Flocculation is a complex characteristic that can provide several benefits, such as cellular defense against stressors and contaminants in a process. However, it also has drawbacks, such as reduced fermentation yield in fermenting tanks due to sedimentation. The impact of this phenomenon on flocculation remains obscure, but its morphology and molecular activity in yeast cells are well understood. Furthermore, it plays a significant role in enhancing bioethanol production in the global industry. In order to achieve optimal efficiency in alcoholic fermentation, it is crucial to have a comprehensive understanding of the entire process, including the key stages and the various degrees of activity that microorganisms undergo over the course of hours and days to maintain their viability and productivity. The research and understanding of the two primary factors that influence fermentation, namely heat and ethanol, are crucial at the onset of the process, whether it is conducted on a small scale, such as in laboratories, or on a big one, such as in industrial operations. The behavioral reactions of fermentation are continuously measured using these factors, with the goal of constantly improving the process. An essential requirement for efficient bioethanol production is the microorganism's capacity to enzymatically degrade carbohydrates and subsequently metabolize them through fermentation. It is crucial to thoroughly analyze and plan this intricate procedure to prevent any harm, particularly when carried out on a large industrial scale. It is imperative to have both an ample quantity of microorganisms and an appropriate number of sugars in order to initiate the production successfully. Considering the present situation, the initial processes underwent modernization and research, leading to the development of second-generation fermentation. This approach strives to optimize the utilization of all remaining leftovers and achieve more sustainable operations. These procedures are continuously advancing and improvementsareconsistentlybeingobservedinordertoachievetheoptimalmethod. Considering this profile, it is crucial to look for yeasts with stronger attributes in industrial settings. Genetic engineering research emphasizes several methodologies involving adaptation and functionality. Progress in the field of process optimization and optimal yeast strains is promising, but there is still a long way to go before achieving perfection. The multitude of secrets and enigmas surrounding fermentation is vast, but ongoing study is continuously transforming our understanding of this process. Gradually, we are unraveling the complexities and intricacies involved, with the key starting point being a comprehensive comprehension of all the procedures and factors involved.

AT A GLANCE

Globally, bioethanol production mostly relies on first-generation technology, which involves fermenting fruits, vegetables, and grains with high sucrose content using yeast. Brazil is one of the top global producers of bioethanol made from sugarcane. The bioethanol production process is intricate and influenced by several environmental conditions, leading to distinct phenotypic profiles of the strain employed. Research has shown that the interplay between environmental conditions and microorganisms can impact the identification of several properties of Saccharomyces cerevisiae. Furthermore, the production of bioethanol is achieved by the utilization of second and third generation technologies. Additionally, the implementation of novel yeast strains could enhance the practicality of bioethanol production. The successful execution of fermentation relies on the yeast's capacity to handle several conditions that arise during the process, including sugar concentration, ethanol levels, nitrogen content, pH levels, resistance to pollutants, stress protein production, temperature fluctuations, and osmotic pressure.

REFERENCES

1. Andrietta MGS, Andrietta SR, Steckelberg C, Stupiello ENA. Bioethanol – Brazil, 30 years of Proálcool. Inernational Sugar Journal. 2007;109:195-200

2. Scholey DV, Burton EJ, Williams PE. The bio refinery; producing feed and fuel from grain. Food Chemistry. 2016;197:937-942. DOI: 10.1016/j.foodchem.2015.11.063

3. Beltrade E, Vandenbergue LPS, Soccol CR, Sigoillot JC, FauldsC. First generation bioethanol. In: Soccol CR, Brar SK, Faulds C, Ramos LP, editors. Green Fuels Technology. Switzerland: Springer International Publishing; 2016. pp. 175-212

4. Amorim HV, Lopes ML, Oliveira JVC, Buckeridge MS, Goldman GH. Scientific challenges of bioethanol production in Brazil. Applied Microbiology and Biotechnology. 2011;91:1267-1275. DOI: 10.1007/s00253-011-3437-6

5. Macedo IC, Seabra JEA, Silva JE. Green house gases emissions in the productions and use of ethanol from sugarcane in Brazil: The 2005/2006 averages and a prediction for 2020. Biomass and Bioenergy. 2008;32:582-595. DOI: 10.1016/j.biombioe.2007.12.006

6. Zanin GM, Santana CC, Bon EP, Giordano RC, Moraes FF, Andrietta SR, Carvalho CC, Macedo IC, Fo DL, Ramos LP, Fontana JD. Brazilian bioethanol program. Applied Biochemistry and Biotechnology. 2000;86:1147-1161. DOI: 10.1385/ABAB:84-86:1-9:1147

7. Goldemberg J. The Brazilian biofuels industry. Biotechnology for Biofuels. 2008;1:6-15. DOI: 10.1186/1754-6834-1-6

8. CONAB. Levantamento da safra 2016/17. 2017. Available from: www.conab.gov.br. [Accessed: 2017/01/20]

9. Amorim HV, Lopes ML. Ethanol production in a petroleum dependent world: The Brazilian experience. Sugar Journal. 2005;67:11-14. DOI: 20053116848

10. Brow NA, Castro PA, Figueiredo BCP, Savoldi M, Buckeridge MS, Lopes ML, Paullilo SCL, Borges EP, Amorim HV, Goldman MHS, Bonatto D, Malavazi I, Goldman GH. Transcriptional profiling of Brazilian Saccharomyces cerevisiae strains selected for semi-continuous fermentation of sugarcane must. FEMS Yeast Research. 2013;13:277-290. DOI: 10.1111/1567-1364.12031

11. Paes LAD, Oliveira MA. Potential trash biomass of the sugar cane plant. 2005; p. 19-23. In: Hassuani SJ, Leal MRLV, Macedo IC. Biomass power generation: sugarcane, bagasse and trash. Piracicaba, SP, Brazil. 2005

12. Canilha L, Chandel AK, Milessi TSS, Antunes FAF, Freitas WLC, Felipe MGA, Silva SS. Bioconversion of sugarcane biomass into ethanol: An overview about composition, pretreatment methods, detoxification of hydrolysates, enzymatic saccharification and ethanol fermentation. Journal of Biomedicine and Biotechnology. 2012;2:1-15. DOI: 10.1155/2012/989572

13. Ensinas AV, Modesto M, Nebra SA, Serra L. Reduction of irreversibility generation in sugar and ethanol production from sugarcane. Energy. 2009;34:680-688. DOI: org/10.1016/j.energy.2008.06.001

14. Amorim HV. Ethanol Production in Brazil: a Successful History. Proceedings of the Sugar Processing Research Conference. 2006;1:44-47.

15. Basílio ACM, Araújo PRL, Morais JOF, Silva-Filho EA, Morais MA, Simões DA. Detection and identification of wild yeast contaminants of the industrial fuel ethanol fermentation process. Current Microbiology. 2008;56:322-326. DOI: 10.1007/s00284-007-9085-5

16. Barbrzadeh F, Jalili R, Wang C, Shokralla S, Pierce S, Robinson-Mosher A, Nyren P, Shafer RW, Basso LC, Amorim HV, Oliveira AJ, Davis RW, Ronaghi M, Gharizadeh B, Stambuk BU. Whole-genome sequencing of the efficient industrial fuel-ethanol fermentative Saccharomyces cerevisiae strain CAT-1. Molecular Genetics and Genomics. 2012;287:485-494. DOI: 10.1007/s00438-012-0695-7

17. Christofoletti CA, Escher JP, Correia JE, Marinho, JF, Fontanetti CS. Sugarcane vinasse: Environmental implications of its use. Waste Management. 2013;33:2752-2761. DOI: 10.1016/j.wasman.2013.09.005

18. Ingledew WD, Austin GD, Kelsall DR, Kluhspies C. The alcohol industry: How has it changed and mature? In: Ingledew WM, Austin GD, Kelsall DR, Kluhspies C, editors. The Alcohol Textbook. A Reference for the Beverage, Fuel and Industrial Alcohol Industries. Nottingham: Nottingham University Press; 2009. pp. 1-6

19. Martinez-Amezcua C, Parsons CM, Singh V, Srinivasan R, Murthy GS. Nutritional characteristics of corn distillers dried grains with solubles as affected by the amounts of grains versus solubles and different processing techniques. Poultry Science. 2007;86:2624-2630. DOI: 10.3382/ps.2007-00137

20. Godoy A, Amorim HV, Lopes ML, Oliveira AJ. Continuous and batch fermentation process: Advantages and disadvantages of these process in the Brazilian ethanol production. International Sugar Journal. 2008;110:175-181

21. Deák T. Methods for the rapid detection and identification of yeasts in foods. Trends in Food Science & Technology. 1995;6:287-292. DOI: 10.1016/S0924-2244(00)89138-4

22. Karabıçak N, Uludağ Altun H, Karatuna O, Hazırolan G, Aksu N, Adiloğlu A, Akyar I. Evaluation of common commercial systems for the identification of yeast isolates in microbiology laboratories: A multicenter study. Mikrobiyol Bul. 2015;49:210-220. DOI: 10.5578/mb.9370

23. Bernardi T, Pereira G, Cardoso P, Dias E, Schwan R. Saccharomyces cerevisiae strains associated with the production of cachaca: Identification and characterization by traditional and molecular methods (PCR, PFGE and mtDNA-RFLP). Journal of Microbiology and Biotechnology. 2008;24:2705-2712. DOI: 10.1007/s11274-008-9799-y

24. Song HT, Liu SH, Gao Y, Yang YM, Xiao WJ, Xia WC, Liu ZL, Li R, Ma XD, Jiang ZB. Simultaneous saccharification and fermentation of corncobs with genetically modified Saccharomyces cerevisiae and characterization of their microstructure during hydrolysis. Bioengineered. 2016;7:198-204. DOI: 10.1080/21655979.2016.1178424

25. Muller H, Annaluru N, Schwerzmann JW, Richardson SM, Dymond JS, Cooper EM, Bader JS, Boeke JD, Chandrasegaran S. Assembling large DNA segments in yeast. Methods in Molecular Biology. 2012;852:133-150. DOI: 10.1007/978-1-61779-564-0_11

26. de Melo Pereira GV, Ramos CL, Galvão C, Souza Dias E, Schwan RF. Use of specific PCR primers to identify three important industrial species of Saccharomyces genus: Saccharomyces cerevisiae, Saccharomyces bayanus and Saccharomyces pastorianus. Letters in Applied Microbiology. 2012;51:131-137. DOI: 10.1111/j.1472-765X.2010.02868.x

27. Guillamón JM, Sabaté J, Barrio E, Cano J, Querol A. Rapid identification of wine yeast species based on RFLP analysis of the ribosomal internal transcribed spacer (ITS) region. Archives of Microbiology. 1998;169:387-392. DOI: 10.1007/s002030050587

28. Oliveira VA, Vicente MA, Fietto LG, Castro IM, Coutrim MX, Schüller D, Alves H, Casal M, Santos JO, Araújo LD, da Silva PH, Brandão RL. Biochemical and molecular characterization of Saccharomyces cerevisiae strains obtained from sugar-cane juice fermentations and their impact in cachaca production. Applied and Environmental Microbiology. 2008;74:693-701. DOI: 10.1128/AEM.01729-07

29. Hector RE, Mertens JA, Bowman MJ, Nichols NN, Cotta MA, Hughes SR. Saccharomyces cerevisiae engineered for xylose requires gluconeogenesis and the oxidate brach of the pentose phosphate pathway for aerobic xylose assimilation. Yeast. 2011;28:645-660. DOI: 10.1002/yea.1893

30. Basso LC, Amorim HV, Oliveira AJ, Lopes ML. Yeast selection for fuel ethanol production in Brazil. FEMS Yeast Research. 2008;8:1155-1163. DOI: 10.1111/j.1567-1364.2008.00428.x

31. Lidzbarrsky GA, Shkolnik T, Nevo, E. Adaptive response to DNA-damaging agents in natural Saccharomyces cerevisiae populations from "Evolution Canyon", Mt. Carmel, Israel. Plos One. 2009;4:1-8. DOI: 10.1371/journal.pone.0005914

32. Liti G, Carter DM, Moses AM, Warringer J, Parts L, James SA, Davey RP, Roberts IN, Burt A, Koufopanou V, Tsai IJ, Bergman CM, Bensasson D, O'Kelly MJ, van Oudenaarden A, Barton DB, Bailes E, Nguyen AN, Jones M, Quail MA, Goodhead I, Sims S, Smith F, Blomberg A, Durbin R, Louis EJ. Populations genomics of domestic and wild yeasts. Nature. 2009;458:337-344. DOI: 10.1038/nature07743

33. Camarasa C, Sanchez I, Brial P, Bigely F, Dequin S. Phenotypic landscape of Saccharomyces cerevisiae during wine fermentation: Evidence for origin-dependent metabolic traits. Plos One. 2011;6:1-12. DOI: 10.1371/journal.pone.0025147

34. Warringer J, Cubillos FA, Zia A, Gjuvsland A, Simpson JD. Trait variation in yeast is defined by population history. Plos Genetics. 2011;6:1-15. DOI: 10.1371/journal.pgen.1002111

35. Bai FW, Anderson WA, Moo-Young M. Ethanol fermentation technologies from sugar and starch feedstocks. Biotechnology. 2008;26:89-105. DOI: 10.1016/j.biotechadv.2007.09.002

36. Gao L, Liu Y, Sun H, Li C, Zha Z, Liu G. Advances in mechanisms and modifications for rendering yeast thermotolerance. Journal of Bioscience and Bioengineering. 2015;121:599-606. DOI: 10.1016/j.jbiosc.2015.11.002

37. Smit G, Straver MH, Lugtenberg BJJ, KIjne JW. Flocculence of Saccharomyces cerevisiae cells is induced by nutrient limitation, with cell surface hydrophobicity as a major determinant. Applied and Environmental Microbiology. 1992;58:3709-3714

38. Alvarez F, CorreA LFM, Araújo TM, Mota BEF, Conceição LEFR, Castro IM, Brandão RL. Variable flocculation profiles of yeast strains isolated from cachaça distilleries. International Journal of Food Microbiology. 2014;190:97-104. DOI: 10.1016/j.ijfoodmicro.2014.08.024

39. Soares EV. Flocculation in Saccharomyces cerevisiae: A review. Journal of Applied Microbiology. 2010;110:1-18. DOI:10.1111/j.1365-2672.2010.04897.x

40. Smukalla S, Caldara M, Pochet N, Beauvais A, Guadagnini S, Yan C, Vinces MD, Jansen A, Prevost MC, Latgé JP, Fink GR, Foster KR, Verstrepen KJ. FLO1 is a variable green beard gene that drives biofilm-like cooperation in budding yeast. Cell. 2008;135:726-737. DOI: 10.1016/j.cell.2008.09.037

41. Kobayashi O, Hayashi N, Kuroki R, Sone H. Region of FLO1 proteins responsible for sugar recognition. Journal of Bacteriology. 1998;180:6503-6510

42. Ma K, Wakisaka M, Sakai K, Shirai Y. Flocculation characteristics of an isolated mutant flocculent Saccharomyces cerevisiae strain and its application for fuel ethanol production from kitchen refuse. Bioresource Technology. 2009;100:2289-2292. DOI: 10.1016/j.biortech.2008.11.010

43. Stratford M, Assinder S. Yeast flocculation: Flo1 and NewFlo phenotypes and receptor structure. Yeast. 1991;7:559-574. DOI: 10.1002/yea.320070604

44. Beavan MJ, Belk DM, Stewart GG, Rose AH. Changes in electrophoretic mobility and lytic enzyme activity associated with development of flocculating ability in Saccharomyces cerevisiae. Canadian Journal of Microbiology. 1979;25:88-95

45. Govender P, Domingo JL, Bester MC, Pretorius IS, Bauer FF. Controlled expression of the dominant flocculation genes FLO1, FLO5, and FLO11 in Saccharomyces cerevisiae. Applied and Environmental Microbiology. 2008;74:6041-6052. DOI: 10.1128/AEM.00394-08

46. Goossens KV, Willaert RG. The N-terminal domain of the Flo11 protein from Saccharomyces cerevisiae is an adhesin without mannose-binding activity. FEMS Yeast Research. 2012;12:78-87. DOI: 10.1111/j.1567-1364.2011.00766.x

47. Goossens KV, Ielasi FS, Nookaew I, Stals I, Alonso-Sarduy L, Daenen L, Van Mulders SE, Stassen C, van Eijsden RG, Siewers V, Delvaux FR, Kasas S, Nielsen J, Devreese B, Willaert RG. Molecular mechanism of flocculation self-recognition in yeast and its role in mating and survival. MBio. 2015;6:1-16. DOI: 10.1128/mBio.00427-15

48. Della-Bianca BE, de Hulster E, Pronk JT, van Maris AJ, Gombert AK. Physiology of the fuel ethanol strain Saccharomyces cerevisiae PE-2 at low pH indicates a context-dependent performance relevant for industrial applications. FEMS Yeast Research. 2014;14:1196-1205. DOI: 10.1111/1567-1364.12217

49. Kitichantaropas Y, Boonchird C, Sugiyama M, Kaneko Y, Harashima S, Auesukaree C. Cellular mechanisms contributing to multiple stress tolerance in Saccharomyces cerevisiae strains with

potential use in high-temperature ethanol fermentation. AMB Express. 2016;6:1-14. DOI: 10.1186/s13568-016-0285-x

50. Henderson CM, Zeno WF, Lerno LA, Longo ML, Block DE. Fermentation temperature modulates phosphatidylethanolamine and phosphatidylinositol levels in the cell membrane of Saccharomyces cerevisiae. Applied and Environmental Microbiology. 2013;79:5345-5356. DOI: 10.1128/AEM.01144-13

51. Ma M, Liu ZL. Mechanisms of ethanol tolerance in Saccharomyces cerevisiae. Applied Microbiology and Biotechnology. 2010;87:829-845. DOI: 10.1007/s00253-010-2594-2593

52. Liu ZH, Qin L, Zhu JQ, Li BZ, Yuan YJ. Simultaneous saccharification and fermentation of steam-exploded corn stover at high glucan loading and high temperature. Biotechnology for Biofuels. 2014;7:1-16. DOI: 10.1186/s13068-014-0167-x

53. Mejía-Barajas JA, Montoya-Pérez R, Salgado-Garciglia R, Aguilera-Aguirre L, Cortés-Rojo C, Mejía-Zepeda R, Arellano-Plaza M, Saavedra-Molina A. Oxidative stress and antioxidant response in a thermotolerant yeast. Brazilian Journal of Microbiology. 2017;205:1-7. DOI: 10.1016/j.bjm.2016.11.005

54. Wallace-Salinas V, Gorwa-Grauslund MF. Adaptive evolution of an industrial strain of Saccharomyces cerevisiae for combined tolerance to inhibitors and temperature. Biotechnology for Biofuels. 2013;6:1-9. DOI: 10.1186/1754-6834-6-151

55. Őhgren K, Bura R, Lesnicki G, Saddler J, Zacchi G. A comparison between simultaneous saccharification and fermentation and separate hydrolysis and fermentation using steam-pretreated corn stover. Process Biochemistry. 2007;42:834-839. DOI:10.1016/j.procbio.2007.02.003

56. Basso LC, Basso TO, Rocha SN. Ethanol production in Brazil: The industrial process and its impact on yeast fermentation. Biofuel Production. 2011;5:85-100. DOI: 10.5772/17047

57. Rep M, Krantz M, Thevelein JM, Hohmann S. The transcriptional response of Saccharomyces cerevisiae to osmotic shock. Hot1p and Msn2p/Msn4p are required for the induction of subsets of high osmolarity glycerol pathway-dependent genes. Journal of Biological Chemistry. 2000;275:290-300. DOI: 10.1074/jbc.275.12.8290

58. Linderholm AL, Findleton CL, Kumar G, Hong Y, Bisson LF. Identification of genes affecting hydrogen sulfide formation in Saccharomyces cerevisiae. Applied and Environmental Microbiology. 2008;74:1418-1427. DOI: 10.1128/AEM.01758-07

59. Stanley D, Bandara A, Fraser S, Chambers PJ, Stanley GA. The ethanol stress response and ethanol tolerance of Saccharomyces cerevisiae. Journal of Applied Microbiology. 2010;109:13-24. DOI: 10.1111/j.1365-2672.2009.04657.x

60. Klis FM, Boorsma A, De Groot PW. Cell wall construction in Saccharomyces cerevisiae. Yeast. 2006;23:85-202. DOI:10.1002/yea.1349

61. Dong SJ, Yi CF, Li H. Changes of Saccharomyces cerevisiae cell membrane components and promotion to ethanol tolerance during the bioethanol fermentation. The International Journal of Biochemistry & Cell Biology. 2015;69:196-203. DOI: 10.1016/j.biocel.2015.10.025

62. Gibson BR, Lawrence SJ, Leclaire JP, Powell CD, Smart KA. Yeast responses to stresses associated with industrial brewery handling. FEMS Microbiology. 2007;31:535-569. DOI: 10.1111/j.1574-6976.2007.00076.x

63. Ding J, Huang X, Zhang L, Zhao N, Yang D, Zhang K. Tolerance and stress response to ethanol in the yeast Saccharomyces cerevisiae. Applied Microbiology and Biotechnology. 2009;85:253-263. DOI: 10.1007/s00253-009-2223-1

64. Zhao XQ, Bai FW. Mechanisms of yeast stress tolerance and its manipulation for efficient fuel ethanol production. Journal of Biotechnology. 2009;144:23-30. DOI: 10.1016/j.jbiotec.2009.05.001

65. Kotarska K, Czupryński B, Kłosowski G. Effect of various activators on the course of alcoholic fermentation.JournalofFoodEngineering.2006;77:967-971.DOI.org/10.1016/j.jfoodeng.2005.08.041

66. Ruyters S, Mukherjee V, Verstrepen KJ, Thevelein JM, Willems KA, Lievens B. Assessing the potential of wild yeasts for bioethanol production. Journal of Industrial Microbiology and Biotechnology. 2015;42:39-48. DOI 10.1007/s10295-014-1544-y

67. Zhang K, Zhang L, Fang Y, Jin X, Qi L, Wu X, Zheng D. Genomic structural variation contributes to phenotypic change of industrial bioethanol yeast Saccharomyces cerevisiae. FEMS. 2016;16:1-12. DOI: 10.1093/femsyr/fov118

68. Santos RM,. Nogueira FCS, Brasil AA, Carvalho PC, Leprevost FV, Domont GB, Eleutherio ECA. Quantitative proteomic analysis of the Saccharomyces cerevisiae industrial strains CAT-1 and PE-2. Journal of Proteomics. 2017;151:114-121. DOI: 10.1016/j.jprot.2016.08.020

69. Della-Bianca BE, Gombert AK. Stress tolerance and growth physiology of yeast strains from the Brazilian fuel ethanol industry. Antonie van Leeuwenhoek. 2013;104:1083-1095. DOI 10.1007/s10482-013-0030-2

70. Breisha GZ. Production of 16% ethanol from 35% sucrose. Biomass Bioenergy. 2010;34:1243-1249. DOI:10.1016/j.biombioe.2010.03.017

71. Jacob P, Hirt H, Bendahmane A. The heat-shock protein/chaperone network and multiple stress resistance. Plant Biotechnology Journal. 2016;10:1-10. DOI: 10.1111/pbi.12659

72. Verghese J, Abrams J, Wang Y, Morano KA. Biology of the heat shock response and protein chaperones: Budding yeast (Saccharomyces cerevisiae) as a model system. Microbiology and Molecular Biology Reviews. 2012;76:115-158. DOI: 10.1128/MMBR.05018-11

73. Zander G, Hackmann A, Bender L, Becker D, Lingner T, Salinas G, Krebber H. mRNA quality control is bypassed for immediate export of stress-responsive transcripts. Nature. 2016;540:593-596. DOI: 10.1038/nature20572

74. Mackenzie RJ, Lawless C, Holman SW, Lanthaler K, Beynon RJ, Grant CM, Hubbard SJ, Eyers CE. Absolute protein quantification of the yeast chaperome under conditions of heat shock. Proteomics. 2016;16:2128-2140. DOI: 10.1002/pmic.201500503

75. Mokry DZ, Abrahão J, Ramos CH. Disaggregases, molecular chaperones that resolubilize protein aggregates. Anais da Academia Brasileira de Ciências. 2015;87:1273-1292. DOI: 10.1590/0001-3765201520140671

76. Aslam K, Hazbun TR. Hsp31, a member of the DJ-1 superfamily, is a multitasking stress responder with chaperone activity. Prion. 2016;10:103-111. DOI: 10.1080/19336896.2016.1141858

77. O'Connell JD, Tsechansky M, Royall A, Boutz DR, Ellington AD, Marcotte EM. A proteomic survey of widespread protein aggregation in yeast. Molecular BioSystems. 2014;10:851-861. DOI: 10.1039/c3mb70508k

78. Needham PG, Patel HJ, Chiosis G, Thibodeau PH, Brodsky JL. Mutations in the yeast Hsp70, Ssa1, at P417 alter ATP cycling, interdomain coupling, and specific chaperone functions. Journal of Molecular Biology. 2015;427:2948-2965. DOI: 10.1016/j.jmb.2015.04.010

79. Mahmud SA, Hirasawa T, Shimizu H. Differential importance of trehalose accumulation in Saccharomyces cerevisiae in response to various environmental stresses. Journal of Bioscience and Bioengineering. 2010;109:262-266. DOI: 10.1016/j.jbiosc.2009.08.500

80. Tapia H, Koshland DE. Trehalose is a versatile and long-lived chaperone for desiccation tolerance. Current Biology. 2014;24:2758-2766. DOI: 10.1016/j.cub.2014.10.005

81. Tapia H, Young L, Fox D, Bertozzi CR, Koshland D. Increasing intracellular trehalose is sufficient to confer desiccation tolerance to Saccharomyces cerevisiae. Proceedings of the National Academy of Sciences of the United States of America. 2015;112:6122-6127. DOI: 10.1073/pnas.1506415112

82. Matsuda F, Kinoshita S, Nishino S, Tomita A, Shimizu H. Targeted proteome analysis of single-gene deletion strains of Saccharomyces cerevisiae lacking enzymes in the central carbon metabolism. PLoS One. 2017;12:1-20. DOI: 10.1371/journal.pone.0172742

83. Inai T, Watanabe D, Zhou Y, Fukada R, Akao T, Shima J, Takagi H, Shimoi H. Rim15p-mediated regulation of sucrose utilization during molasses fermentation using Saccharomyces cerevisiae strain PE-2. Journal of Bioscience and Bioengineering. 2013;16:591-594. DOI: 10.1016/j.jbiosc.2013.05.015

84. Souza CS, Thomaz D, Cides ER, Oliveira KF, Tognolli JO, Laluce C. Genetic and physiological alterations occurring in a yeast population continuously propagated at increasing temperatures with cell recycling. World Journal of Microbiology and Biotechnology. 2007;23:1667. DOI: 10.1007/s11274-007-9414-7

85. Koschwanez JH, Foster KR, Murray AW. Sucrose utilization in budding yeast as a model for the origin of undifferentiated multicellularity. PLoS Biology. 2011;9:1-10. DOI: 10.1371/journal.pbio.1001122

86. Lopes ML, Paulillo SC, Godoy A, Cherubin RA, Lorenzi MS, Giometti FH, Bernardino CD, Amorim Neto HB, Amorim HV. Ethanol production in Brazil: A bridge between science and industry. Brazilian Journal of Microbiology. 2016;47:64-76. DOI: 10.1016/j.bjm.2016.10.003

87. Naumov GI, Naumova ES. Invertase overproduction may provide for inulin fermentation by selection strains of Saccharomyces cerevisiae. Microbiology. 2015;84:130-134. DOI: 10.1134/S0026261715020095

88. Ivey M, Massel M, Phister TG. Microbial interactions in food fermentations. Annual Review of Food Science and Technology. 2013;4:141-162. DOI: 10.1146/annurev-food-022811-101219

89. Favaro L, Viktor MJ, Rose SH, Viljoen-Bloom M, van Zyl WH, Basaglia M, Cagnin L, Casella S. Consolidated bioprocessing of starchy substrates into ethanol by industrial Saccharomyces cerevisiae strains secreting fungal amylases. Biotechnology and Bioengineering. 2015;12:1751-1760. DOI: 10.1002/bit.25591

90. Fang Z, Deng H, Zhang X, Zhang J, Bao J. Evaluation of the cellulase cost during the cassava cellulose ethanol fermentation process. Sheng Wu Gong Cheng Xue Bao. 2013;29:312-324

91. Li H, Shen Y, Wu M, Hou J, Jiao C, Li Z, Liu X, Bao X. Engineering a wild-type diploid Saccharomyces cerevisiae strain for second-generation bioethanol production. Bioresources and Bioprocess. 2016;3:1-17. DOI: 10.1186/s40643-016-0126-4

92. Grimaldi MP, Marques MP, Laluce C, Cilli EM, Pombeiro Sponchiado SR. Evaluation of lime and hydrothermal pretreatments for efficient enzymatic hydrolysis of raw sugarcane bagasse. Biotechnology for Biofuels. 2015;8:205. DOI: 10.1186/s13068.015.0384.y

93. Djajadi DT, Hansen AR, Jensen A, Thygesen LG, Pinelo M, Meyer AS, Jørgensen H. Surface properties correlate to the digestibility of hydrothermally pretreated lignocellulosic Poaceae biomass feedstocks. Biotechnology for Biofuels. 2017;23:1-15. DOI: 10.1186/s13068-017-0730-3

94. Choudhary J, Singh S, Nain L. Bioprospecting thermotolerant ethanologenic yeasts for simultaneous saccharification and fermentation from diverse environments. Journal of Bioscience and Bioengineering. 2017;123:34-36. DOI: 10.1016/j.jbiosc.2016.10.007

95. Miranda I, Masiero M. O, Zamai T, Capella M, and Laluce C. Improved pretreatments applied to the sugarcane bagasse and release of lignin and hemicellulose from the cellulose-enriched fractions by sulfuric acid hydrolysis. Journal of Chemical Technology and Biotechnology. 2016;91:476-482. DOI: 10.1002/jctb.4601

96. Raghavi S, Sindhu R, Binod P, Gnansounou E, Pandey A. Development of a novel sequential pretreatment strategy for the production of bioethanol from sugarcane trash. Bioresource Technology. 2016;63:495-409. DOI: 10.1016/j.biortech.2015.08.062

97. Alff-Tuomala S, Salusjärvi L, Barth D, Oja M, Penttilä M, Pitkänen JP, Ruohonen L, Jouhten P. Xylose-induced dynamic effects on metabolism and gene expression in engineered Saccharomyces cerevisiae in anaerobic glucose-xylose cultures. Applied Microbiology and Biotechnology. 2016;100:969-985. DOI: 10.1007/s00253-015-7038-7

98. Jeffries TW, Jin YS. Metabolic engineering for improved fermentation of pentoses by yeasts. Applied Microbiology and Biotechnology. 2004;63:495-409. DOI: 10.1007/s00253-003-1450-0

99. Ko JK, Um Y, Lee SM. Effect of manganese ions on ethanol fermentation by xylose isomerase expressing Saccharomyces cerevisiae under acetic acid stress. Bioresource Technology. 2016;222:422-430. DOI: 10.1016/j.biortech.2016.09.130

100. Moysés DN, Reis VC, de Almeida JR, de Moraes LM, Torres FA. Xylose fermentation by Saccharomyces cerevisiae: Challenges and prospects. International Journal of Molecular Sciences. 2016;17:1-18. DOI: 10.3390/ijms1703020

CHAPTER 10

CHARACTERIZATION OF THE PHYSICAL AND CHEMICAL PROPERTIES OF YEAST CELLS AND LIGNOCELLULOSIC WASTE USED IN IMMOBILIZING CELLS FOR ETHANOL PRODUCTION

1. INTRODUCTION: AN OVERVIEW

The volatility of global petroleum prices and availability, coupled with the imperative to mitigate the environmental consequences of fossil fuel combustion, has sparked a renewed focus on renewable alternative fuels. Ethanol is a highly significant biofuel that may be used either independently or in conjunction with gasoline in the existing transportation system. As a result, there has been a rise in initiatives aimed at enhancing the efficiency and cost-effectiveness of ethanol production technology. An enhancement suggested for alcoholic fermentation is the implementation of a continuous process utilizing immobilized cells. This application has significant benefits, including the simplicity of separating products, the ability to reuse biocatalysts, and achieving high volumetric productivity [1]. The investigated immobilization matrices consist of hydrophilic polymer gels, including alginate, carrageenan, agarose, and others. The cells in these matrices are fixed in place by being trapped within the gel [2, 3]. Nevertheless, using this immobilization technique on a large-scale industrial level is not feasible due to the high expenses associated with acquiring raw materials, the intricate process of preparing biocatalysts, and the limited duration of their effectiveness in fermentations [4]. Over the past ten years, researchers have examined agro-industrial leftovers as potential carriers for immobilizing cells. On these medium, cell immobilization primarily occurs through adsorption, which is a straightforward and cost-effective strategy. The cellulosic materials possess the qualities of being renewable, reusable, sterilizable, heat-resistant, physiologically and chemically stable under various fermentation conditions, and having adequate mechanical strength. These characteristics make them suitable for possible use in industrial settings [5-8]. The investigation of cell-cell interactions has been undertaken by numerous researchers. However, these studies have been limited in their scope and simplifications of the phenomenon, as it is impractical to consider all the components and forces involved in this type of contact. A comprehensive analysis should encompass factors such as the chemical composition and structure of cells, the presence of positive and negative surface charges and their non-uniform distribution, the enduring alterations occurring in cellular membrane components, the ionic permeability of the membrane, the changes in cellular topology throughout the process, the potential for active movement or locomotion exhibited by the cell, and the occurrence of random movement. The text mentions various factors such as Brownian motion, ionic force, pH, temperature, and viscosity. The user's text is "[9]." The physico-chemical treatment of the cellular adhesion problem involves analyzing the adhesion event by assuming the cell as a colloidal particle with consistent surface properties. However, it should be noted that the cell may alter its shape and redistribute its membrane components. This methodology yields valuable insights into the initial phases of the adhesion process; yet, it may not accurately forecast the occurrence of adhesion in certain instances. Microbial adhesion is commonly understood as the interaction of Van der Waals long-range forces, electrostatic forces, and numerous short-range interactions from a physicochemical perspective [10]. This study examined the physical and chemical properties of four lignocellulosic residues: wood shavings, bagasse, corncobs, and corn leaves. These residues were utilized as carriers to immobilize yeast cells for ethanol generation in packed bed bioreactors. The characterization was also conducted on yeast cells that were both free and immobilized in the four lignocellulosic materials during the initial stages of immobilization. The hypothesis suggests that the alteration of yeast's surface qualities is a result of the unique conditions of the process, which are characteristics that influence the process of immobilization on lignocellulosic substrates. In order to enhance the cell immobilization process, it is crucial to comprehend the specific interactions between the support and cells that take place during the process. Additionally, it is vital to determine the primary surface properties of both the media and the cells, since these factors significantly impact the immobilization process.

2. METHODS

2.1. The organism and culture media

In this work, we used the commercial yeast Saccharomyces cerevisiae Ethanol

Red supplied by Fermentis. The strain was kept lyophilized and stored at 4°C. The medium for inoculum and fermentation is composed of: glucose 100 g/L, peptone 3.6 g/L, (NH4)2SO4 3 g/L, yeast extract 4 g /L, KH2PO4 2 g/L, and MgSO4.7H2O 1 g/L. The pH was adjusted to 5.0.

2.2. Composition material determination

To determine the content of the lignin, cellulose, hemicellulose, ash, and cell wall of lignocellulosic waste, we followed the methodology used by Van Soest [11–15]. This methodology consists primarily in the digestion of the materials in a sequence of detergent solutions, with changes in acidity to achieve solubilization of components due to their resistance or susceptibility to chemical reagents.

2.3. Surface area determination

The technique of Brunauer–Emmett–Teller (BET) adsorption isotherms was used for establishing the surface area for each material. It was also performed to determine the type of pore and pore-size distribution, using the methodology of adsorption isotherms barrett-joyner-halenda (BJH) desorption.

2.4. Sedimentation test

The technique consists in the measurement of cell sedimentation after contact cell suspension with solution of CaCl2, which promotes the aggregates cells formation. This methodology is used to establish the flocculating ability of cells. And their performance taking into account the work of Domingues et al. [16] and Branyik et al. [17].

2.5. Microbial Adhesion To Hydrocarbons test (MATH tests)

The technique consists of determining the ability of cell to adhere to several solvents and establish their electron-donor/electron acceptor or Lewis acid-base properties. This method is based on the comparison between microbial cell affinity to a polar solvent and to a nonpolar solvent by simply measuring the fraction of cell removal from the aqueous phase in the presence of these solvents. To perform the test MATH, we used two pairs of substances, chloroform is an electron acceptor solvent and hexadecane is a nonpolar solvent. We also used ethyl acetate as a solvent strongly electron donor and decane as a nonpolar solvent. Due to surface tension properties of these solvents, the differences between the results obtained with chloroform and hexadecane, and the results obtained with ethyl acetate and decane indicate that cell surface interactions are of type electron donor or electron acceptor and reveal hydrophobic and hydrophilic properties. Strains adhering well to the hydrocarbon are considered to be "hydrophobic" and strains adhering poorly are considered "hydrophilic." The removal is greatest when you have a pH value where the zeta potential of the organisms and/or hydrocarbons is zero. Under this condition, there is no static repulsion [10]. This methodology was developed taking into account the work of Domingues et al. [16], Branyik et al. [17], and Bellon-Fontaine et al. [18].

2.6. Zeta potential measuresment

Zeta potential describes the static electric field strength of the double layer at the boundary between a particle and the fluid in which is immersed. Measurements are performed on equipment Zeta meter Malvern Instruments that employed folded capillary cells (zetasizer nano series Malvern).

2.6.1. For cells

Cells recovered from fermentation are centrifuged and resuspended in solution 10 mM KNO3. They are washed at least twice with this solution. Prepare a cell suspension of concentration 0.85 g/L dry cell weight with solution 10 mM KNO3. Adjust the pH with HNO3. We can prepare three different cells suspension with pH values of 3.0, 4.0, and 5.0. The cell suspension is filled into the electrophoresis cell, and after at least 40 electrophoretic mobility readings, the average zeta potential is calculated.

2.6.2. For lignocellulosic materials

An amount of 0.5 g in dry state was triturated and then suspended in 100 mL of 10 mM KNO3. The suspension of carrier particles was filtered through a polyester mesh (PE 15 mm Seidengazefabrik AG Thal, Switzerland) with mesh openings of 15 × 15 mm and the pH is adjusted with HNO3. The cell suspension is filled into the electrophoresis cell and after at least 40 electrophoretic mobility readings the average zeta potential was calculated.

2.7. Contact angles

The technique consist in the measurement of contact angles, formed by sessile drops of three different liquids (two polar and one apolar), to enable the calculation of the surface free energy and the degree of hydrophobicity of cells and lignocellulosic materials. This methodology was developed taking into account the work of Henriques et al. [19] and Branyik et al. [17]. The apparatus used was a model OCA 15 PLUS, DATAPHYSICS.

2.7.1. Contact angle measurement for cells

Cells were harvested by centrifugation at 6000 g and 4°C for 10 min and washed with increasing concentrations of ethanol in water (10, 20, and 50% v/v). The resulting pellet was resuspended in 50% (v/v) ethanol. The final concentration was adjusted to 1.2 g/L dry cell weight. An aliquot of 1 mL of cell suspension is spread over the solidified agar layer in order to cover the entire surface. This layer is let to dry; this step is repeated four times. Contact angles are measured by the sessile drop technique at room temperature using water, formamide, and α-bromonaphtalene. Each assay is performed in triplicate and at least 20 contact angles, per sample, are measured.

2.7.2. Contact angle measurement for carriers

To contact angle measurements, carrier particles are fixed on a microscopic slide by an adhesive tape. Contact angles are measured by the sessile drop technique, it is employed a drop volume of 3 μL and the same procedure for cells determination.

2.8. Scanning electron microscope (SEM)

For the realization of the images, we used the scanning electron microscope (FEG/SEM) FEI Nova 200 with EDAX, EDS/EBSD, and STEM. The images were performed on samples of materials before and after basal immobilization.

2.9. FT-IR and FT-Raman measurements

Fourier Transform Infrared Perkin Elmer, Spectrum one model, deuterated triglycine sulfate (DTGS) detector was employed to obtain the spectra. The samples were dried at 80°C for 30 min, dispersed in KBr to form a tablet, which is analyzed in the spectrometer. The analysis conditions were temperature 24°C, the number of sweeps of 8 with a resolution of 4 cm-1 and a range of wavelengths (ν) 4000–400 cm-1.

2.10. Basal immobilization process

We used pieces of each material: wood shavings, sugarcane bagasse, corncobs, and corn leaves. A cell suspension prepared in isotonic solution with 15 g/L of biomass was continuously recirculated. The process was realized by a period of 12 hours at 30°C, and we used column bioreactors with a volume of 150 mL.

2.11. Dry weight technique modified

Immobilized biomass was determined by the difference in dry weight of support before and after cell removal protocol. The removal was performed with a solution of NaOH 0.1% by mechanical agitation at 150 rpm for 24 hours at room temperature. The drying was performed at 105°C by 12 hours [20].

2.12. Free energy determinations

The total surface tension (γtot) and its components (γLW, γ+, γ−, γAB), the values of the free energy of interaction between cells and water and the components (ΔGcwctot) (ΔGcwcLW, and ΔGcwcAB) are calculated according to van Oss and co-workers [21].

3. RESULTS AND DISCUSSION

Table 1 shows the results for determining the composition of lignocellulosic waste. The content of lignin, cellulose, hemicellulose, ash, and cell wall presented significant differences for each material. Wood shavings, corn leaves, and bagasse have between 20 and 24% cellulose content more that lignin, however, the corncob has only 8%. The corn leaves in the material had higher content of cell wall and hemicellulose, whose main function is to provide the bond between the cellulose and lignin [22]. Lignin is the substance that gives structural rigidity to the material because it is responsible for holding the fibers harden and polysaccharide [23]. The corn leaf contains a low amount of lignin, a characteristic that may influence the ability of immobilization and the stability of the material in fermentation. We can experimentally verify that this material in the operation presents less stability. It tends to gather and pile up by the top of the bioreactor, resulting in not homogeneous distribution of the bed and its obstruction. The corn leaf was the material that showed less stability during cell removal protocol; it is easily disintegrated when placed in contact with NaOH at high concentration.

Table 1. Material's composition obtained using the Van Soest Method.

	Lignin (%)	Cellulose (%)	Hemicellulose (%)	Ash (%)	Cell wall (%)	Total
Wood shavings	27.09	46.79	16.62	0.45	9.05	100
Bagasse	29.68	52.18	7.24	0.01	10.89	100
Corncobs	38.15	45.95	9.95	0.34	5.61	100
Corn leaves	5.44	29.19	24.63	1.71	39.03	100

Natural materials typically have very low porosities and surfaces, so the amount of nitrogen adsorbed at 77 K was very low, this corresponds to materials with a poorly developed mesoporosity and no micropores. The size distributions of mesopores determined by the BJH method (desorption isotherm) resulted in none of the samples has a significant amount of mesopores than 20 nm; most of them are between 3 and 12 nm. The corn leaves and wood shaving materials presented mesopores between 3.1 and 3.4 nm. The values are summarized in Table 2. There is good agreement between the values of the areas calculated by the BET method and BJH method from the desorption isotherm, with the exception of wood shaving material. It is important that the BJH method counts only the pores with diameters greater than 3.1 and it is possible that this sample has a significant amount (in relative terms) of these pores. The corresponding size distribution suggests this. We can establish a small difference between the samples of bagasse, with the corncob, bagasse can absorb a little more nitrogen than the cob, and this is consistent with the parameters. The bagasse has a higher adsorption at low relative pressure zone.

Also size distribution of present mesopores is less than 20 nm, while most are distributed in sizes 3 to 12 nm. The top curves correspond to diameters between about 3.5 and 4.5 nm, values that do not differ significantly from the other samples. We can say with some certainty that all samples have a very low porosity, mainly relating to mesopores with diameters below 12 nm, with specific areas between 1 and 4 surface m2/g and pore specific volume between 0.002 and 0.006 cm3/g. The advantages of cell adhesion to nonporous carriers consist in lower mass transfer limitation of substrates and products due to direct contact between cells and bulk liquid and also in the simplicity of the immobilization. The main risk of this method is the biofilm detachment induced by changes in cell environment [24].

Table 2. Specific surface areas and volumes for lignocellulosic wastes, determined by the methods of BET and BJH.

Materials	S_{BET} $(m^2/g)^a$	S_{BJH} $(m^2/g)^b$	S_{BJH} $(m^2/g)^c$	V_P $(cm^3/g)^d$	V_P $(cm^3/g)^b$	V_P $(cm^3/g)^c$
Wood shavings	2.6	2.5	2.1	0.005	0.005	0.005
Corn leaves	2.8	2.2	2.0	0.004	0.004	0.004
Bagasse	5.3	3.7	3.7	0.006	0.005	0.006
Corncobs	4.8	3.1	3.0	0.005	0.005	0.005

[a] Calculated by the BET method (adsorption isotherm).

[b] Calculated by the BJH (desorption isotherm).

[c] Calculated by the BJH (adsorption isotherm).

[d] Calculated by the maximum amount adsorbed (P/P0 = 0.99).

All the spectra show similar absorption features, although the intensities of the absorption bands are different. The assignment of the following absorbance bands is in accordance with the literature [25]. The methylene (CH2) structural unit has symmetric and asymmetric vibrations. The asymmetric C–H stretching vibrations for CH2 involve one C–H bond contracting, while the other bond is lengthening. This vibration is observed at 2926 ± 10 cm−1. The symmetric methylene stretch involves both the C–H bonds lengthening or contracting at the same time. This band typically appears at 2855 ± 10. These bands are present in both blank cells and immobilized cells. The presence of band at 1375 ± 10 cm−1 is a strong indication of the presence of a metyl group in a sample. These bands correspond to the symmetric C–H bending vibrations of the methyl (CH3) group and are observed in the wood shaving sample. The spectral signature bands due to the C=O stretching vibration, which appears as an intense band between 1800 and 1600 cm-1. This vibration is observed in both blank cells and immobilized cells. The overriding spectral feature of a carboxylic acid is the broad, intense O–H stretching band typically found from 3500 to 2500 cm−1, and often centered around 3000 cm−1. This band almost by itself, tells you that sample contains a carboxylic acid. The sharper C–H stretching vibrations are superimposed upon the broad O–H stretching band. However, some times the O–H stretch masks the C–H stretching bands. Note that on the low wavenumber side of the O–H stretching band in the blank and immobilized cells of both samples, between 2500 and 2800 cm−1, there are some broad features of medium intensity. The in-plane O–H bending band is found from 1440 to 1395 cm−1. In the spectrum, this band is found at 1400 cm−1. Carboxylic acids contain a C–O bond, and the C–O stretch of carboxylic acids appears between 1320 and 1210 cm−1. The C=O stretch of carboxylic acids is in the same regions as the C=O stretch of ketones, aldehydes and esters. The C=O stretching band by itself is not sufficient evidence to identify a carboxylic acid. Other bands, such as the O–H stretch must be found to confirm the assignment.

The NH2 group in a primary amide can bend in addition to stretching. The in-plane bending vibration involves the H–N–H bond angle getting bigger and smaller, like the opening and closing of a pair of scissors. This vibration is called the NH2 scissors mode, and is found from 1650 to 1620 cm−1. Note that the wavenumber range for the C=O stretch and the NH2 scissors overlap. The NH2 part of a primary amide can also bend out of the plane defined by the functional group. This out-of-plane bending band is typically very broad and is found from 750 to 600 cm−1. This band is observed in all spectra for the materials. Primary amides contain one C–N bond, and the stretching of this bond gives rise to a band from 1430 to 1390 cm−1. It is found well-defined in the spectrums of corncobs, corn leaves, and wood shaving, and it is less evident in the bagasse spectrum. The secondary amides are probably the most common and important type of amide. Proteins, nylons, and other polymers contain secondary amide linkages. This band appears between 3370 and 3170 cm−1. In the cells spectrum, this band is overlapped by the O–H stretching band. The carbonyl stretch of secondary amides appears in the same range as other amides, from 1680 to 1630 cm−1. The N–H moiety can bend as well as stretch and, like many other functional groups, there is an in-plane and out-of-plane bending vibration in secondary amides. The in-plane N–H bending vibration usually appears from 1570 to 1515 cm−1. It is found in the spectrum of both cells samples in 1542 cm−1. The primary amine C–N stretching vibration for saturated molecules occurs from 1250 and 1020 cm−1. It is observed at 1088 cm−1 for cells spectrum and 1062 cm−1 for the materials spectrum. In summary, the FT-IR spectra obtained showed the presence of CH, CH2, CH3, NH, NH2, NH3, COOH, and CONH groups on the surface of all samples cells and materials as shown in Figure 1).

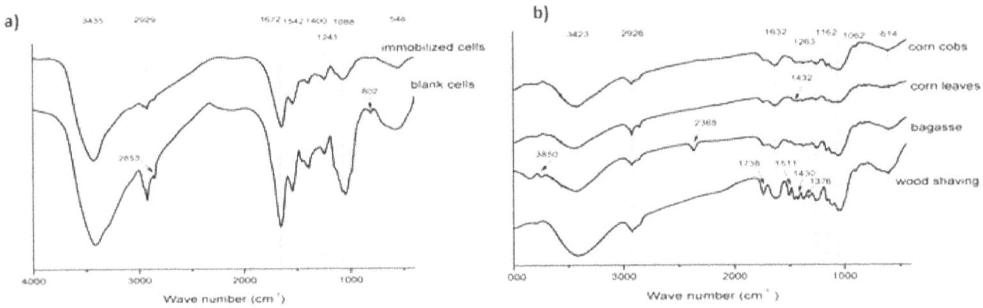

Figure 1. FT-IR results for (a) yeast cells and (b) lignocellulosic wastes.

Images obtained by SEM for the materials before and after basal immobilization are presented in Figure 2(a) and (b). In Figure 2(a), we can see the superficial differences between the materials prior to immobilization. The corncob is the material that has a rough, irregular surface, and multiple pores are observed, corresponding to embryo sacs that make up the fruit of the corn. The wood shaving has a surface with two types of arrays, one consisting of a flat, smooth region, equivalent to the wood fibers and is the main constituent. The other region is composed of pores corresponding to the tracheae in the body secondary woody plants. The corn leaf is the material having a smooth and homogeneous surface; there is complete cellular structure and the presence of trichomes or hairinesses characteristic of this tissue. The bagasse has an ordered structure in the form of overlapping plates; equivalent to longitudinal segment of cane stalk and due to mechanical treatment for the extraction of juice in sugar cane cannot be seen defined cellular structure.

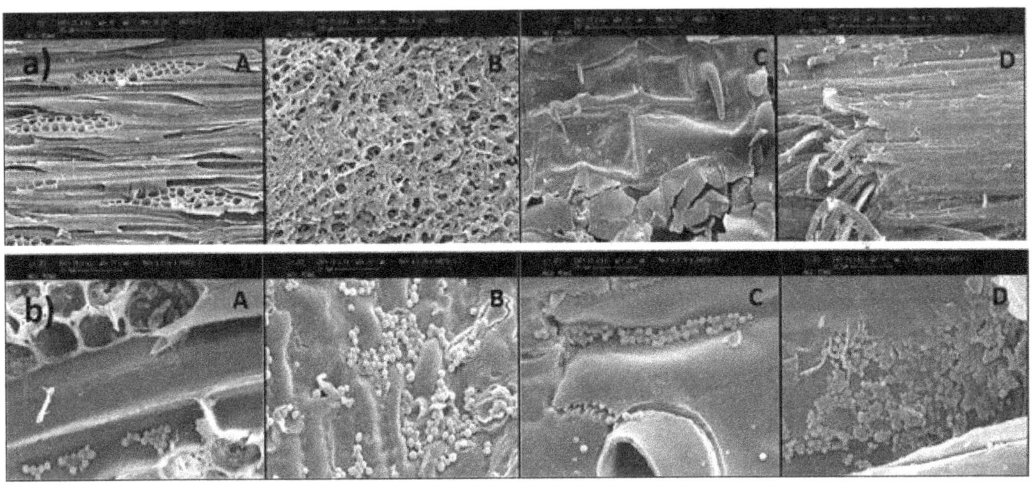

Figure 2. (a) Scanning Electronic Micrograph of lignocellulosic wasted, (b) Scanning Electronic Micrograph of lignocellulosic wasted after basal immobilization; (A) wood shavings, (B) corncobs, (C) corn leaves and (D) bagasse.

Figure 2(b) shows the images of the material after basal immobilization. We can see that the yeast cells adhered to all surfaces of different materials. In the case of the wood shaving, immobilization was performed in both the pore region and the smooth region. In the corncob, the cells display more uniform adhesion across the surface of the material; however, in the corn leaf, the cells were immobilized preferentially in regions where there were creases. The bagasse laminar surface was covered uniformly by the cells. It can be seen that the cells are attached to chains and forming group. The highest number of immobilized cells was bagasse and corncob.

The superficial characteristic of cells before basal immobilization process, as shown in Figure 3, indicate an affinity for chloroform greater than hexadecane. The differences in affinity were due to interactions of Lewis acid-base, i.e., interactions donor / acceptor of electrons resulting from the electron donor nature of the yeast. These results are consistent with those obtained by Mercier-Bonin et al. [26]. From the results obtained with the other pair of solvents, we can establish that there is a higher affinity of the cells by the ethyl acetate solvent than the nonpolar solvent decane. It reveals an electron acceptor nature even slightly higher than its electron donor nature. The low affinity of cells to the nonpolar solvents indicates hydrophilic properties [10, 27]. We can conclude that yeast cells used in this study have an electron acceptor nature and hydrophilic properties.

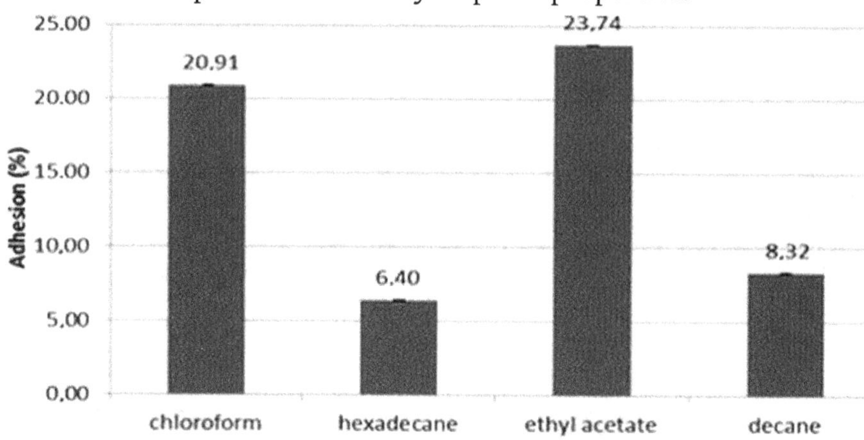

Figure 3. Test MATHS for free cells.

Adhesion analysis for cellular samples obtained from the basal immobilization process on wood shaving, corncobs, and corn leaves carriers revealed an increase in the adhesion of immobilized cells over no immobilizes cells. Both free and immobilizes cells exhibited an electron donor nature, a result contrary to what was obtained with blank cells (free cells suspended in isotonic solution not subjected to immobilization). However, the results of adhesion to cells in the process with bagasse like carrier showed differences in this trend, i.e., retained their electron acceptor nature. It is important to note that all cell samples retained the hydrophilic character; it was evidenced by the poor adherence to the nonpolar solvents. These results suggest that the immobilization process modifies the surface characteristics of the cell and the immobilization by adhesion is influenced also by the surface characteristics of the carrier used. The results of cell adhesion after the basal immobilization process with each of the carriers are presented in Figure 4.

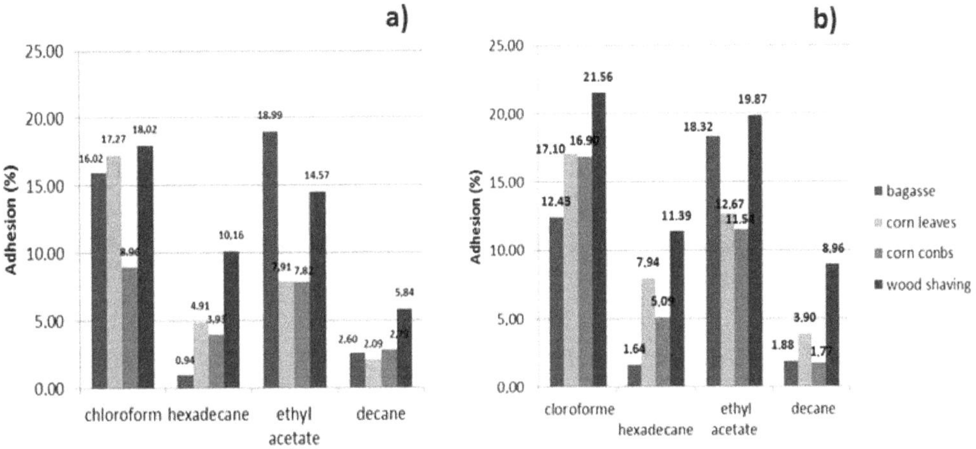

Figure 4. Test MATH for cells in basal immobilization process. (a) Free cells and (b) immobilized cells.

The results obtained for the sedimentation test performed on the cells used as blank and the free cells taken from the basal immobilization process with bagasse are presented in Figure 5. As shown no increased capacity flocculating or settling of the cells after being subjected to the immobilization process conducted by 12 hours was observed. It is possible that a change occurs with the time in cell physiology, allowing an increase in the capacity of sedimentation. This behavior has been reported by other researchers [17]; however, the period of time of immobilization process that the cells were subjected in these studies was superior to 100 hours; it was much higher than the evaluated in this study.

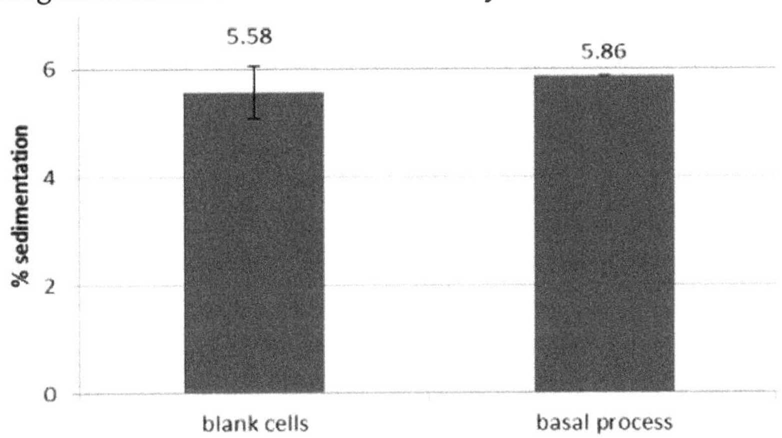

Figure 5. Sedimentation test for free cells obtained in different process.

An approach based on a balance of free energy of interaction between cell-liquid, liquid-medium, and medium-cell interface was used to estimate the physicochemical properties of the surface of cells and media, which could lead to adhesion. Electrodynamic forces (Lifshitz-Van der Waals) and hydrophobic forces (acids/Lewis bases) were determined using contact angle measurements and tests of electrophoretic mobility (zeta potential). The contact angle measurements were obtained from the formation of sessile drops of three different liquids; water and formamide as polar liquids and liquid α-bromonaphthalene as apolar. The calculation of surface free energy and the degree of hydrophobicity can be defined in terms of the change in free energy of interaction between two particles of the same material immersed in water (w). The free energy comprises a polar component (AB) and an apolar component (LW), ΔG1w1TOT = ΔG1w1AB + ΔG1w1LW. When the value of de ΔG1w1TOT is negative, the interaction energy between molecules is attractive, which means that cells have less affinity for water than for themselves giving them a hydrophobic character. On the other hand, the cells are hydrophilic when this value is positive [27].

The value of contact angle with water can provide preliminary information on the hydrophobicity of the cells. If the value is greater than 50°, the surface is considered hydrophobic; if the value is less than 50°, the surface is hydrophilic [28]. In the results presented in Table 3, we can see values of contact angles above 50° for all lignocellulosic waste, indicating their hydrophobic character. In the case of blank cells, the value of contact angle with water is less than 50°, indicating a hydrophilic character. These results are consistent with those previously obtained in the MATH test.

Table 3. Contact angle for lignocellulosic waste materials and yeast cells.

	Wood savings	Corncobs	Corn leaves	Bagasse	Free cells
Water (θ_w (°))	84.50 ± 06.40	84.40 ± 06.10	115.67 ± 15.28	104.20 ± 11.10	19.88 ± 1.87
α-Bromonaphthalene (θ_B (°))	32.27 ± 07.84	33.19 ± 09.63	44.63 ± 11.49	28.60 ± 10.00	76.50 ± 2.67
Formamide (θ_F (°))	74.25 ± 11.53	67.57 ± 10.15	75.72 ± 12.47	75.11 ± 11.37	13.80 ± 1.04

The cell surface is the site of the physicochemical interactions with the support leading to adhesion. Previous studies have suggested the relationship between the ability of adhesion and cell surface hydrophobicity [29]. The hydrophobicity of yeast cells is affected by temperature, nutrition, and growth phase [19]. While the nature of the substrate surface can be considered temporarily unchanging, the nature of the yeast surface is a function of their physiological state. In the Table

4, we shows the values of contact angles obtained for cell samples of immobilized process carried out with the four materials. The properties of the cell surface have varied. There is an increase in the value of contact angle with water for all samples of cells that were immobilized; this indicates an increase in cell surface hydrophobicity generated by changing the physiological state of the cell and the process conditions. The only material that did not show this behavior was the corn leaf. In this material, the contact angle value increased for the cells that remained free during the process, however, the immobilized cells showed a similar value to the blank cells.

Table 4. Contact angles obtained for free cells and immobilized cells in basal immobilization process with each carried.

		Wood shaving		Corncobs		Corn leaf		Bagasse	
	Cells blank	Free cells	inmobilized cells	Free cells	Inmobilized cells	Free cells	inmobilized cells	Free cells	Inmobilized cells
θ_W (°)	19.88 ± 1.9	22.7 ± 3.4	27.3 ± 2.7	23.9 ± 3.7	19.5 ± 1.5	19.4 ± 1.7	29.5 ± 3.0	22.8 ± 2.0	29.0 ± 2.6
θ_B (°)	76.50 ± 2.7	75.6 ± 5.2	81.0 ± 2.9	80.1 ± 1.0	85.4 ± 1.3	85.6 ± 2.3	86.8 ± 2.1	84.5 ± 2.3	86.5 ± 3.2
θ_F (°)	13.80 ± 1.0	27.9 ± 4.0	29.4 ± 1.7	24.5 ± 2.6	22.6 ± 2.8	25.7 ± 1.9	27.2 ± 2.7	25.5 ± 2.1	27.1 ± 2.9

In Tables 5 and 6, the results for surface tension and interaction energies are calculated. The high values obtained for Y's- are consistent with measurements of zeta potential, and also showed a negative surface charge to the cells (see results in Figure 6(a). The result showed a pH of 3.0–5.0 and an ionic strength of 10 mM; the zeta potential has a range of -13 to 16 mV. These values reveal an electron donor nature of the yeast cell surface. Some authors suggest that it may be due to presence of carboxyl and phosphate groups [17]. Figure 6(b) shows that the zeta potential is negative for all lignocellulosic materials and there is a marked increase in negativity with the pH increase. For a pH range of 3.0 –5.0, the zeta potential range is between -7 and -21 mV.

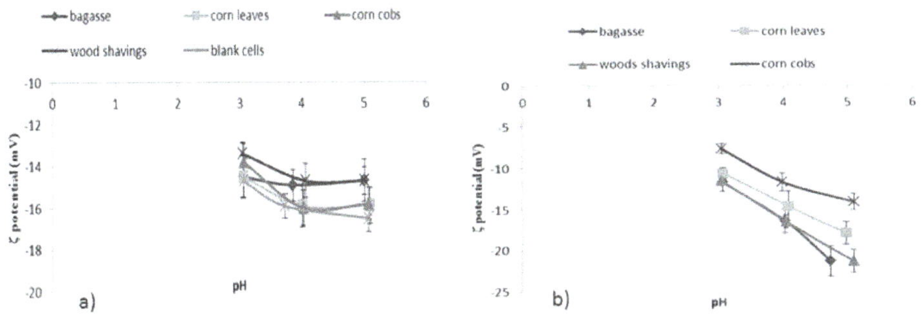

Figure 6. Zeta pontential (a) immobilized cells on basal process and (b) lignocellulosic wasted.

Table 5. Surface tensions and free energies of interaction between cells and water (1w1). Free energy of interaction between cells and substrates in water (1w2).

	Cells blank	Wood shaving		Corncobs		Corn leaf		Bagasse	
		Free cells	inmobilized cells	Free cells	Inmobilized cells	Free cells	Inmobilized cells	Free cells	inmobilized cells
Y's^{LW} (mJ/m^2)	16.898	17.345	14.860	15.247	12.955	12.880	12.383	13.341	12.518
Y's$^+$ (mJ/m^2)	11.368	7.992	9.826	10.656	12.865	12.014	13.183	11.965	13.017
Y's$^-$ (mJ/m^2)	46.504	51.619	47.823	47.869	50.270	52.211	43.519	49.271	44.039
s^{AB} (mJ/m^2)	45.985	40.622	43.355	45.170	50.862	50.090	47.905	48.560	47.886
s (mJ/m^2)	62.883	57.967	58.215	60.417	63.817	62.970	60.288	61.901	60.404
ΔG_{1w1}^{AB} (mJ/m^2)	11.929	19.022	14.292	13.343	11.942	13.791	8.795	12.543	9.175
ΔG_{1w1}^{LW} (mJ/m^2)	-0.658	-0.637	-1.371	-1.174	-2.298	-2.366	-2.672	-2.097	-2.619
ΔG_{1w1}^{IF} (mJ/m^2)	11.271	18.386	12.921	12.169	9.643	11.426	6.124	10.456	6.557
ΔG_{1w2}^{AB} (mJ/m^2)	-11.842	-13.574	-11.960	-10.128	-13.439	-3.734	0.304	-5.319	-2.813
ΔG_{1w2}^{LW} (mJ/m^2)	1.636	1.478	2.386	2.185	3.059	2.192	2.334	3.176	3.534
ΔG_{1w2}^{IF} (mJ/m^2)	-10.206	-12.096	-9.574	-7.943	-10.380	-1.541	2.638	-2.143	0.721

Table 6. Superficial tensions and free energy of interaction between the carriers and water (2w2).

	Wood shaving	Corncobs	Corn leaves	Bagasse
Y's^{LW} (mJ/m^2)	37.631	37.195	32.305	38.829
Y's$^+$ (mJ/m^2)	0.000	0.000	0.267	0.000
Y's$^-$ (mJ/m^2)	10.582	6.760	0.000	0.220
s^{AB} (mJ/m^2)	0.000	0.000	0.000	0.000
s (mJ/m^2)	37.631	37.195	32.305	38.829
ΔG_{2w2}^{AB} (mJ/m^2)	-36.315	-49.113	-92.455	-95.837
ΔG_{2w2}^{LW} (mJ/m^2)	-4.354	-4.184	-2.299	-4.959
ΔG_{2w2}^{IF} (mJ/m^2)	-40.666	-53.297	-94.754	-100.796

Positive values obtained for ΔG1w1IF (interaction free energy of cells) are positive, indicating the hydrophilic character of the cells. There were no changes to this feature during the process of cell immobilization, although there were differences between cells that were free and those that were immobilized. Cells that were immobilized had a value less hydrophilic, indicating a variation in the surface properties of yeast cells. For lignocellulosic material was obtained ΔG2w2IF negative value, indicating its hydrophobicity. Also the high value of Y'sLW confirms its apolar character. The interaction energy between the carrier and the cell when immersed in water is presented at the end of the table ($\Delta G1w2IF$), this value reveal the possibility of cell immobilization. Positive values of $\Delta G1w1AB$ involve repulsion between the particles or molecules of the material 1 immersed or dissolved in the liquid (w). It is also a major factor in the stability of particles in suspensions.

It is theoretically possible for a given material to only have a value of $\gamma s+$ or presence of only one value for $\gamma s-$. So your γsAB is zero and its surface tension (γs) is equal to total γsLW. However, such substances, which are designated as single-pole, can interact strongly with bipolar materials and mono-polar materials of opposite polarity, despite the apparently non-polar nature of its surface tension [27]. This feature shows the lignocellulosic materials.

When the interfacial free energy of interaction $\Delta G1w2$ is greater than zero, it indicates that there will be a net repulsion between particles or molecules of materials 1 and 2 when they are immersed or dissolved in the liquid w. This repulsion occurs unless there is a cancellation of the electrostatic attraction between materials 1 and 2. Conversely, when $\Delta G1w2$ is less than zero, it signifies an attraction between the materials. The energy of interaction between the carriers and cells is mutually favorable for both wood shaving and corncobs free cells when they are immobilized. The interaction energy (attraction) between the free cells and carriers of bagasse and maize leaves is negative. However, positive values are obtained for the energy of contact with immobilized cells. Notwithstanding these findings, the cell immobilization takes place on the supports, as evidenced by the images obtained by scanning electron microscopy (SEM).

4. CONCISE SUMMARY

The characterisation of both lignocellulosic materials utilized as carriers and yeast cells provides significant insights into the properties of their surfaces. Four lignocellulosic wastes exhibit variations in both their surface characteristics and content. The presence of these variants can have a substantial impact on the capacity of cells to adhere and remain stable as biocatalysts in the fermentation conditions employed for ethanol generation. It emphasizes that the cells were immobilized on each of the four carriers that were assessed. Yeast cells in an isotonic solution were attached to a support when they were brought into contact for a period of 12 hours, which is the standard immobilization technique. The yeast cells used in this investigation were found to have a hydrophilic surface and possess electron acceptor properties. Following the basal immobilization process, the characteristics of the cell surface undergo modification. The cells exhibit reduced hydrophilicity and undergo a shift from being electron acceptors to electron donors. Furthermore, there exist distinctions between free cells and the cells that attach to the surface of the supports. The results validate the variability of surface characteristics in cells, which is influenced by their physiological condition and the specific conditions of the immobilization procedure. In order to assess the capabilities of these materials in creating efficient biocatalysts for continuous packed-bed reactors in ethanol production, it is imperative to evaluate their performance under fermentative conditions and ensure that the yeast cells remain attached to the carriers throughout the process. It is imperative to determine the variances that exist in their surface properties. Currently, these studies are being conducted within the organization.

AT A GLANCE

Ethanol is a prominent substitute for traditional fuels. Efforts have been intensified to further the advancement of technology aimed at the efficient and cost-effective production of ethanol. An excellent solution is offered in the form of continuous fermentation employing yeast cells that are immobilized in low-cost materials. Four lignocellulosic wastes were utilized in the immobilization process. The materials underwent physicochemical characterization. The composition was ascertained using the Van Soest technique. The zeta potential was measured in order to determine the hydrophobic or hydrophilic nature of the material surfaces. The contact angle measurements were utilized to verify the hydrophobic or hydrophilic nature, while the free energy of interaction was determined. The images were acquired using a scanning electron microscope, and the calculation of surface areas and volumes was carried out by adsorption and desorption isotherms. The immobilization method to which cell surface properties are treated has been found to modify them. The relationship between cell immobilization and the carrier's properties, as well as the features of the cell surface, was clearly apparent. Therefore, to enhance the cell immobilization process, it is crucial to comprehend the interactions between the carrier and cells that take place during immobilization. This requires understanding the key surface characteristics of both the media and the cells that can influence the process.

REFERENCES

1. Wendhausen R, Fregonesi A, Moran PJS, Joekes ISJ, Rodrigues AR, Tonella E, Althoff K. Continuous fermentation of sugar cane syrup using immobilized yeast. Journal of Bioscience and Bioengineering. 2001;91(1):48-52

2. Kurillová L, Gemeiner P, Vikartovská A, Mikov H, Rosenberg M, Ilavsky M. Calcium pectate gel beads for cell entrapment. 6. Morphology of stabilized and hardened calcium pectate gel beads with Immobilized cells for biotechnology. Journal of Microencapsulation. 2000;17:279-296

3. Martins dos Santos VAP, Leenen EJTM, Rippoll MM, Sluis CVD, Vliet TV, Tramper J, Wijffels RH. Relevance of rheological properties of gel beads for mechanical their stability in bioreactors. Biotechnology and Bioengineering. 1997;56:517-529

4. Serp D. From Whole Cell Immobilization: Practical Insight To Tools Applications. [thesis]. Switzerland: École Polytechnique Fédérale De Lausanne; 2002. p. 50

5. Vasconcelos JN, Lopes CE, França FP. Continuous ethanol production using immobilized yeast on sugar-cane stalks. Brazilian Journal of Chemical Engineering. 2004;21(3):357-365

6. Kourkoutas Y, Psarianos C, Koutinas AA, Kanellaki M, Banat IM, Marchant R. Continuous whey fermentation using kefir yeast immobilized on delignified cellulosic material. Journal of Agricultural of Food Chemistry. 2002;50:2543-2547

7. Plessas S, Kourkoutas Y, Psarianos C, Kanellaki M, Koutinas AA. Continuous baker's yeast production using orange peel as promoting support in the bioreactor. Journal of the Science of Food and Agriculture. 2005;86(3):407-414

8. Brányik T, Vicente AA, Machado JM, Teixeira JA. Spent grains - a new support for brewing yeast Immobilization. Biotechnology Letters. 2001 ;23:1073-1078

9. Popescu AI. Cell-cell interactions. A physical approach. Bioelectrochemistry and Bioenergetics. 1996;40:153-157

10. Van der Mei HC, Meinders JM, Busscher HJ. The influence of ionic strength and pH on diffusion of micro-organisms with different structural surface features. Microbiology. 1994;140:3413-3419

11. Van Soest PJ. Use of detergents in the analysis of fibrous feeds. I. Preparation of fiber residues of low nitrogen content. Journal - Association of Official Analytical Chemists. 1963;46:825-829

12. Van Soest PJ. Use of detergents in the analysis of fibrous feeds. II. A rapid method for the determination of fiber and lignin. Journal - Association of Official Analytical Chemists. 1963;46:829-835

13. Van Soest PJ. Wine RH. Determination of lignin and cellulose in acid detergent fiber with permanganate. Journal - Association of Official Analytical Chemists. 1968;51:780-785

14. Van Soest PJ. Collaborative study of acid-detergent fiber and lignin. Journal - Association of Official Analytical Chemists. 1973;56:781-784

15. Van Soest PJ, Robertson JB, Lewis BA. Methods for dietary fiber, neutral detergent fiber and non-starch polysaccharides in relation to animal nutrition. Journal of Dairy Science. 1991;74:3583-3597

16. Domingues L, Dantas M, Lima N, Teixeira J. Continuous ethanol fermentation of lactose by a recombinant flocculating Saccharomyces cerevisiae strain. Biotechnology and Bioengineering. 1999;64(6):692-697

17. Branyik T, Vicente A, Oliveira R, Teixeira J. Physicochemical surface properties of brewing yeast influencing their immobilization onto spent grains in a continuous reactor. Biotecnolology and Bioengineering. 2004;88:84-83

18. Bellon-Fontaine M-N, Rault J, van Oss CJ. Microbial adhesion to solvents: A novel method to determ ine the electron-donor/electron-acceptor or Lewis acid-base properties of microbial cells. Colloids and Surfaces B: Biointerfaces. 1996;7:147-143

19. Henriques M, Gasparetto K, Azevedo J, Oliveira R. Experimental methodology to quantify Candida albicans cell surface hydrophobicity. Biotechnology Letters. 2002;24:1111-1115

20. Branyik T, Vicente A, Dosta P, Teixeira A. Continuous beer fermentation using immobilized yeast cell bioreactor systems. Biotechnology Progress. 2005;21:653-663

21. Van Oss CJ. Hydrophobicity of biosurfaces: Origin, quantitative determination and interaction energies. Colloids Surfaces: B Biointerfaces. 1995;5:91-110

22. Sun RC, Fang JM, Rowlands P, Bolton J. Physicochemical and thermal characterization of wheat straw hemicelluloses and cellulose. Journal of Wood Chemistry and Technology. 1998;46:2804-2809

23. Howard RL, Abotsi E, Rensburg J, Howard S. Lignocellulose biotechnology: Issues of bioconversion and enzyme production. African Journal of Biotechnology. 2003;2:602-619

24. Mozes N, Rouxhet PG. Microbial hydrophobicity and fermentation technology. In: Doyle RJ, Rosenberg M, editors. Microbial cell Surface Hydrophobicity. Washington: American Society for Microbiology; 1990. pp. 76-105

25. Smith B, editor. Infrared Spectral Interpretation. 1st ed. Boca Raton Florida: CRC Press LLC; 1999. p. 269

26. Mercier-Bonin M, Ouazzani K, Schmitz P, Lorthois S. Study of bioadhesion on a flat plate with a yeast/glass model system. Journal of Colloid and Interface Science. 2004;271:342-350

27. Van Oss CJ. Hydrophobicity and hydrophilicity of biosurfaces. Current Opinion in Colloid & Interface Science. 1997;2:503-512

28. Van Oss CJ, Giese RF. The hydrophilicity and hydrophobicity of clay minerals. Clay Minerals. 1995;43:474-477

29. Panagoda GJ, Ellepola AN, Samaranayake LP. Adhesion to denture acrylic surfaces and relative cell surface hydrophobicity of Candida parapsilosis and Candida albicans. APMIS. 1998;106:736-742

CHAPTER 11

POTENTIAL ETHANOL PRODUCTION BY IMMOBILIZED AND COIMMOBILIZED SACCHAROMYCES CEREVISIAE AND ZYMOMONAS MOBILIS: A VIABLE OPTION FOR RECYCLING ORGANIC WASTE

1. INTRODUCTION: AN OVERVIEW

1.1. Utilization of agroindustrial waste in fermentation processes

Alcoholic fermentation is a metabolic process in which bacteria enzymatically transform hexoses, such as glucose, fructose, mannose, and galactose, into alcohol (ethanol), carbon dioxide, and adenosine triphosphate (ATP) molecules in the absence of oxygen. Around 70% of the energy is emitted as heat, while the remaining portion is stored in two terminal phosphate bonds of ATP. This stored energy is utilized in transfer reactions, such as the activation of glucose (phosphorylation) and amino acids prior to polymerization. Fermentation refers to a series of chemical processes performed by microbes, where an organic compound is partially oxidized in the absence of oxygen to produce chemical energy. This process can be seen as a partial oxidation when all the carbon atoms in the substance are oxidized to make CO2. This procedure, which has been known since ancient times, is today the sole industrial method used worldwide for producing ethyl alcohol. Glucose is utilized as a raw material, however, there are other sorts of raw materials that are significantly more cost-effective. Alcoholic fermentation is a natural process that is initiated by certain microorganisms through their anaerobic energy cell metabolism. However, in order to achieve large-scale production, it is necessary to use microorganisms such as bacteria, fungi, and yeasts to speed up the process and enhance the conversion rate [1]. From the 20th century until the early 21st century, the primary focus of alcoholic fermentation has been on enhancing fermentation processes. Specifically, the emphasis has been on optimizing industrial performance by carefully selecting yeast strains. Yeast is the most commonly used microorganism for ethanol production through fermentation due to its high efficiency in converting sugars and facilitating the separation of biomass after fermentation. Yeasts are single-celled microorganisms, often spherical in shape, with a size ranging from 2 to 4 μm. They occur naturally in certain items including fruits, grains, and vegetables. Various types of fermentative bacteria have been found, mostly include Saccharomyces cerevisiae, Kluyveromyces fragilis, Torulaspora, and Zymomonas mobilis [2].

S. cerevisiae is a single-celled organism that may utilize two metabolic pathways, alcoholic fermentation and aerobic respiration, to acquire the energy required for its essential functions. The first phase is marked by the release of carbon dioxide (CO2) and the synthesis of ethanol in the absence of oxygen. This allows the organism to receive the energy required for its essential functions through the breakdown of carbohydrates. The yeast necessitates glucose for catalysis through the glycolysis or Embden-Meyerhof route, resulting in the production of pyruvate. Subsequently, pyruvate is transformed into ethanol and CO2 using specialized enzymes under anaerobic conditions. The organism's preferred temperature range for growth ranges from 22 to 29°C, and it cannot withstand temperatures over 53°C. The organism undergoes fermentation when exposed to a sugar solution with a concentration below 12%. However, its activity is halted when the sugar concentration surpasses 15% due to the osmotic pressure exerted by the medium on the cell. Z. mobilis is a gram-negative bacteria that can ferment specific carbohydrates anaerobically to produce bioethanol. In some cases, it can do so more efficiently than yeasts. The organism possesses an unfinished Krebs cycle, yet it exhibits traits that enable it to carry out the pyruvic synthesis pathways from glucose or glyceraldehyde-3-phosphate. This organism exhibits a significant level of sugar absorption and achieves a 97% production of ethanol as a fuel source [3]. The utilization of agroindustrial products in alcoholic fermentation poses significant challenges due to potential issues arising from using raw materials meant for human consumption or edible crops for ethanol production. Additionally, the shift in land use from vegetable cultivation to ethanol and bioethanol production can result in deforestation, food scarcity, expansion of desert regions, and reduced soil water retention capacity, thereby disrupting the hydrological cycle's equilibrium [4]. At a worldwide level, the utilization of energy raw resources for energy purposes and in the manufacturing of ethanol has resulted in increased prices for commodities like maize or barley, while also rendering ethanol production economically impractical. Hence, it is crucial to employ raw materials that do not compete with food goods and are economically viable in the manufacturing of biofuels. Additionally, these materials must guarantee both profitability and environmental sustainability in the projects. It has been projected that utilizing all global agricultural surpluses in the energy sector would only be sufficient to generate bioethanol capable of replacing 1% of the current oil consumption. It can be inferred that utilizing food crops for ethanol production will create a series of imbalances in the food chain, leading to an unsustainable situation [5].

Another option for ethanol production involves utilizing unconventional raw materials, which are derived as by-products or trash from industrial processes. Suggest innovative technologies for the production of ethanol from cane leftovers, solid waste, and materials rich in cellulose and hemicellulose. These technologies enable the transformation of waste from different industries into valuable raw materials for ethanol production.

Currently, the focus has primarily been on finding inexpensive alternative raw materials to replace old ones, with the aim of improving the efficiency of alcohol production through fermentation, recovery, and purification procedures. The primary objective of bioethanol production is to provide a viable alternative to fossil fuels. Bioethanol, which may be used as fuel for motor vehicles, has several advantages such as increased octane number, reduced consumption, and lower levels of carbon monoxide and hydrocarbon emissions (10-15% less). Ethanol can be blended with unleaded gasoline in a range of 10 to 25% without any issues. However, several engines have been designed to run solely on 100% ethanol as fuel. Therefore, ethanol has the potential to replace methyl tert-butyl ether (MTBE), which is a substance used to oxygenate gasoline in Mexico since 1989 and has resulted in decreased CO2 emissions. This effect is crucial due to the high stability and low degradation of MTBE, as well as its high solubility in water. The production of bioethanol declined in significance by the end of the first half of the 20th century, as it was substituted by the manufacture of synthetic ethanol derived from petroleum. Although synthetic ethanol is cheaper, it is unsuitable for use in food preparation, alcoholic beverages, or pharmaceuticals. The increase in oil costs has prompted us to focus on the fermentation method for producing ethanol. Currently, our main objective is to find inexpensive raw materials to replace the conventional sweet substances. Various studies conducted by researchers indicate that the by-products of mango juice, cane juice, and molasses are a highly effective substitute for producing ethanol. This alternative does not have any negative impact on the food supply and actually enhances the productivity and concentration of ethanol in the fermentation process. As a result, it significantly reduces the costs associated with ethanol production [6]. The sugar industry in Mexico has a significant historical importance. It is known for its year-round sugarcane harvests, producing a total of 46,231,229 tons of cane per year. Additionally, the byproduct of sugarcane, known as residue, is utilized as energy biomass and for the production of various biotechnological products through fermentation. Additional options for raw materials include mango juice and its byproducts. Mangoes are cultivated in all Latin American countries, with Mexico being the primary exporter, producing around 1.452 million tons of mango annually. Over 60% of this production is allocated to the South-Southeast region of our country. Exploring the use of alternative residues or substitutes for the traditional raw materials in ethanol production is a highly promising option. This is because the cost of producing ethanol is closely tied to the cost, volume, and composition of the raw materials. In Mexico, the current economy is centered around cane cultivation, which has a long history and tradition in the production of sugar. Additionally, the export of various types of mangoes is also a significant part of the economy. Technological advancements have provided alternatives for the fermentation of cane juice, molasses, and mango juice using S. cerevisiae and Z. Mobilis. These microorganisms serve as viable sources for the production of ethanol, which can be used in the manufacturing of alcoholic beverages or as a biofuel.

1.2. Tecnología de inmovilización

Research has been developed in order to increase the productivity of alcoholic fermentation processes. The productivity, expressed as grams of ethanol produced per hour per unit of fermentation volume, can be increased by optimizing the composition of the culture medium, by the selection of an appropriate microorganism strain or through the adaptation of the design of reactors [7]. One challenge today is to reduce ethanol production costs, and an alternative is to reduce the cost of the culture media, which can represent about 30% of the final production costs of ethanol [8].

Some fermentation technologies have been developed to improve the production of ethanol and its concentration in the culture media [9, 10]. Among these, the immobilization technology offers advantages in contrast to free cell cultures, such as increased retention time in bioreactors, high cellular metabolic activity, high cell load and protection for cells from stress [10, 11]. The immobilization cell technologies have been applied for different purposes as for the production of hydrogen [12] and compounds commercially used in the food industry [13]. Other studies have been developed with immobilized algal cells to remove nutrients (N and P) from wastewater, phenol and hexavalent chromium [14–17]. Similarly, the immobilization of Zymomonas and Saccharomyces have been used for the bio-ethanol production from waste materials [7–10, 18–20].

On the other hand, the immobilization technology provides the possibility of efficiently incorporating symbiotic bacteria [21, 22]. The interaction between two microorganisms in the same matrix is called coimmobilization, and this association can be positive with higher growth and production. However, there are relatively less applications in the ethanol production involving the immobilization of mixed-culture systems and/or coimmobilized cultures.

In a petroleum deficiency situation, bioethanol from yeast and bacterial fermentation has become a promising alternative source for fuel. Agricultural and industrial waste containing sugar, starch and cellulose, such as cassava peels, fruit bunches, and the effluents from sugar and pineapple cannery productions have been successfully applied for the bioethanol production [23, 24]. In this context, the municipality of Ciudad del Carmen, Campeche, Mexico, has an annual production of about 2.868 ha mango (Mangifera indica), obtained through various forms of cultivation and orchard-based technology, but the lack of local market and the poor product distribution to other locations cause much of the product be wasted, with significant losses in the locality. Hence, the need to seek alternatives to use these wastes and generate added value in the economy of the region.

This study was to determine whether the association between S. cerevisiae coimmobilized with Z. mobilis improved growth, and ethanol production using a culture medium equivalent to mango juice (M. indica) creates an opportunity for a regional fruit for exploitation in the production of ethanol. In this study, both microorganisms were confined in small alginate beads, a practical means of using microorganisms for environmental applications.

2. MATERIALS AND METHODS

2.1. Microorganism and medium

The yeast strain S. cerevisiae (ATCC® 2601) and bacteria Z. mobilis (ATCC® 8938) were obtained from the laboratory Microbiologis® and used for fermentation in coimmobilized and immobilized systems. Both microorganisms were cultured in a medium containing composition (g L-1), as described by Demirci et al. [25]: 20 g glucose, 6 g yeast extract, 0.23 g $CaCl_2 \cdot 2H_2O$, 4g $(NH_4)_2SO_4$, 1 g $MgSO_4 \cdot 7H_2O$ and 1.5 g of KH_2PO_4, previously sterilized by autoclave. Strains were maintained in 250 mL of culture at 30°C and pH 4.5 with manual shaking three times a day. Transfers of fresh medium were made every 24 h for three consecutive days prior to use in experiments.

2.2. Preparation of immobilized and coimmobilized cells

For the preparation of immobilized cells, we used the technique described by Tam and Wong [26]. Both microorganisms were harvested by centrifugation at 3500 rpm for 10 min. The bacteria and yeast cells were resuspended in 50 mL of distilled water to form a concentrated cell suspension. The suspension was then mixed with a 4% sodium alginate solution in 1:1 volume ratio to obtain a mixture of 2% microorganism–alginate suspension. The mixture was transferred to a 50-mL burette, and drops were formed when "titrated" into a calcium chloride solution (2%). This method produced approximately 6500 uniform algal beads of approximately 2.5 mm in diameter with biomass content for Z. mobilis-alginate of 0.0055 g bead-1 and for S. cerevisiae of 0.00317 g bead-1 for every 100 mL of the microorganism–alginate mixture (Figure 1).

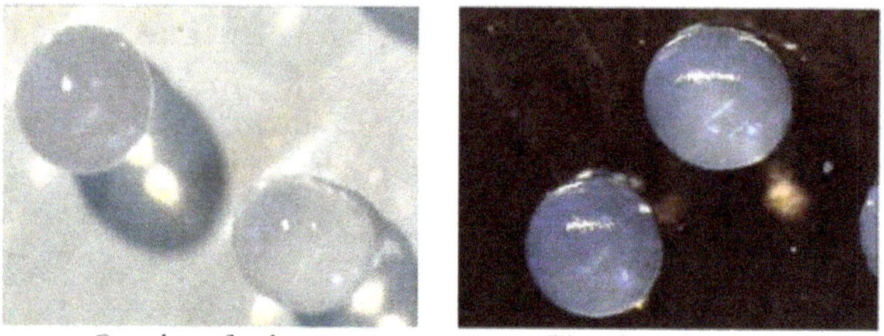

One day of culture Five day of culture

Figure 1. Microorganism–alginate beads suspension.

The beads were kept for hardening in the CaCl2 solution for 4 h at 25 ± 2°C and then rinsed with sterile saline solution (0.85% NaCl) and subsequently with distilled water. A concentration of 2.6 beads mL-1 of medium (equivalent to 1:25 bead: medium v/v) was placed in a Chemostat Ommi Culture Plus (Virtis) containing 2 L of culture medium. The reactor was maintained under stirring at 120 rpm and 30°C (Figure 2). A similar procedure was used for coimmobilization, with the difference that the concentrate of bacteria (25 mL) and yeast (25 mL) was mixed and then mixed with 50 mL of alginate; this procedure allowed retaining the same concentration of cells in all experiments.

Figure 2. Chemostat Ommi Culture Plus (Virtis) used for fermentation experimental process.

2.3. Experimental setup and procedure

This study was divided into two parts: (1) the batch experiment consisted of evaluating the growth and ethanol production in a medium equivalent to mango juice in cultures with free cells, immobilized and coimmobilized, and (2) evaluate the effect of glucose concentration in the production of ethanol in the system previously selected in the first experimental part based on the ethanol productivity obtained. Fermentation was performed in a Chemostat Ommi Culture Plus (Virtis) with a volume of 2 L operation, adjusting stirring at 120 rpm and maintaining a temperature of 30°C. The medium equivalent to mango juice was similar to that described by Demirci et al. [25] by adjusting the composition of the medium to a concentration of 200 g L−1 glucose, equivalent to that observed in the mango juice (M. indica).

The experimental design consisted of triplicate cultures in a Chemostat reactor Ommi Culture Plus for S. cerevisiae and Z. mobilis in free cell culture, immobilized and coimmobilized. For each experiment, the biomass was collected, as well as samples of the culture medium to the end of the logarithmic phase every 20 h. For the determination of ash-free dry weights, five beads were dissolved and filtered through a GF-C glass fiber filter (2.5 cm diameter), previously rinsed with distilled water, and incinerated at 470°C for 4 h. The samples were dried at 120°C and put to constant weight for 2 h in a conventional oven and then in a muffle furnace at 450°C for 3 h. The soluble solids of each fermenting medium were determined every 20 h by taking 1 mL aliquot from each reactor and testing for the Brix level in an refractometer.

Ethanol content (% v/v) was obtained using the Anton Paar DMA 4100M instrument, which determines the density of the mixture in relation to the standard OIML-STD-90, which can determine the content of distillate ethanol (% v/v); according to the ethanol density recorded, it was possible to obtain an approximate of ethanol content (grams of ethanol per liter of culture) produced for each experiment. Prior to the determination of the ethanol content, a distillation of cultures was conducted with a plate column distiller PS-DA-005/PE of four plates, at small scale. The cooling water flow was 3 L h−1 at 15°C. An aliquot of 3 L was distilled for 4 h, maintaining the operating conditions at atmospheric pressure, without reflux and with a temperature ramp in the heating jacket of 30°C up to 80°C.

The STATISTICA 7.0 software for statistical analysis and calculated mean and standard deviation for each treatment were used. The covariance analysis (ANCOVA) with $P \leq 0.05$ was used to evaluate the growth in free cell cultures, immobilized and coimmobilized. The Tukey test ($P \leq 0.05$) was used when significant differences were observed.

3. RESULTS

3.1. Growth

In free cell cultures, the growth was observed immediately after being inoculated in the reactor of 2 L. Growth kinetics shows an exponential phase for S. cerevisiae and Z. mobilis of 120 h. After this period of cultivation, both species showed a decline in the production of biomass, finalizing treatment after 200 h of culture. The maximum values of biomass concentration were 14.18 and 11.80 g L-1 dry weight for S. cerevisiae and Z. mobilis, respectively. Both microorganisms grew satisfactorily under the culture conditions used in this study (Figure 3A), with a higher growth rate (μ) for S. cerevisiae (0.0547 d-1) with respect to Z. mobilis (0.0418 d-1). Growth rates in free cell cultures for both microorganisms S. cerevisiae and Z. mobilis were not significantly different ($P \geq 0.05$).

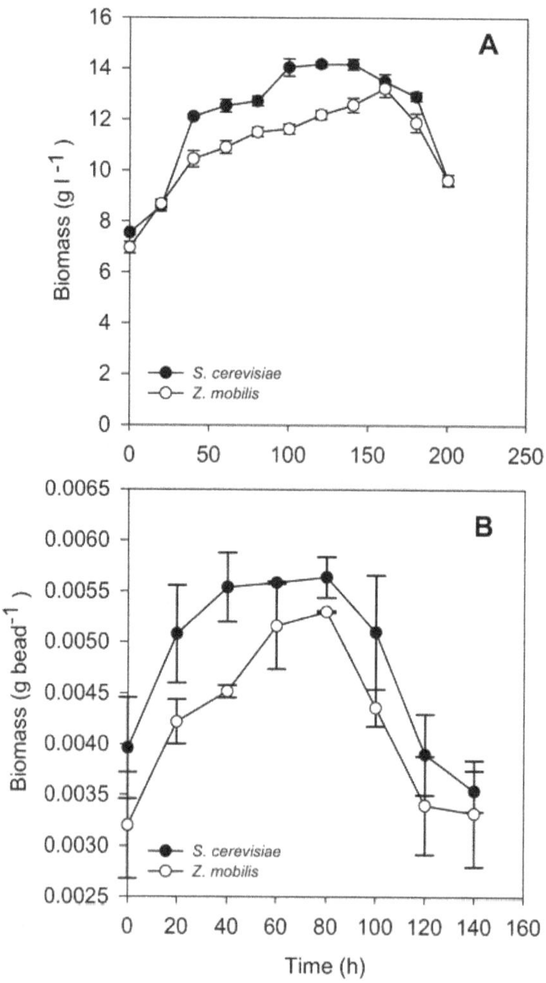

Figure 3. Average increase of biomass for Saccharomyces cerevisiae and Zymomonas mobilis in free culture (A) and immobilized cells (B).

For immobilized cells, both yeast and bacteria presented immediate growth after adding the beads to the culture medium; in both treatments, the exponential phase of growth reached a maximum of 80 h. It is noteworthy that although both microorganisms were immobilized under the same procedure, the content of biomass per bead at the beginning of treatment was lower for Z. mobilis (0.0031 g bead-1) compared to S. cerevisiae (0.0039 g bead-1). Despite these differences, both microorganisms were able to tolerate immobilization (Figure 3B), reaching maximum biomass content values of 0.0055 and 0.0047 g bead-1 for S. cerevisiae and Z. mobilis, respectively. In relation to growth, Z. mobilis showed a higher growth rate (0.142 d-1) with respect to S. cerevisiae (0.106 d-1), but there were no significant differences ($P \leq 0.0001$).

3.2. Glucose-substrate removal

The decrease of substrate showed significant differences ($P \leq 0.0001$) between treatments with free and immobilized cells for both species (Figure 4). However, the Tukey test analysis showed that the two species in free culture were not significantly different ($P > 0.05$) in 200 h of treatment. While for the immobilized and coimmobilized cell cultures, only the immobilized Z. mobilis bacteria showed no significant differences ($P = 0.245$) during removal of the substrate with the coimmobilized system during 140 h of culture (Figure 4B).

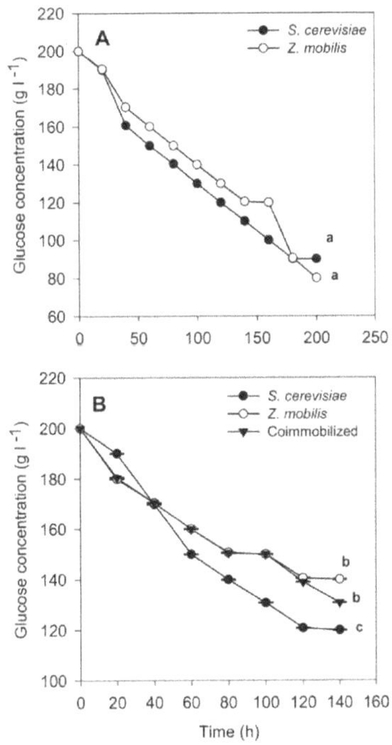

Figure 4. Average reduction of glucose (g L-1) for Saccharomyces cerevisiae and Zymomonas mobilis in free culture (A) and immobilized culture (B). Different letters indicate significant differences ($P \leq 0.05$).

It is a fact that consumed substrate was greater in free culture for S. cerevisiae and Z. mobilis from 200 to 80 g L−1 (60% removal) after the 200-h treatment period (Figure 4A), compared to the immobilized system with 40% removal for S. cerevisiae (from 200 to 120 g L−1) and 30% removal for Z. mobilis (from 200 to 140 g L−1), while in those cultures of coimmobilized cells consumption ranged from 200 to 130 g L-1 (35% removal) (Figure 4B).

The average consumption analysis based on removal rates determined during the exponential growth for both species showed that free culture S. cerevisiae and Z. mobilis reached removal rates of 2.0 and 2.7 g-substrate per g-biomass d, respectively. This suggested greater productivity for the bacteria (5.76 g h−1) with respect to yeast (5.29 g h−1) (Table 1).

Table 1. Uptake rate, productivity (Y) and ethanol mole produced per glucose mole for Saccharomyces cerevisiae and Zymomonas mobilis in free culture, immobilized and coimmobilized.

Culture	Cells	Ethanol formed (% v/v)	Y (g L^{-1} h^{-1})	Glucose consumed (g L^{-1})	Uptake rate (g-substrate removed per g-biomass d)	moles Ethmole Glc
Free cells	*S. cerevisiae*	80.6	5.29a	90	2.71a	2.76
	Z. mobilis	87.7	5.76a	80	2.0a	3.38
Immobilized cells	*S. cerevisiae*	86.0	8.45b	60	0.165b	4.42
	Z. mobilis	87.7	8.70b	43	0.056b	6.29
Coimmobilized cells	*Z. mobilis* and *S. cerevisiae*	88.7	8.80c	47	0.235c	5.82

[a,b,c] Indicate significant differences ($P \leq 0.05$).

In cultures with immobilized cells, the removal rate in the exponential phase (80 h) was greater for S. cerevisiae (0.165 g-substrate per g-biomass d) with respect to Z. mobilis (0.056 g-substrate per g-biomass d), but in coimmobilized culture it was greater (0.235 g-substrate per g-biomass d) since both species contribute to reducing glucose and increasing the removal rate. Similar results were observed in the productivity, where the coimmobilized cell culture showed higher values (8.80 g L−1 h−1) with respect to the immobilized cells of S. cerevisiae (8.45 g L−1 h−1) and Z. mobilis (8.70 g L-1 h-1 (Table 1). In general, the highest productivity levels were recorded in coimmobilized and immobilized cultures with respect to free cell cultures because shorter ethanol production time (80 h) compared to free cultures (120 h).

3.3. Effect of initial concentration of glucose on ethanol production

It is a fact that most cultures from fruits may contain a high concentration of fiber solids that cause problems of mixture in the reactor, and consequently a low contact between cells and substrate. Mango juice is no exception. In this study, we evaluated the growth and alcohol production of Z. mobilis coimmobilized with S. cerevisiae in cultures with dilutions of 200 and 50 g L-1 of substrate in equivalent medium.

For the coimmobilized of Z. mobilis and S. cerevisiae within alginate beads, an immediate increase in biomass content was observed. Although the biomass content for both treatment showed significant differences ($P \leq 0.0024$), the results suggest that the concentration of substrate was not a limiting factor for the growth for bacteria and yeast (Figure 5). The maximum biomass content in the treatment of glucose to 50 g L-1 (Gl50) was obtained in the first 100 h of culture with about 0.0063 g beads-1; while for the treatment of 200 g L-1 glucose (Gl200) was of 0.053 g beads-1 during a period of 80 h (Figure 5).

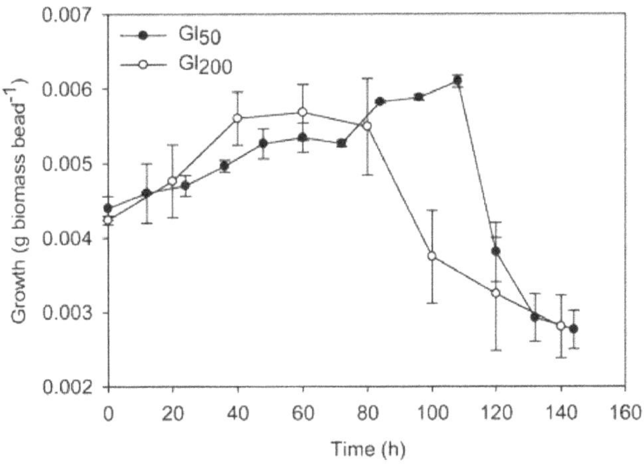

Figure 5. Growth (g biomass bead-1) in coimmobilized system at different substrate concentrations.

The content of alcohol produced had no significant differences ($P \leq 0.05$) with respect to glucose concentration. However, uptake rates exhibit a decline as the glucose content in the reactor decreases (Table 2). The highest uptake rate occurred at a concentration of 200 g L-1 glucose (0.235 g-substrate per g-biomass d) with a 76.5% removal, compared to 50 g L-1 glucose (0.08 g-substrate per g-biomass d). Although the production of alcohol was similar in both treatments, the ratio mol-ethanol produced per consumed mol-glucose was higher in cultures of 50 g L-1 glucose with a value of 6.91, with respect to 200 g L -1 glucose with a ratio of 5.82 mol-ethanol produced per consumed mol-glucose (Table 2). Similarly, higher productivity was obtained (8.85 g L-1 h-1) at a lower glucose concentration compared to a medium with high glucose content (8.80 g L-1 h-1).

Table 2. Productivity (Y), uptake rate and ethanol mole produced per glucose with minimum content of glucose for coimmobilized Z. mobilis and S. cerevisiase.

C_0	Ethanol formed (% v/v)	Y (g L^{-1} h^{-1})	Glucose consumed (g L^{-1})	Uptake rate (g-substrate removed per g-biomass d)	moles EthmoleGlc
50	89.63	8.85	40.0	0.08	6.91
200	88.70	8.80	47.0	0.235	5.82

C_0: Initial concentration of glucose (g L^{-1}).

4. GROWTH AND PRODUCTIVITY

To evaluate the capacity of growth in free and immobilized culture, two microorganisms, yeast and bacteria, were subjected to the same culture conditions (Figure 3A). In free culture, the yeast S. cerevisiae showed a higher cell density and specific growth rate (0.0547 d-1) with respect to bacteria Z. mobilis (0.0418 d-1) in a treatment time of 120 h. The immobilized systems are known to have a greater capacity of cell growth and high metabolic activity [27, 28], which is consistent with the results obtained in this study. The result showed a high growth rates for immobilized Z. mobilis (0.142 d-1) and S. cerevisiae (0.106 d-1) with respect to free cell cultures, suggesting that immobilization did not affect growth in both microorganisms and increased biomass content favorably. Furthermore, the high activity in immobilized cell was observed with in a decrease of substrate in a shorter time of treatment (80 h) compared to free cell cultures (120 h). The short time of treatment for immobilized cell could be attributed to the increase of biomass within the beads and consequently an immediate decay of the substrate; however, this indicates that increasing cell population within the beads can cause a limited effects of nutrients on the cells located at the center of the beads, causing a decrease in cellular activity [28, 29].

Another factor that probably favors the rapid decline in cell density is attributed to the production of CO2 as result of fermentation activity. Studies suggest the adverse effect of CO2 gas, because if the diffusion of CO2 is lower compared to its production, it will accumulate inside of alginate bead [30]. In this study, the CO2 is observed in the reactor as bubbles attached on the surface of the beads, suggesting that the spread of CO2 gas in the first 80 h was not a factor that inhibited growth and the production of alcohol; however, after this time, gas saturation in the reactor was probably high, affecting the diffusion of CO2. This coupled with a limitation in the transport of nutrients and subsequent inhibition of microorganisms and may have caused glucose consumption to be lower compared to free cells (Table 1).

In particular, in free cell culture, the lower percentage of alcohol obtained by yeast during the increasing fermentation culture commonly relates to the fact that this is affected by the high concentration of ethanol in the solution, which may inhibit metabolism and decrease efficiency [31], unlike bacteria Z. mobilis [30]. In the present study, the lowest biomass produced by the bacteria (0.0047 g L-1) with respect to yeast (0.0055 g L-1) may be practical from the standpoint of waste generation. Similarly, observations were reported by Amin and Verachtert [9] for Z. mobilis and Saccharomyces bayanus immobilized in carrageenan with 5.6 and 9.9 g L-1, respectively.

It is evident that ethanol production was not inhibited in immobilized or coimmobilized systems, and even showed higher productivity with respect to free cells (Table 1), suggesting that they are more efficient in the conversion of sugar with respect to time. Krishnan et al. [19] reported lower productivity for Z. mobilis immobilized in carrageenan (1.6 g L-1 h-1) compared to that obtained in this study (8.7 g L-1 h-1); this difference may be attributed to the lower amount of glucose content in the culture medium of 32 g L-1 with respect to that used in the present study of 200 g L-1.

Interestingly, the immobilized systems showed a higher conversion of substrate of 4.42 and 6.29 mole of ethanol per mole of glucose for yeast and bacteria, respectively, compared to the obtained by free cells, from 2.7 to 3.3 mole of ethanol per mole of glucose. In general, treatments with immobilized cells showed a higher output of ethanol per mole of glucose with respect to that reported by Amin and Verachtert [9] for Z. mobilis and S. bayanus immobilized in carrageenan with values of 1.8–1.9 mole of ethanol produced per mole of consumed glucose. Gunasekaran et al. [32] and Krishnan et al. [19] suggest that Z. mobilis is a good candidate to obtain alcohol with approximately 1.9 mole ethanol per mole of glucose; similarly, Rogers et al. [33] reported that specific productivity of ethanol (g ethanol g-1 biomass dry weight) is greater for Zymomonas than for Saccharomyces uvarum.

According to the results, immobilization and coimmobilization exhibited a lower uptake rate compared to free cells; this shows that there was less consumption of substrate (Table 1). Nevertheless, there was greater productivity, which indicates that it is possible to obtain high alcohol content with a lower requirement of substrate, but with the disadvantage of residual glucose in the medium; this problem can be solved with sequenced systems, as suggested by Demirci et al. [25]. Another alternative of solving this problem is to increase the cell number or inoculum size within the reactor. This is reasonable because a high number of cells could create a greater sorption of substrate (glucose) into the cell and eventually consumed substrate. However, Siripattanakul-Ratpukdi [10] suggested that with different cell yeast loads, the same reduction (>90%) of substrate is obtained at the end of a treatment period of 10 h.

The low glucose reduction observed in this study in alginate beads can be attributed to the decline in cell density, but it is likely that the diffusion of substrate could have been prevented. Studies have reported that the adsorption of substrate by the matrix was observed in the first hours of treatment, with a possible decrease of substrate diffusion within the matrix in a continuous process [34].

On the other hand, Robinson et al. [35] suggest that the diffusion rate within the alginate matrix depends on the concentration gradient between the culture medium and matrix; this is, when the nutrient concentration in the culture medium decreases, the diffusion rate occurs within the matrix and therefore the removal rate. In this study, during the first hours of treatment, the matrix is probable a partial saturation with substrate (glucose), because the substrate is decreased during the culture time, and cell growth for both microorganisms was continuous. Clearly, the immobilized cell system successfully decreased glucose by adsorption of the matrix (immobilized glucose) and biodegradation (bioconversion of glucose), being the main process the biodegradation. This suggests that the main factor that could limit glucose removal may have been the high concentration of CO2 in the reactor.

4.1. Coimmobilization of *Z. mobilis* and *S. cerevisiae* at different glucose concentrations

The analysis of biomass content in alginate beads indicates that a high concentration of glucose (200 g L^{-1}) results in a significant and quick decline, in contrast to cultures with a low glucose concentration (50 g L^{-1}). This observation demonstrates that a higher concentration of glucose leads to a faster saturation of beads, which in turn reduces the diffusion between the beads and the culture medium. As a result, the diffusion of CO2 produced can be hindered, leading it to remain trapped inside the bead. This trapped CO2 then leads to a decrease in growth and substrate consumption. On the other hand, a low concentration of substrate in the culture media suggests that there is an efficient transport system and accumulation of substrate inside the matrix. This enables the bacteria and yeast to consume and grow properly. Thus, a decrease in substrate concentration might actually enhance alcohol production while minimizing the presence of glucose residues. This can result in the production of 6.91 mol of ethanol per mole of glucose, as compared to a high concentration of glucose (Table 2). Our previous research has demonstrated that when using a coimmobilized system, the fermentation process of mango juice yields a production ratio of 1.4 liters of alcohol (with an ethanol concentration of 79% v/v) for every 3 liters of mango juice.

5. CONCISE SUMMARY

The current study has demonstrated the inherent capacity of utilizing coimmobilized systems for ethanol generation. The co-cultivation of Z. mobilis and S. cerevisiae resulted in a synergistic effect, leading to increased ethanol production and efficient conversion of the substrate, surpassing the performance of both free and immobilized cells individually. Overall, immobilization technology provides an alternate method that enhances productivity and substrate conversion compared to culture systems that use free cells. The immobilized systems in this investigation demonstrated a high capacity for conversion, resulting in a higher alcohol content while requiring less substrate. Substrate inhibition did not impact cell growth in either organism. It is evident that the immobilized cell system effectively decreased glucose levels through matrix adsorption (immobilized glucose) and biodegradation (conversion of glucose into other substances), with biodegradation being the primary process. These findings indicate that the primary reason impeding further expansion was the elevated CO2 levels in the reactor. In addition, while there were no notable variations in the alcohol level in immobilized culture in a diluted medium, the conversion from glucose to ethanol is higher in media with a glucose concentration of 50 g L-1. In order to achieve practical goals, it is preferable to carry out the fermentation of waste organic matter by diluting it. This will help to enhance the uniformity of alginate beads in the reactor, hence facilitating the diffusion of CO2 and substrate through the beads.

AT A GLANCE

Advancements in fermentation technologies have been made to enhance the efficiency of ethanol production. One such alternative is immobilization technology, which allows for the effective inclusion of symbiotic bacteria inside the same matrix. This study examines the possible application of immobilized and coinmobilized systems on calcium alginate beads for ethanol production using mango waste (Mangifera indica). The study compares the performance of Zymomonas mobilis and Saccharomyces cerevisiae in these systems with that of free cells culture. Additionally, the study evaluates the impact of glucose concentration on productivity in the coimmobilized system using a Chemostat reactor called Ommi Culture Plus. In the case of free cell culture, Z. mobilis exhibited a greater productivity rate of 5.76 g L-1 h-1 compared to S. cerevisiae's productivity rate of 5.29 g L-1 h-1. However, in the coimmobilized culture, a higher productivity rate of 8.80 g L-1 h-1 was achieved, surpassing the productivity rates of immobilized cultures which ranged from 8.45 g L-1 h-1 to 8.70 g L-1 h-1. The conversion of glucose to ethanol in the coimmobilized system was more efficient, resulting in a higher yield of ethanol (6.91 mol) when using 50 g L-1 of glucose, compared to 200 g L-1 of glucose (5.82 mol). This suggests that the use of immobilized and coimmobilized cultures, as opposed to free cells, presents an opportunity for both the recycling of organic residues and the production of a larger quantity of alcohol.

REFERENCES

1. Mathews KC, Van Holde KE, Kevin GA. Biochemistry, San Francisco, California, editorial Addison Wesley, 3er edition, 2002: P.1368
2. Beltran G, Torija MJ, Novo M, Ferrer N, Poblet M, Guillamon JM, Rozes N, Mas A. Analysis of yeast populations during alcoholic fermentation: A six year follow-up study. Systematic and Applied Microbiology. 2002;25(2):287-293
3. Doelle HW, Kirk L, Crittenden R, Toh H, Doelle MB. Zymomonas mobilis—science and industrial application. Critical Reviews in Biotechnology. 1993;13(1):57-98
4. Sheehan J, Himmel M. Enzymes, energy, and the environment: A strategic perspective on the U.S. Department of Energy's research and development activities for bioethanol. Biotechnology Progress. 1999;15(5):817-827
5. Christen P, Minier M, Renon H. Ethanol extraction by supported liquid membrane fermentation. Biotechnology & Bioengineering. 1990;36:116-123
6. Mejía-Giraldo LF, Martínez-Correa HA, Betancourt-Gutiérrez JE, Castrillón Castaño CE. Usage of the common mango agroindustrial waste (Mangifera indica L.) in the destruction of fermentable sugars. Ingeniería y Ciencia. 2007;3(6):41-62
7. Amin GV, Eynde D, Verachtert E. Determination of by-products formed during the ethanolic fermentation, using batch and immobilized cell systems of Zymomonas mobilis and Saccharomyces bayanus. European Journal of Applied Microbiology and Biotechnology. 1983;18:1-5
8. Sivers MV, Zacchi G, Olsson L, Hahn-Hagerdal B. Cost analysis of ethanol production from willow using recombinant Escherichia coli. Biotechnology Progress. 1994;10:555-560
9. Amin G, Verachtert H. Comparative study of ethanol production by immobilized-cell systems using Zymomonas mobilis or Saccharomyces bayanus. European Journal of Applied Microbiology and Biotechnology. 1982;14:59-63
10. Siripattanakul-Ratpukdi S. Ethanol production potential from fermented rice noodle wastewater treatment using entrapped yeast cell sequencing batch reactor. Applied Water Science. 2012. DOI: 10.1007/s13201-011-0024-z
11. Tam NFY, Lau PS, Wong YS. Wastewater inorganic N and P removal by immobilized Chlorella vulgaris. Water Science & Technology. 1994;30:369-374
12. Wu SY, Lin CN, Chang JS. Hydrogen production with immobilized sewage sludge in three-phase fluidized-bed bioreactors. Biotechnology Progress. 2003;19:828-832
13. Kawaguti HY, Manrich E, Sato HH. Production of isomaltulose using Erwinia sp. D12 cells: Culture medium optimization and cell immobilization in alginate. Biochemical Engineering Journal. 2006;29:270-277
14. Bandhyopadhyay K, Das D, Bhattacharyya P, Maiti BR. Reaction engineering studies on biodegradation of phenol by Pseudomonas putida MTCC 1194 immobilized on calcium alginate. Biochemical Engineering Journal. 2001;8:179-186
15. Park DH, Chab JM, Ryua HW, Lee GW, Yu EY, Rheea JI, Park JJ, Kimd SW, Joe YI, Ryu YW, Hur BK, Park JK, Park K. Hydrogen sulfide removal utilizing immobilized Thiobacillus sp. IW with Ca-alginate bead. Biochemical Engineering Journal. 2002;11:167-173
16. Humphries AC, Nott KP, Hall LD, Macaskie LE. Reduction of Cr (VI) by immobilized cells of Desulfovibrio vulgaris NCIMB 8303 and Microbacterium sp. NCIMB 13776. Biotechnology & Bioengineering. 2005;90(5):597-605
17. Ruiz-Marin A, Mendoza-Espinosa GL, Stephenson T. Growth and nutrient removal in free and immobilized green algae in batch and semi-continuous cultures treating real wastewater. Bioresource Technology. 2010;101:58-59

18. Vallejo-Becerra V, Vazquez-Bahena JM, Santiago-Hernández JA, Hidalgo-Lara ME. Immobilization of the recombinant invertase INVB from Zymomonas mobilis on nylon-6. Journal of Industrial Microbiology and Biotechnology. 2008;35:1289-1295

19. Krishnan MS, Blanco M, Shattuck KC, Nghiem PN, Davison BH. Ethanol production from glucose and xylose by immobilized Zymomonas mobilis CP4 (pZB5). Applied Biochemistry and Biotechnology. 2000;84-86:525-541

20. Yu J, Yue G, Zhong J, Zhang X, Tan T. Immobilization of Saccharomyces cerevisiae to modified bagasse for ethanol production. Renewable Energy. 2010;35:1130-1134

21. Travieso L, Benitez F, Weiland P, Sánchez E, Dupeyrón R, Dominguez AR. Experiments on immobilization of microalgae for nutrient removal in wastewater treatments. Bioresource Technology. 1996;55:181-186

22. De-Bashan LE, Moreno M, Hernández JP, Bashan Y. Removal of ammonium and phosphorus ions from synthetic wastewater by the microalgae Chlorella vulgaris coimmobilized in alginate beads with the microalgae growth promoting bacterium Azospirillum brasilense. Water Research. 2002;36:2941-2948

23. Nigan JN. Continuous ethanol production from pineapple cannery waste using immobilized yeast cells. Journal of Biotechnology. 2000;80:189-193

24. Kassim MA, Kheang LS, Bakar NA, Aziz AA, Som RM. Bioethanol production from enzymatically saccharified empty fruit bunches hydrolysate using Saccharomyces cerevisiae. Research Journal of Environmental Sciences. 2011;5:573-586

25. Demirci A, Pometto AL, Ho KLG. Ethanol production by Saccharomyces cerevisiae in biofilm reactors. Journal of Industrial Microbiology & Biotechnology. 1997;19:299-304

26. Tam NFY, Wong YS. Effect of immobilized microalgal bead concentrations on wastewater nutrient removal. Environmental Pollution. 2000;107(1):145-151

27. Cohen Y. Biofiltration—the treatment of fluids by microorganisms immobilized into the filter bedding material. Bioresource Technology. 2000;77:257-274

28. Mallick N. Biotechnological potential of immobilized algae for wastewater N, P and metal removal: A review. BioMetals. 2002;15:377-390

29. Uemoto H, Saiki H. Distribution of Nitrosomonas europaea and Paracoccus denitrificans immobilized in tubular polymeric gel for nitrogen removal. Applied and Environmental Microbiology. 2000;66(2):816-819

30. Kim BG, Choi CY. A study on the ethanol production by immobilized cells of Zymomonas mobilis. Korean Journal of Chemical Engineering. 1984;1(1):13-19

31. Nuwamanya E, Chiwona-Karltun L, Robert SK, Yona B. Bio-ethanol production from non-food parts of Cassava (Manihot esculenta Crantz). Ambio. 2011;41(3):262-270

32. Gunasekaran P, Karunakaran T, Kasthuribai M. Fermentation pattern of Zymomonas mobilis strains on different substrate: A comparative study. Journal of Biosciences. 1986;10(2):181-186

33. Rogers PL, Jon JL, Tribe DE. Kinetics of alcohol production by Zymomonas mobilis at high sugar concentration. Biotechnology Letters. 1979;1:165

34. Siripattanakul S, Khan E. Fundamentals and applications of entrapped cell bioaugmentation for contaminant removal. In: Shah V, editor. Emerging Environmental Technologies. Vol. 2. Berlin: Springer; 2010. pp. 147-169

35. Robinson PK, Reeve JO, Goulding KH. Phosphorus uptake kinetics of immobilized Chlorella in bath and continuous-flow culture. Enzyme and Microbial Technology. 1989;11:590-596

SECTION 4

BIOREMEDIATION

CHAPTER 12

UTILIZING YEAST BIOMASS AS AN ALTERNATIVE METHOD FOR BIOREMEDIATION OF HEAVY METALS

1. INTRODUCTION: AN OVERVIEW

Industrialization has been widely recognized as a defining characteristic of civilization. Nevertheless, millions of tons of hazardous substances, including toxic heavy metals such as Cadmium (Cd), Copper (Cu), Mercury (Hg), Lead (Pb), Manganese (Mn), Arsenic (As), Nickel (Ni), and Zinc (Zn), are generated and subsequently discharged into the environment, either directly or indirectly [1]. In contrast to organic contaminants, these pollutants are not capable of being broken down by natural processes and can be passed throughout the food chain by the accumulation of these substances in living organisms [2]. The accumulation of hazardous levels of toxic metals in water sources and in crops cultivated in polluted soils is a matter of great concern due to the detrimental impact of metals on human health and aquatic organisms [1]. The primary objective is to create inventive and economical methods for purifying contaminated surroundings and safeguarding the functionality of ecosystems. Volesky [3] and Domenech [4] identified several typical ways for eliminating dissolved heavy metals, including as chemical precipitation, filtration, ion exchange, oxidation or reduction, reverse osmosis, evaporation, membrane technology, and electrochemical treatment. However, the majority of these methods lose their effectiveness when the levels of heavy metals are below 100 mg/L [5]. In addition, potent and polluting substances are employed for desorption, leading to the formation of toxic sludge and subsequent environmental contamination [1].

Bioremediation, a biological approach to metal removal, was proposed as a means to mitigate the impacts of environmental contamination. Various species obtained from polluted soil, wastewater, compost, and harsh conditions have been demonstrated to be beneficial for bioremediation, ranging from plants to bacteria [6, 7]. Their ability to thrive in a challenging environment can be ascribed to metabolic capabilities that enable biological organisms to investigate, cleanse, and endure in unusual and intricate substances. Microbes are well-suited for cleanup due to their diverse and adaptable metabolism, which sets them apart from other species [8]. Microorganisms have developed many strategies to deal with heavy-metal stress. These include transporting the metals across the cell membrane, binding them to cell walls and trapping them in extracellular capsules, causing precipitation, forming complexes, and engaging in oxidation-reduction reactions [9–14]. Yeast cells are a highly prevalent kind of biomass that can be used for bioremediation of heavy metals. These cells have the capacity to accumulate a wide range of heavy metals to different extents, even in diverse external settings [15]. Yeast is an appropriate model for investigating the physiological and molecular mechanisms of how eukaryotic cells interact with heavy metals. Research conducted on bacteria, yeast, fungi, and microalgae has demonstrated that yeasts are more effective at absorbing heavy-metal ions from wastewater. This is attributed to their rapid growth rate and the structure of their cell walls [16]. The presence of phosphates and carboxyl groups in the cell wall is known to be the primary factor responsible for the negative charge on the surface of yeast cells. This negative charge enables yeast cells to effectively bind heavy-metal cations. The most probable cause of this phenomenon is the electrostatic interactions [17]. The polymers of the cell wall of fungi contain different metal-binding groups, including amine, imidazole, phosphate, sulfate, sulfhydryl, and hydroxyl, as documented [18]. Moreover, yeasts possess a bigger size compared to bacteria and have the ability to physically interweave the mycelia/pseudomycelia, resulting in the formation of floes. The net-like structure of the yeast floc enhances the diffusion of oxygen and avoids the need for excessive oxygen dissolution in water. Thus, the utilization of yeasts can result in a highly efficient oxygen supply and decrease energy consumption by reducing the amount of air flow provided [19]. However, yeast biomass is a cost-effective and easily accessible form of biomass. It can be readily grown in inexpensive growth media and can be obtained in large quantities at a low cost from fermentation enterprises for the purpose of wastewater cleanup [20, 21]. Several types of industrial waste-biomass were examined to determine their potential for biosorption. These include two types of fungi: Saccharomyces cerevisiae, which is commonly found in the food and beverage business, and Candida albicans, which is a strain isolated from clinical settings [14].

2. BIOREMEDIATION

Bioremediation is regarded as an alternate approach for reducing environmental contaminants by transforming them into less harmful substances [2]. Biodegradation refers to the biological breakdown or transformation of organic or inorganic wastes into harmless substances [22]. Mueller et al. provided a definition of bioremediation as the regulated biological degradation of organic wastes to a harmless state or to concentrations below the limits set by regulatory authorities [23]. Two primary approaches for carrying out bioremediation processes are biostimulation and bioaugmentation [22]. Biostimulation refers to the process of enhancing bioremediation by introducing an electron acceptor, nutrition, or other elements to a contaminated site. The goal is to stimulate the expansion of the existing microbial population [2]. Bioaugmentation refers to the process of introducing microorganisms to a polluted area in order to increase the breakdown of contaminants. Therefore, the microorganisms utilized in bioremediation can either originate from the contaminated area itself or be obtained from another location. Recent research indicates that microorganisms found in polluted areas demonstrate a remarkable ability to adapt to various environmental conditions and effectively remove substantial quantities of metals [2, 24–26]. Yeasts and fungi have been found to thrive in matrices with elevated amounts of metal compounds, surpassing other microbes in their ability to grow under such conditions. Furthermore, Ksheminska et al. [27] observed that yeast strains possess the ability to effectively eliminate substantial amounts of these contaminants. Additional research conducted on yeasts has demonstrated that when exposed to metals, the primary objective of yeast cells is to safeguard and cleanse the environment by rendering the metal ions inactive, hence preventing any harmful impacts on cell viability. In contrast to bacteria, which rely solely on active metabolizing capabilities, yeasts possess the unique trait of being able to eliminate pollutants whether they are living (metabolically active) or dead (metabolically inactive/passive) [24, 28, 29].

3. YEAST BIOREMEDIATION MECHANISMS

There is a significant lack of information regarding metal detoxification processes in yeasts in comparison to prokaryotes. The yeasts that have been extensively researched are primarily from the ascomycetous group, including S. cerevisiae, Schizosaccharomyces pombe, and Candida sp. [30]. The findings demonstrated that yeasts have developed multiple distinct methods to detoxify metals, enabling them to mobilize, immobilize, or convert them. The mechanisms of immobilization involve (i) biosorption, where metals interact with the cell membrane through processes such as ion exchange, complexation, crystallization, adsorption, and precipitation; (ii) biotransformation, where toxic metals are converted into less toxic forms; and (iii) bioaccumulation, which refers to the uptake of metal ions by living microorganisms within their cells [2]. The mobilization methods mostly involve bioleaching, which occurs through the synthesis and excretion of certain acids that interact with metal ions, resulting in the formation of insoluble complexes. Therefore, in most cases, immobilization and mobilization are the primary methods employed for the bioremediation of metals using yeast and fungi (Figure 1) [31–33].

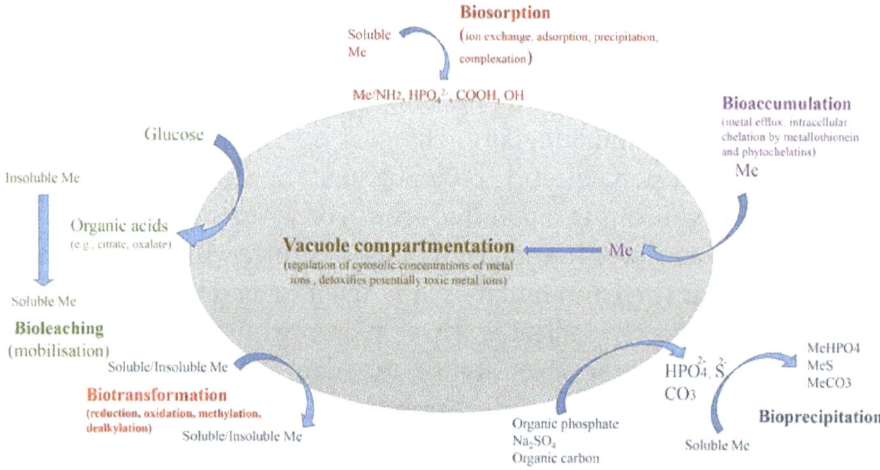

Figure 1. Interactions of metal and fungi cells (adapted from Refs. [31–33]).

3.1. Biosorption

Biosorption is an indiscriminate physico-chemical interaction that can happen between metals and cellular constituents of biological organisms [34]. The phenomenon described involves the capacity of biological substances to attach to and accumulate heavy metals through either metabolic or physico-chemical processes [35]. Yeast and fungal biomass have been identified as attractive biosorbents due to their high capacity and selectivity for heavy metals [36].

3.1.1. Yeast cell wall properties

The first stage of metal ion binding in microorganism cells does not depend on their metabolism and consists in ion chemisorption into cell wall components. Thus, the biosorption efficiency of heavy metals by microbial biomass is mainly connected with the structure of the microorganism cell wall and consequently with cell surface properties in which structure determines the interaction nature between micro-organism and metal cation [36–38].

Yeast cell walls are negatively charged, and the ability of yeast cells to bind heavy-metal cations is likely due to electrostatic interactions [39]. Indeed, heavy metals can be biosorbed by microbes at binding sites present in cellular structure without the involvement of energy. Among the various reactive compounds associated with cell walls, the extracellular polymeric substances such as exopolysaccharide (EPS) are well known to have a considerable effect on acid-base properties and a great ability to complex heavy metals [40]. The structure and the distribution of homopolysaccharides (mannans and glucans), single saccharides and acid components, which are good binding agents, also dictate the cell wall's biosorption capacity [41].

3.1.2. Mechanisms of biosorption

It is likely that various mechanisms can be involved in biosorption and can operate simultaneously to various degrees. Interaction of metal with yeast cell wall involves a complex mechanism that includes several processes such as ion exchange, complexation, adsorption and precipitation [2, 42]. Many evidences have proved that ion exchange mechanism exists in biosorption system [38, 43, 44]. However, it was suggested by many researchers that ion exchange is neither the sole nor the main mechanism for metal biosorption [45]. Ion exchange is the replacement of an ion in a solid phase in contact with a solution by another ion. More specifically, it is the replacement of an absorbed, readily exchangeable ion by another [46]. Rapid release of 70% of cellular K+, followed by a slower release of approximately 60% of cellular Mg2+, but little loss of Ca2+, was observed in Cu2+ removal by S. cerevisiae [47], indicating the existence of an ion exchange mechanism. Chen and Wang [48] also reported that S. cerevisiae acts as a biosorbent for the removal of Zn (II) and Cd (II) through the ion exchange mechanism. According to Vasudevan et al. [49], the release of Ca2+, Mg2+ or H+ was also observed in the biosorption process of heavy-metal ions (Sr2+, Mn2+, Zn2+, Cd2+, Cu2+, Ti+) by living, non-metabolizing cells of S. cerevisiae, which also confirmed the existence of ion exchange. Although it is a simple concept, in reality, ion exchange can be a mechanistically highly complex process depending on the system [50, 51].

Metal precipitation is also involved in biosorption. The term precipitation in most cases refers to the formation of insoluble inorganic metal precipitates [52]. This may be more easily understood when metals are bound to extracellular polymeric substances excreted by eukaryotic microorganisms such as yeast and fungi. The precipitates may be formed and remain in contact with or inside the microbial cells or may be independent from the solid phase of the microbial cell. Some researches proved that purified biomolecule products from isolated cells such as glucan, mannan, and chitin accumulate greater quantities of cations than the intact cells and can form metal precipitates [53].

Several yeast species such as S. cerevisiae, Pichia anomala, Candida tropicalis, C. albicans, and Cunninghamella elegans emerged as a promising sorbents against heavy metals [54]. Table 1 summarizes some of the important results of metal biosorption using these yeast biomasses.

Table 1. Some data on the biosorptive capacities of yeasts for different metal ions reported in literatures.

Biomass type	Metal studied	Initial concentration of metal studied (mg/L)	Biosorption capacity (mg/g) or % removal	Biomass concentration (g/L)	References
S. cerevisiae	Hg	25–200	76.20	2	[55]
S. cerevisiae	Pb		67–82%		[56]
	Cd		73–79%		[56]
S. cerevisiae	Cd	169	5.96		[57]
Schizosac charomyces pombe	Ni	400	33.8		[58]
S. cerevisiae	Cr(III)	200	35.00		[59]
S. cerevisiae	Cr(VI)	150	120	0.5	[60]
S. cerevisiae	Pb		270.30	10	[61]
S. cerevisiae	Zn		23.40	1	[62]
S. cerevisiae	Hg		64.20	1	[62]
S. cerevisiae	Cd		15.60		[63]
S. cerevisiae	Pb		17.50		[63]
S. cerevisiae	Pd		40.60		[64]
S. cerevisiae	Cr(VI)		32.60	10	[61]
S. cerevisiae	Ni		46.30	10	[61]
S. cerevisiae sp.	Zn		100%	0.4	[65]
S. cerevisiae	Cd		95%	0.4	[65]
C. pelliculosa	Cu	100	95.04%	13.39	[66]
C. utilis	Cd	50	81.46	3	[67]
C. tropicalis	Cr	100	29.10	2	[24]
W. anomalus	Cr	100	28.14	2	[24]
C. fabianii	Cr	100	18.90	2	[24]

3.1.3. Factors affecting yeast biosorption capacity

It is noted that the same yeasts species have different biosorption capacities for the same metal ion. It was shown that different yeasts species present different cell surface properties and cell wall compositions, which brings about a differentiation in biosorption ability, affinity and interaction specificity. Moreover, biosorption depends on many factors that are related to the biomass or to environmental conditions. Indeed, Nguyen et al. [68] studied the polysaccharide composition of the cell walls of several yeast species, such as Debaryomyces hansenii, Zygosaccharomyces bailii and S. cerevisiae, and results indicated that the cell wall composition varied over the species and strains. Growth, nutrition and age of the biomasscan also influence the variability of cell size, cell wall composition and extracellular product formation [69]. It was found that the cell wall composition and polysaccharide content could vary by more than 50% with the nature of the carbon source, nitrogen limitation, temperature and aeration and pH and with the mode of cell cultivation [70]. Temperature is considered as one of the important factors in the biosorption process. It was reported that adsorption reactions are mostly exothermic and the extent of adsorption augment with decreasing temperature [32]. Thus, a better biosorption capacity for Ni and Pb by S. cerevisiae was observed at a low temperature (25°C) and found to diminish as the temperature was increased to 40°C [71].

Metal biosorption was also frequently shown to be strongly pH dependent. Sulaymon et al. [35] indicated that pH of solution affects the solution chemistry of the metals, the activity of the functional groups in the biomass and the competition of metallic ions. Some studies showed that yeast cells of S. cerevisiae are able to remove heavy-metals between pH 5.0 and 9.0, being pH close to 5–6 as the optimal for Cu2+, Cd2+, Pb2+, Ni2+, Zn2+ and Cr3+ biosorption by yeast cells [47, 72]. However, other studies demonstrated that biosorption of metals like Cu, Cd, Ni, Co and Zn is often reduced at low pH values [73, 74]. Generally, the heavy metal uptake for most of the biomass types declines significantly when pH of the metal solutions is decreased from pH 6.0 to 2.5 [72]. The cell surface hydrophobicity may also affect biosorption capacity, facilitating hydrophobic bonds. Edyta [75] has reported that a lower relative hydrophobicity and a higher negative surface charge can be related to better availability of polar/charged groups such as carboxyls and mannosylphosphates on yeast cell surface.

3.1.4. Industrial application

Biosorption process has not only been used in laboratory scale (batch and column) but also in pilot plant-scale studies. Tigini et al. [76] tested this process using C. elegans biomass in 200 L of pilot plant installation for dye removal. Obtained results showed that biosorption was an effective process for the removal of different pollutants from spent baths and wastewaters [76].

The apparent exploitation potential of biosorption is often cited in the literature, and the commercial applications of biosorption are currently available on the market. The biological materials can be commercialized as powdered biosorbents, e.g. BIO-FIX® (sphagnum, peat moss, algae, yeast, bacteria and aquatic flora immobilized in polysulfone), MetaGeneR and RAHCO Bio-Beads [77]. All the preparations are suitable for the treatment of wastewaters from metallurgical industry or mining operations. The future possible applications may concern using biosorption in the separation and purification of high value molecules, e.g. high-value proteins, steroids, pharmaceuticals or antibodies [77].

3.2. Bioaccumulation

Bioaccumulation is defined as the uptake of toxicant by living cells and their transport into the cell [78]. It is a growth-dependent process mediated only by living biomass [79]. The mechanism of intracellular uptake is more complex than biosorption itself and is not fully understood yet. Generally, the process is regarded as a two-step process [80]. The first step called passive biosorption proceeds rapidly within several minutes by any one or a combination of the following metal-binding mechanisms: coordination, complexation, ion exchange or physical adsorption (e.g. electrostatic). The metal ions are adsorbed to the surface of cells by interactions between metal-functional groups displayed on the surface of cells. The second step which slowly takes place is that metal ions penetrate the cell membrane and enter into the cells. Raspor et al. [81] pointed out that after initial rapid biosorption on cell walls, active transport mechanisms into the cells take place and metal ions penetrate the cell membrane and enter into the cells. Metal ions transported across the cell membrane, are transformed to other species or precipitated within the cell by active cells, including transportation.

Metal accumulation strategies for essential and non-essential metal ions may be different. Literature in model yeasts indicates that upon non-essential metal exposure, cytoplasmic detoxification is the major strategy [30]. This cytoplasmic detoxification can be achieved by metal transport to the outside of the cell or to less sensitive cellular compartments, making the metal ions unavailable to promote cytotoxic effects [82]. Thus, microorganisms developed different mechanisms including cell membrane metal efflux [83], intracellular chelation by metallothionein proteins and glutathione-derived-peptides called phytochelatins [84, 85] as well as metal compartmentalization in vacuoles [3], but the exact mechanism of intracellular accumulation is not elucidated.

In general, metals enter yeast cells by dedicated transporters, which are often target of negative regulation when the specific metal is present in excess intracellularly. In the past decades, a large number of references on metal ion transport in the baker's yeast S. cerevisiae were published [86, 87]. Baker's yeast was found to possess two or more substrate-specific transport systems to accumulate any single metal ion, and a large number of yeast genes that function in metal ion transport or its regulation were confirmed [87]. After entering into the cell, the metal ions react with thiol groups present in Cys residues such thiolated peptides include the glutathione (GSH), phytochelatins (PC), and the metallothioneins (MT) [82]. The resulting metal-thiolated peptide complexes may be used as a substrate for metal(loid) extrusion to the outside of the cell, or for accumulation in cellular compartments such as the vacuole. However, thiolated peptides can be produced to chelate metals, reducing their reactivity and availability to the cells [30].

The most described yeast MT is the Cup1 of S. cerevisiae, which is mainly associated with Cu detoxification [88]. Several authors attributed Cd, in addition to Cu, detoxification to chelation by this MT [88, 89]. In S. cerevisiae, GSH synthesis was also described. In comparison with a wild-type strain of S. cerevisiae, the mutant strain displayed a high sensitivity to arsenite, selenite and cadmium which confirms the role of GSH on detoxification of these metals.

The role of the vacuole in the detoxification of metal ions was also investigated and a large number of researches signaled that fungal vacuole plays an important role in molecular degradation, storage of metabolites and regulation of cytosolic concentrations of metal ions and detoxifies potentially toxic metal ions. The results showed that vacuole-deficient strain displayed much higher sensitivity and decreased large uptake of As, Zn, Mn, Co and Ni [90, 91]. Avery and Tobin [92] also confirmed that Sr2+ accumulated mainly stays in the vacuole of the living yeast cell of S. cerevisiae.

The active mode of metal accumulation by living cells is dependent on structural properties, physiological and genetic adaptation, environmental modification of metal specification, availability and toxicity [93]. The process also depends of several factors (which are almost identical to the factors influencing the cultivation of an organism): the composition of the growth medium, pH, temperature, the presence of other pollutants or other inhibitors, surfactants, etc. [94]. Metabolic activities such as respiration, nutrient uptake, and metabolite release will alter the microenvironment around the cells which, in turn, may affect mechanisms involved in bioaccumulation (adsorption, ion exchange, complexation and precipitation) [69].

In yeasts, heavy metals can be accumulated by bioaccumulation process more than biosorption. But the biosorption process seems to be more feasible for large scale application compared to the bioaccumulation process, because microbes will require addition of nutrients for their active uptake of heavy metals [27]. Thus, the biosorptive capacity of yeast was studied extensively in comparison to bioaccumulation.

3.3. Bioreduction

The detoxification of metal ions can also be realized by oxidation or reduction. When reduction of a metal to a lower redox state occurs, mobility and toxicity can be reduced, thus offering potential bioremediation applications [2].

Information available for metal detoxification, such as reduction mechanisms in yeasts, only considers neutrophilic yeasts and there is considerably less information available when compared to prokaryotes. Indeed, for eukaryotic microbial cells and primarily yeasts, the data on the metal-reducing systems are more ambiguous. It is generally unknown what system enzymatic or non-enzymatic and intracellular or extra-cellular plays a leading role in the chromate detoxification process [9].

Tamás and Wysocki [82] proved that one mechanism of detoxification of As(V) was the reduction of As(V) to As(III), a process catalysed by arsenate reductase enzymes. The removal of toxic hexavalent chromium from aqueous solution by biosorption by different biomass types was as well extensively reported [9]. Cr(VI) can be reduced as a powerful oxidative agent to Cr(III) by cellular-reducing systems that can include enzymatic and non-enzymatic pathways.

The intracellular reduction of Cr(VI) to Cr(III) is known to be the main detoxification mechanism of chromium. In aerobic condition, microbial reduction of Cr6+ is catalysed by soluble enzymes (chromate reductase) [95]. Many yeasts like Cyberlindnera fabianii, P. anomala, Rhodotorula pilimanae D-76, and Pichia guilliermondii ATCC 201911 were known for their enzymatic reduction ability of Cr6+ to Cr3+ [27, 28, 96, 97].

Cr(VI) removal may also be associated with its simultaneous reduction to Cr(III). It was reported that Cr(VI) can be reduced to Cr(III) through a redox reaction unrelated to any enzyme activity [98, 99]. Thus inactivated biomass, e.g. S. cerevisiae, C. tropicalis, P. anomala and Penicillium chrysogenum, removes Cr(VI) from aqueous solutions by reduction to Cr(III) when contacted with the biomass [96, 100]. Glutathione and cysteine can be considered as the most powerful non-enzymatic chromate reductants for microbial cells and ascorbate for higher organisms [2, 101]. Therefore, the removal of Cr(VI) from aqueous solution by dead cells may involve two mechanisms: (i) direct reduction, in the aqueous phase, Cr(VI) is directly reduced to Cr(III) by contact with the electron-donor groups of the biomass which has lower reduction potential values than that of Cr(VI) (+1.3 V), and (ii) indirect reduction, which includes three steps—(a) adsorption of Cr(VI) anionic species to the biomass surface containing the positively charged groups, (b) reduction of Cr (VI) to Cr(III) by adjacent electron-donor groups and (c) release of Cr (III) ions in the aqueous phase due to the electronic repulsion between the Cr (III) ions and positively charged groups, or adsorption of Cr(III) with adjacent groups.

3.4. Bioleaching

In the context of bioremediation, immobilization processes such as biosorption, accumulation and precipitation may enable metals to be transformed in situ into insoluble forms. They are particularly applicable to remove metals from mobile phases such as ground waters and leachates. In contrast, the process of metal solubilization provides a way to remove metals from soils and sediments by leaching. Generally bioleaching is a process described as being "dissolution of metals from their mineral source by certainly and naturally occurring microorganisms". However, there are some slight differences in definition: Usually, "bioleaching" is described as the conversion of solid metal values into their water soluble forms using microorganisms [102].

In the soil environment, metals can be held on inorganic soil constituents through various sorption or ion exchange reactions, complexed with soil organic materials or precipitated as pure or mixed solids [103]. However, in most acidic soils, metals may be speciated into more mobile forms [104]. It was reported that in such locations, fungi are the most predominating and often comprise the largest biomass suggesting its intervention in metal solubilization [33, 104]. Gadd [33] confirmed that microbes interact with metal and mineral in natural and synthetic environments, altering their physical and chemical state. Therefore, the biochemical activity of fungi and other microorganisms can affect metal speciation and mobility in the soil, modifying their biogeochemical cycles. The most important mechanisms of metal and minerals solubilization by fungi are acidolysis and complexolysis. It was revealed that some excreted metabolites with metal-complexing properties, e.g. phenolic compounds, and organic acids may be involved in metals solubilization [105]. Gadd [104] also indicated that low molecular weight organic acids, such as citric and oxalic acid, are the most important chemical fungal and yeast products and they were used for heavy metal solubilization. In fact, previous studies showed that many metal citrates are highly mobile. They also proved that oxalic acid can act as a leaching agent for those metals, which forms soluble oxalate complexes, including Al and Fe [106]. It was shown that organic acids provide both source of protons for solubilization and metal-chelating anion to complex the metal cation [107]. They have the double function: (i) to acidify the substrate, thus enhancing ions solubility, and (ii) to form complexes with solubilized ions, which leads into mobilizing them [104, 105].

Yeast doesn't show a broad area of examples in the field of bioleaching and the available information about the metal extraction from solid substrates using yeasts is limited. Some yeasts associated with bioleaching are Rhodotorula rubra and Rhodotorula mucilaginosa. Studies of yeast R. mucilaginosa sp. lm9 isolated from Kupferschiefer black shale showed that organic acids (malic and oxalic) produced by this strain can effectively mobilize copper from sedimentary rock [108]. Marcinčáková et al. [109] also demonstrated that 0.17% of lithium can be recovered from lepidolite by microbial leaching using the heterotrophic microorganism of R. rubra.

The bioleaching process is carried out at the mine site. It is indicated that the bioleaching process can be used to process low-grade ores and arsenic-containing ores that could not be processed effectively by high temperature smelting. Two types of bioleaching processes exist: bioleaching in stirred tank reactors (STR), which is dedicated to high-grade ores due to the relatively high costs of investments, and heap leaching, which is dedicated to low-grade ores. Bench scale columns were also used with ore in Australia in the period from 1964 to 1968 [110].

There are many and various applications of bioleaching. It can be applied to a wide variety of base-metal sulfides, mainly in large operations located in many countries and several interesting projects were conducted in order to develop this technology. The earliest commercial applications of the process involved in situ leaching of uranium in Canada, heap bioleaching of copper in Toromocho and dump leaching of copper in the United States (bioleach, Chili). Nowadays, the production of copper from low-grade ores is the most important industrial application. Early application of bioleaching to copper mining was centered in its recovery from heap or dump/stockpile [111, 112]. In these operations, copper is extracted from ores containing minerals: secondary copper sulfides such as covellite (CuS), chalcocite (Cu2S) and bornite (Cu5FeS4) and the primary copper sulfide, chalcopyrite (CuFeS2).

As well as detoxification of pollutant metals and copper production, the recovery of precious metals such as gold is also a potential area for exploitation. Industrial-scale bioleaching of refractory gold concentrates was practiced in South Africa, Brazil, Australia, Ghana, Peru, China and Kazakhstan. In this case, the process is used to leach sulfide minerals such as pyrite (FeS2) and arsenopyrite (FeAsS), which encapsulate microscopic and sub-microscopic gold particles. By dissolving these sulfide minerals, the gold particles are exposed and can be recovered by further treatment [113].

Although yeasts have a high potential for bioleaching, there are no studies on the use of these microorganisms in bioleaching projects. So, it would be interesting to develop this technology using yeasts by testing their capacity on a large scale.

4. BIOTECHNOLOGICAL METHODS: BIOAUGMENTATION

Bioaugmentation is a potential bioremediation technology that involves introducing specific microbial strains or mixed cultures to biological waste treatments or contaminated areas. This process efficiently enhances the removal of certain contaminants [114]. The justification for this method is that native microbial communities may lack the ability to break down the many substances found in complicated combinations [115], especially when the number of pollutant-degrading microbes is small. Hence, the expedited pace of decontamination is the principal benefit derived from the implementation of microorganisms. This technology utilizes microbial consortia that are specifically engineered to exploit the physico-chemical features of the bioprocess [116]. This repair method has been proven to be more effective than utilizing unspecified inocula [114]. Nevertheless, a challenge in bioaugmentation processes arises from the abundance of uncharacterized species that engage in competition with the transplanted populations. According to reports, certain foreign microorganisms (namely those in inocula) were effectively utilized in laboratory settings. However, their effectiveness is contingent upon their capacity to outcompete native microorganisms, predators, and various non-living variables. Therefore, the effectiveness of bioaugmentation treatments relies on the utilization of inocula comprising of microbial strains or microbial consortia that are highly suited to the specific region that needs to be decontaminated [117]. Various microorganisms were employed for bioaugmentation. The utilization of yeast-based bioaugmentation has been demonstrated to be a beneficial method for remediation of polluted areas, which are contaminated with heavy metals and/or organic contaminants [118–120]. A comprehensive evaluation was conducted on numerous microorganisms that were isolated and qualified in laboratory settings. However, they have not yet been proven to be valid in their natural environment. C. fabianii bioaugmentation effectively reduced over 60% of the chromium pollution in the soil, which initially had a concentration of 40 mg.Kg-1. It has decreased its harmful effects on Phaseolus vulgaris L. and enhanced its development under chromium stress [121]. In a separate study, it was found that both living and dead C. tropicalis biomass had a remarkable capacity to effectively decrease the bioavailability of 40 mg.Kg-1 of Cr(VI) in soils, with reductions of up to 58.7% and 72.25% respectively [120]. The study utilized clover plants as bioindicators to discover a notable enhancement in seed germination and seedling growth in soils that were treated with C. tropicalis cells. In subsequent research, the introduction of the particular yeast strain C. tropicalis SK21 through bioaugmentation proved to be highly effective in the process of bioremediating soil contaminated with petroleum. The strain demonstrated impressive results, degrading 96% and 42% of the total petroleum hydrocarbons (TPH) when exposed to initial diesel oil concentrations of 0.5% and 5% (v/v) respectively. Candida digboiensis shown significant proficiency in breaking down acidic petroleum hydrocarbons in laboratory and field settings when used to bioaugment acidic oily sludge-contaminated soil [118]. From an application standpoint, utilizing microbial consortia for bioaugmentation instead of pure cultures is undoubtedly more beneficial [122–125]. It offers various metabolic pathways and the necessary resilience needed for practical use in the field. Bacteria were widely utilized in many places worldwide for the implementation of bioaugmentation. Several studies have proposed the utilization of yeast strains as promising agents for the bioaugmentation process [117]. However, a forthcoming disparity between experiments conducted in the laboratory and those carried out in real-world settings was identified. Therefore, the scientific community should make additional efforts to implement this on a larger scale. In order to achieve a successful industrial application of bioaugmentation, it is necessary to implement tailored techniques that are specifically designed to address the unique environmental characteristics of the contaminated sites [124]. Therefore, numerous research have been conducted to exploit the benefits of this treatment technique for industrial use, as it offers a cost-effective and environmentally benign method of remediation [126].

AT A GLANCE

The issue of heavy metal pollution has emerged as a highly significant global environmental concern. Bioremediation, which utilizes biological materials such as living or deceased microbes, is an encouraging, secure, and cost-effective approach for treating polluted water and soil. Yeast is a highly prevalent kind of biomass that is commonly used for bioremediation of heavy metals. Yeast cells are a cost-effective and easily accessible biomass source that can effectively remove a wide spectrum of heavy metals to different extents. In addition, yeasts demonstrate the capacity to adjust to harsh environments, including severe temperatures, pH levels, and elevated concentrations of both organic and inorganic pollutants. This paper provides an overview of the role that yeasts play in the immobilization and mobilization of harmful metals in the environment. It aims to explain the many methods of interaction between metals and yeast strains, as well as the factors that influence these processes. The discussion will also include the use of biotechnological applications, namely bioaugmentation using the degrading abilities of yeasts, in the bioremediation of heavy metals.

REFERENCES

1. Anushree M. Metal bioremediation through growing cells. Environmental International. 2004;30:261-278

2. Siddique S, Rovina K, Al Azad S, Naher L, Suryani S, Chaikaew P. Heavy metal contaminants removal from wastewater using the potential filamentous fungi biomass: A review. Journal of Microbial and Biochemical Technology. 2015;7:384-393. DOI: 10.4172/1948-5948.1000243

3. Volesky B. Advances in biosorption of metals: Selection of biomass types. FEMS Microbiology Review. 1994;14:291-302.

4. Domenech X. Quimica Ambiental, El Impacto Ambiental de los Residous. Madrid: Miraguano; 1998. p. 127

5. Ahluwalia SS, Goyal D. Microbial and plant derived biomass for removal of heavy metals from wastewater. Bioresource Technology. 2007;98:2243-2257

6. Kolvenbach BA, Helbling DE, Kohler HP, Corvini PF. Emerging chemicals and the evolution of biodegradation capacities and pathways in bacteria. Current Opinion in Biotechnology. 2014;27:8-14

7. McGenity TJ. Hydrocarbon biodegradation in intertidal wetland sediments. Current Opinion in Biotechnology. 2014;27:46-54

8. Trama B, Fernandes JDS, Labuto G, Franco de Oliveira JC, Viana-Niero C, Pascon RC, Vallim MA. The evaluation of bioremediation potential of a yeast collection isolated from composting. Advances in Microbiology. 2014;4:796-807

9. Tahri Joutey N, Sayel H, Bahafid W, El Ghachtouli N. Mechanisms of hexavalent chromium resistance and removal by microorganisms. Reviews of Environmental Contamination and Toxicology. 2015;233:45-69. DOI: 10.1007/978-3-319-10479-9_2

10. Avery SV, Tobin JM. Mechanism of adsorption of hard and soft metal ions to Saccharomyces cerevisiae and influence of hard and soft anions. Applied and Environmental Microbiology. 1993;59:2851-2856

11. Brady D, Duncan JR. Bioaccumulation of metal cations by Saccharomyces cerevisiae. Applied Microbiology and Biotechnology. 1994;41:149-154

12. Brady D, Stoll A, Duncan JR. Biosorption of heavy metal cations by non-viable yeast biomass. Environmental Biotechnology. 1994;15:429-438

13. Krauter P, Martinelli R, William K, Martins S. Removal of Cr4+ from ground water by S. cervisiae. Biodegradation. 1996;7:277-286

14. Veglio F, Beolchini F. Removal of metals by biosorption: A review. Hydrometallurgy. 1997;44:301-316

15. Abbas BA, Badr SQ. Bioremediation of some types of heavy metals by Candida spp. International Journal of Engineering and Technical Research. 2015;3:2454-4698

16. Ram CB, Nelly G, Nevena L. Bioremediation of chromium ions with filamentous yeast Trichosporon cutaneum R57. Journal of Biology and Earth Sciences. 2012;2:70-75

17. Kordialik-Bogacka E. Studies on surface properties of yeast cells during heavy metal biosorption. Central European Journal of Chemistry. 2011;9:348-351

18. Crist RH, Oberholser K, Shank K, Nguyen M. Nature of bonding between metallic ions and algal cell walls. Environmental Science and Technology. 1981;15:1212-1217

19. Wongkarnka M. The application of aerobic yeast for treatment of high strength food processing wastewater containing furfural. [thesis]. Iowa State University Ames; 2005

20. Ertugrul S, San NO, Dönmez G. Treatment of dye (Remazol blue) and heavy metals using yeast cells with the purpose of managing polluted textile wastewaters. Ecological Engineering. 2009;35:128-134

21. Sanghi R, Sankararamakrishnan N, Dave BC. Fungal bioremediation of chromates: Conformational changes of biomass during sequestration, binding, and reduction of hexavalent chromium ions. Journal of Hazardous Materials. 2009;169:1074-1080

22. Huang JP, Huang CP, Morehart AL. Removal of heavy metals by fungal (Aspergillus oryzae) adsorption. In: Vernet JP, editor. Heavy Metals in the Environment. London: Elsevier; 1991

23. Mueller JG, Cerniglia CE, Pritchard PH. Bioremediation of environments contaminated by polycyclic aromatic hydrocarbons. In: Bioremediation: Principles and Applications. Cambridge University Press: UK; 1996. p 125-194.

24. Bahafid W, Tahri Joutey N, Sayel H, Iraqui-Houssaini M, El Ghachtouli N. Chromium adsorption by three yeast strains isolated from sediments in Morocco. Geomicrobiology Journal. 2013;30:422-429

25. Sayel H, Bahafid W, Tahri Joutey N, Derraz K, Fikri Benbrahim K, Ibnsouda Koraichi S, El Ghachtouli N. Cr(VI) reduction by enterococcus gallinarum isolated from tannery waste-contaminated soil. Annals of Microbiology. 2012;62:1269-1277

26. Tahri Joutey N, Bahafid W, Sayel H, El Abed S, El Ghachtouli N. Remediation of hexavalent chromium by consortia of indigenous bacteria from tannery waste contaminated biotopes in Fez, Morocco. International Journal of Environmental Studies. 2011;686:901-991

27. Ksheminska H, Fedorovych D, Honchar T, Ivash M, Gonchar M. Yeast tolerance to chromium depends on extracellular chromate reduction and Cr(III) chelation. Food Technology and Biotechnology. 2008;46:419-426

28. Bahafid W, Tahri Joutey N, Sayel H, EL Ghachtouli N. Mechanism of hexavalent chromium detoxification using Cyberlindnera fabianii yeast isolated from contaminated site in Fez Morocco. Journal of Materials and Environmental Science. 2013;4:840-847

29. Brierley CL. Bioremediation of metal-contaminated surface and groundwater. Geomicrobiology Journal. 1990;8:201-223

30. Assis Fidalgo CI. Heavy metal resistance in extremophilic yeasts: A molecular and physiological approach. [thesis]. Faculdade de Ciências: Universidade de lisboa; 2011

31. Nancharaiah YV, VenkataMohan S, Lens PNL. Biological and bioelectrochemical recovery of critical and scarce metals. Trends in Biotechnology. 2016;34:137-155

32. Valls M, de Lorenzo V. Exploiting the genetic and biochemical capacities of bacteria for the remediation of heavy metal pollution. FEMS Microbiology Review 2002;26:327-338.

33. Gadd GM. Interactions of fungi with toxic metals. The New Phytologist. 1993;124:25-60

34. Shumate ES, Strandberg WG. Accumulation of metals by microbial cells. Comprehensive Biotechnology. 1985;13:235-247

35. Abbas SH, Ismail IM, Mostafa TM, Sulaymon AH. Biosorption of heavy metals: A review. Journal of Chemical Science and Technology. 2014;3:74-102

36. Wang JL, Chen C. Biosorption of heavy metals by Saccharomyces cerevisiae: A review. Biotechnology Advances. 2006;24:427-451

37. Wang JL, Chen C. Biosorbents for heavy metals removal and their future. Biotechnology Advances. 2009;27:195-226. DOI: 10.1016/j.biotechadv.2008.11.002

38. Ahluwalia SS, Goyal D. Microbial and plant derived biomass for removal of heavy metals from wastewater. Bioresource Technolology. 2007;98:2243-2257

39. Fukudome K, Sato M, Takata Y, Kuroda H, Watari J, Takashio M. Evaluation of yeast physiological state by Alcian blue retention. Journal of the American Society of Brewing Chemists. 2002;60:149-152

40. Comte S, Guibaud G, Baudu M. Biosorption properties of extracellular polymeric substances (EPS) towards Cd, Cu and Pb for different pH values. Journal of Hazardous Materials. 2008;151:185-193

41. Raspor P, Zupan J. Yeasts in extreme environments. Yeast Handbook, Biodiversity and Ecophysiology of Yeasts. Springer-Verlag: Berlin; 2006. 371-417p

42. Ruchita D, Wasiullah DM, Kuppusamy P, Udai BS, Asha S, Renu S, Bhanu PS, Jai PR, Pawan KS, Harshad L, Diby P. Bioremediation of heavy metals from soil and aquatic environment: An overview of principles and criteria of fundamental processes. Sustainability. 2015;7:2189-2212. DOI: 10.3390/su7022189

43. Schneider IAH, Rubio J, Smith RW. Biosorption of heavy metals onto plant biomass: Exchange adsorption or surface precipitation? International Journal of Mineral Processing. 2001;62:111-120

44. Davis TA, Volesky B, Mucci A. A review of the biochemistry of heavy metal biosorption by brown algae. Water Research. 2003;37:4311-4330

45. Brady JM, Tobin JM. Binding of hard and soft metal ions to Rhizopus arrhizus biomass. Enzyme and Microbial Technology. 1995;17:791-796

46. Sposito G. The Chemistry of Soils. New York: Oxford University Press; 1989

47. Brady D, Duncan JR. Cation loss during accumulation of heavy metal cations by Saccharomyces cerevisiae. Biotechnology Letters. 1994;16:543-548

48. Chen C, Wang JL. Characteristics of Zn2+ biosorption by Saccharomyces cerevisiae. Biomedical and Environmental Sciences. 2007;20:478-482

49. Vasudevan P, Padmavathy V, Dhingra SC. Biosorption of monovalent and divalent ions on baker's yeast. Bioresource Technology. 2002;82:285-289

50. Stumm W, Morgan JJ. Aquatic Chemistry. Chemical Equilibria and Ratesin Natural Waters. NewYork: Wiley; 1996

51. Crist DR, Crist RH, Martin JR, Watson JR. Ion exchange systems in proton-metal reactions with algal cell walls. FEMS Microbiology Review. 1994;14:309-314

52. Macaskie LE, Bonthrone KM, Rauch DA. Phosphatase-mediated heavy metal accumulation by Citrobacter sp. and related enterobacteria. FEMS Microbiology Letters. 1994;121:141-146

53. Remoundaki E, Hatzikioseyian A, Kousi P, Tsezos M. The mechanism of metals precipitation by biologically generated alkalinity in biofilm reactors. Water Research. 2003;37:3843-3854

54. Tigini V, Prigione V, Giansanti P, Mangiavillano A, Pannocchia A, Varese GC. Fungal biosorption, an innovative treatment for the decolourisation and detoxification of textile effluents. Water. 2010;2:550-565

55. Anaemene IA. The use of Candida sp. in the biosorption of heavy metals from industrial effluent. European Journal of Experimental Biology. 2012;2:484-488

56. Damodaran D, Suresh G, Mohan BR. Bioremediation of soil by removing heavy metals using Saccharomyces cerevisiae. In: . 2nd International Conference on Environmental Science and Technology. IPCBEE; 2011

57. Breierova E, Vajczikova I, Sasinkova V, Stratilova E, Fiserac M, Gregorc T, Sajbidor J. Biosorption of cadmium ions by different yeast species. Zeitschrift für Naturforschung. 2002;57:634-639

58. Selcen DS, Alpagu SA, Topal-Sarikaya A, Alp S. Biosorption of Ni (II) by Schizosaccharomyces pombe: Kinetic and thermodynamic studies. Bioprocess and Biosystems Engineering. 2011;34:997-1005

59. Abd-Elsalam IS. Factorial design for some parameters affecting on chromium (iii) uptake by Saccharomyces cerevisiae. International Journal of Applied Biology and Pharmaceutical Technology. 2011;2:33-40

60. Bingol A, Ucun H, Bayhan YK, Karagunduz A, Cakici A, Keskinler B. Removal of chromate anions from aqueous stream by a cationic surfactant-modified yeast. Bioresource Technology. 2004;94:245-249

61. Ozer A, Ozer D. Comparative study of the biosorption of Pb(II), Ni(II) and Cr(VI) ions onto S. cerevisiae: Determination of biosorption heats. Journal of Hazardouz Material. 2003;27(1-3):219-229

62. Al-Saraj MR, Monzir A, El-Nahhal IM, Baraka R. Bioaccumulation of some hazardous metals by sol–gel entrapped microorganisms. Journal of Non-Crystalline Solids. 1999;248(2-3):137-140. DOI: 10.1016/S0022-3093(99)00306-3

63. Göksungur Y, Uren S, Güvenç U. Biosorption of cadmium and lead ions by ethanol treated waste baker's yeast biomass. Bioresource Technology. 2005;96(1):103-109

64. Xie DD, Liu YY, CL W, JK F, Xue R. Studies on biosorption of Pd2+ by the immobilized Saccharomyces cerevisiae waste biomass. Microbiology. 2003;30:29-34

65. Mapolelo M, Torto N. Trace enrichment of metal ions in aquatic environments by Saccharomyces cerevisiae. Talanta. 2004;64:39-47

66. Andréa HA, Luciana VP, Vanete ST, Masaki R, Ernani FF, Alexsandro G, Carlo SR. Optimization of biomass production with copper bioaccumulation by yeasts in submerged fermentation. Brazilian Archives of Biology and Technology. 2011;54:1027-1034

67. Zu Y, Zhao XH, Hu M, Ren Y, Xiao P, Zhu L, Cao Y, Zhang Y. Biosorption effects of copper ions on Candida utilis under negative pressure cavitation. Journal of Environmental Sciences. 2006;18:1254-1259

68. Nguyen T, Fleet G, Rogers P. Composition of the cell walls of several yeast species. Applied Microbiology and Biotechnology. 1998;50:206-212

69. Gadd GM. Biosorption: Critical review of scientific rationale, environmental importance and significance for pollution treatment. Journal of Chemical Technology and Biotechnology. 2009;84:13-28

70. Aguilar-Uscanga B, François JM. A study of the yeast cell wall composition and structure in response to growth conditions and mode of cultivation. Letters in Applied Microbiology. 2003;37:268-274

71. White C, Sayer JA, Gadd GM. Microbial solubilization and immobilization of toxic metals: Key biogeochemical processes for treatment of contamination. FEMS Microbiology Reviews. 1997;20:503-516

72. Machado MD, Soares EV, Soares HMV. Removal of heavy metals using a brewer's yeast strain of Saccharomyces cerevisiae: Application to the treatment of real electroplating effluents containing multielements. Chemical Technology and Biotechnology. 2010;85:1353-1360

73. Greene B, Darnall DW. Microbial oxygenic photoautotrophs (cyanobacteria and algae) for metal-ion binding. In: Ehrlich HL, Brierley CL, editors. Microbial Mineral Recovery. New York: McGraw-Hill; 1990. p. 227-302

74. Deng X, Wang P. Isolation of marine bacteria highly resistant to mercury and their bioaccumulation process. Bioresource Technology. 2012;121:342-347

75. Edyta K-B. Surface properties of yeast cells during heavy metal biosorption. Central European Journal of Chemistry. 2011;9(2):348-351. DOI: 10.2478/s11532-011-0008-8

76. Tigini V, Prigione V, Anastasi A, Spina F, Varese GC. Scale up of biosorption process for the treatment of the textile wastewaters with fungal biomasses. In: 7th International Conference on Polymer and Textile Biotechnology; 2011; Milano. Italy

77. Volesky B. Biosorption and me. Water Research. 2007;41:4017-4029

78. Malik A. Metal bioremediation through growing cells. Environment International. 2004;30:261-278

79. Karna RR, Sajani LS, Mohan PM. Bioaccumulation and biosorption of CO2+ by Neurospora crassa. Biotechnology Letters. 1996;18:1205-1208

80. Chojnacka K. Biosorption and Bioaccumulation in Practice. New York: Nova Science Publishers; 2009 137 p

81. Raspor P, Batic M, Jamnik P, Josic D, Milacic R, Pas M, Recek M, Rezic-Dereani V, Skrt M. The influence of chromium compounds on yeast physiology (a review). Acta Microbiologica et Immunologica Hungarica. 2000;47:143-173

82. Tamás MJ, Wysocki R. How Saccharomyces cerevisiae copes with toxic metals and metalloids. FEMS Microbiology Reviews. 2010;34:925-951

83. Kamizono A, Nishizawa M, Teranishi Y, Murata K, Kimura A. Identification of a gene conferring resistance to zinc and cadmium ions in the yeast Saccharomyces cerevisiae. Molecular & General Genetics. 1989;219:161-167

84. Kneer R, Kutchan TM, Hochberger A, Zenk MH. Saccharomyces cerevisiae and Neurospora crassa contain heavy metal sequestering phytochelatin. Archives of Microbiology. 1992;157:305-310

85. Presta A, Stillman MJ. Incorporation of copper into the yeast Saccharomyces cerevisiae. Identification of cu(I)-metallothionein in intact yeast cells. Journal of Inorganic Biochemistry. 1997;66:231-240

86. Portnoy ME, Schmidt PJ, Rogers RS, Culotta VC. Metal transporters that contribute copper to metallochaperones in Saccharomyces cerevisiae. Molecular Genetics and Genomics. 2001;265(5):873-882

87. Eide DJ. The molecular biology of metal ion transport in Saccharomyces cerevisiae. Annual Review of Nutrition. 1998;18:441-469

88. Tamás MJ, Wysocki R. Mechanisms involved in metalloid transport and tolerance acquisition. Current Genetics. 2001;40:212

89. Perego P, Howell SB. Molecular mechanisms controlling sensitivity to toxic metal ions in yeast. 1997;147:312-318.

90. Ramsay LM, Gadd GM. Mutants of Saccharomyces cerevisiae defective invacuolar function confirm a role for the vacuole intoxic metalion detoxification. FEMS Microbiology Letters. 1997;152(2):293-298

91. Thorsen M, Jacobson T, Vooijs R, Navarrete C, Bliek T, Schat H, Tamás MJ. Glutathione serves an extracellular defence function to decrease arsenite accumulation and toxicity in yeast. Molecular Microbiology. 2012;84(6):1177-1188

92. Avery SV, Tobin JM. Mechanisms of strontium uptake by laboratory and brewing strains of Saccharomyces cerevisiae. Applied and Environmental Microbiology. 1992;58:3883-3889

93. Blackwell KJ, Singleton I, Tobin JM. Metal cation uptake by yeast: A review. Applied Microbiology and Biotechnology. 1995;43:579-584

94. Kujan P, Votruba J, Kamenik V. Substrate-dependent bioaccumulation of cadmium by growing yeast Candida utilis. Folia Microbiologica. 1995;40(3):288-292

95. Cheung KH, Gu J. Mechanism of hexavalent chromium detoxification by microorganisms and bioremediation application potential: A review. International Biodeterioration and Biodegradation. 2007;59:8-15

96. Bahafid W, Sayel H, Tahri Joutey N, EL Ghachtouli N. Removal mechanism of hexavalent chromium by a novel strain of Pichia anomala isolated from industrial effluents of Fez (Morocco). Journal of Environmental Science & Engineering. 2011;5:980-991

97. Muneer B, Rehman A, Shakoori FR, Shakoori AR. Evaluation of consortia of microorganisms for efficient removal of hexavalent chromium from industrial wastewater. Bulletin of Environmental Contamination and Toxicology. 2009;82(5):597-600

98. Ackerley DF, Gonzalez CF, Park CH, Blake R II, Keyhan M, Matin A. Chromate-reducing properties of soluble flavoproteins from Pseudomonas putida and Escherichia coli. Applied and Environmental Microbiology. 2004;70:873-882

99. Cardenas-Gonzalez JF, Acosta-Rodriguez I. Hexavalent chromium removal by a Paecilomyces sp. fungal strain isolated from environment. Bioinorganic Chemistry and Applications. 2010;1-6

100. Park D, Yun YS, Park JM. Use of dead fungal biomass for the detoxification of hexavalent chromium: Screening and kinetics. Process Biochemistry. 2005;40:2559-2565

101. Elangovan R, Abhipsa S, Rohit B, Ligy P, Chandraraj K. Reduction of Cr(VI) by a Bacillus sp. Biotechnology Letters. 2006;28:247-252

102. Hamidian H. Microbial leaching of uranium ore. In: Tsvetkov P, editor. Nuclear Power – Deployment, Operation and Sustainablity. InTech; 2011. p. 291-304

103. Knox AS, Seaman JC, Mench MJ, Vangronsveld J. Remediation of metal- and radionuclide-contaminated soils by in situ stabilization techniques. In: Iskandar IK, editor. Environmental Restoration of Metals-Contaminated Soil. BocaRaton, FL: Lewis Publishers; 2000. pp. 21-61

104. Gadd GM. Fungal production of citric and oxalic acid: Importance in metal speciation, physiology and biogeochemical processes. Advances in Microbial Physiology. 1999;41:47-92

105. Gadd GM. Mycotransformation of organic and inorganic substrates. Mycologist. 2004;18:60-70. DOI: 10.1017/S0269915XO4002022

106. Strasser H, Burgstaller W, Schinner F. High yield production of oxalic acid for metal leaching purposes by Aspergillus niger. FEMS Microbiology Letters. 1994;119:365-370

107. Devevre O, Garbaye J, Botton B. Release of complexing organic acids by rhizosphere fungi as a factor in Norway spruce yellowing in acidic soils. Mycological Research. 1996;100:1367-1374

108. Rajpert L, Sklodowska A, Matlakowska R. Biotransformation of copper from Kupferschiefer black shale (fore-Sudetic monocline, Poland) by yeast Rhodotorula mucilaginosa LM9. Chemosphere. 2013;91:1257-1265

109. Marcinčáková R, Kaduková J, Mražíková A, Velgosová O, Vojtko M. Lithium bioleaching from lepidolite using the yeast Rhodotorula rubra. Journal of the Polish Mineral Engineering Society. 2015;16:1-6

110. Allman MB, Harris JA. An experiment in heap leaching. Mining Congress Journal. 1969;28-31

111. Dreisinger D. Copper leaching from primary sulfides: Options for biological and chemical extraction of copper. Hydrometallurgy. 2006;83:10-20

112. Watling HR. The bioleaching of sulphide minerals with emphasis on copper sulphides - a review. Hydrometallurgy. 2006;84:81-108

113. Neale J. Bioleaching Technology in Minerals Processing. Randburg, South Africa: Mintek; 2006

114. Herrero M, Stuckey DC. Bioaugmentation and its application in wastewater treatment: A review. Chemosphere. 2015;140:119-128

115. Leahy JG, Colwell RR. Microbial degradation of hydrocarbons in the environment. Microbial Reviews. 1990;53(3):305-315

116. Van der Gast CJ, Whiteley AS, Thompson IP. Temporal dynamics and degradation activity of a bacterial inoculum for treating waste metal-working fluid. Environmental Microbiology. 2004;6:254-263

117. Godleads OA, Prekeyi TF, Samson EO, Igelenyah E. Bioremediation, biostimulation and bioaugmention: A review. International Journal of Environmental Bioremediation & Biodegradation. 2015;3:28-39

118. Sood N, Sonali P, Banwari L. Bioremediation of acidic oily sludge-contaminated soil by the novel yeast strain Candida digboiensis TERI ASN6. Environmental Science and Pollution Research. 2010;17:603-610

119. Fan M, Rui-Jie X, Gang Q. Bioremediation of petroleum contaminated soil by a combined system of biostimulation-bioaugmentation with yeast. Environmental Technology. 2013;35:391-399

120. Bahafid W, TahriJoutey N, Sayel H, Boularad I, El Ghachtouli N. Bioaugmentation of chromium-poluted soil of microcosms with Candida tropicalis diminishes phytoavailable chromium. Journal of Applied Microbiology. 2013;115:727-734

121. Bahafid W, Tahri Joutey N, Sayel H, Asri M, Laachari F, EL Ghachtouli N. Soil bioaugmentation with Cyberlindnera fabianii diminish phytotoxic effects of chromium (VI) on Phaseolus vulgaris L. Journal of Materials and Environmental Sciences. 2017;8:438-443

122. Gallego JL, García-Martínez MJ, Llamas JF, Belloch C, Peláez AI, Sánchez J. Biodegradation of oil tank bottom sludge using microbial consortia. Biodegradation. 2007;18:269-281

123. Li X, Xin L, Peijun L, Wan L, Li W, Fang M, Chukwula KS. Biodegradation of the low concentration of polycyclic aromatic hydrocarbons in soil by microbial consortitum during incubation. Journal of Hazardous Materials. 2009;172:601-605

124. Tyagi M, Da Fonseca MR, Carvalho CCCR. Bioaugmentation and biostimulation strategies to improve the effectiveness of bioremediation processes. Biodegradation. 2011;22:231-241

125. He Z, Yao Y, Lu Z, Ye Y. Dynamic metabolic and transcriptional profiling of Rhodococcus sp. strain YYL during the degradation of tetrahydrofuran. Applied and Environmental Microbiology. 2014;80:2656-2664. DOI: 10.1128/AEM.04131-13

126. Chai L, Huang S, Yang Z, Peng B, Huang Y, Chen Y. Cr(VI) remediation by indigenous bacteria in soils contaminated by chromium-containing slag. Journal of Hazardous Materials. 2009;167:516-522

INDEX

www.ingramcontent.com/pod-product-compliance
Lightning Source LLC
LaVergne TN
LVHW060055220926
848484LV00006BA/1471